Routledge Revivals

Chemical Discovery and Invention in the Twentieth Century

Chemical Discovery and Invention in the Twentieth Century

Sir William A. Tilden,
F.R.S.

Routledge
Taylor & Francis Group

First published by George Routledge & Sons Ltd.

This edition first published in 2018 by Routledge
2 Park Square, Milton Park, Abingdon, Oxon, OX14 4RN
and by Routledge
605 Third Avenue, New York, NY 10017

Routledge is an imprint of the Taylor & Francis Group, an informa business

Publisher's Note
The publisher has gone to great lengths to ensure the quality of this reprint but points out that some imperfections in the original copies may be apparent.

Disclaimer
The publisher has made every effort to trace copyright holders and welcomes correspondence from those they have been unable to contact.

A Library of Congress record exists under ISBN: 17021524

ISBN 13: 978-1-138-61737-7 (hbk)

ISBN 13: 978-0-429-46175-0 (ebk)

CHEMICAL DISCOVERY AND INVENTION

IN THE

TWENTIETH CENTURY

UNIFORM WITH THIS WORK

DISCOVERIES AND INVENTIONS OF THE TWENTIETH CENTURY

By EDWARD CRESSY

With 281 large Illustrations

The Revival of Water Power; Coal, Gas, and Petroleum; Gas, Petrol, and Oil Engines; Generation and Transmission of Electricity; Electric Lighting and Heating; Speed and Economy in Factory and Workshop; Foundry and Forge; Electric Furnace; Artificial Production of Cold; Soil and Crops; Railways; Electric Traction; Motor-Cars; Modern Ships; Conquest of the Air; Wireless Telegraphy; Ships of War and their Weapons; Applications of Photography; Radium, Electricity, and Matter. Index

William Crookes.

Frontispiece.

CHEMICAL DISCOVERY AND INVENTION

IN THE

TWENTIETH CENTURY

BY

SIR WILLIAM A. TILDEN, F.R.S.

D.Sc., LL.D., Sc.D.

Professor Emeritus of Chemistry in the Imperial College of Science and Technology

"For wee are borne to quest and seeke after trueth; to possesse it belongs to a greater power."—Florio's *Montaigne*.

LONDON

GEORGE ROUTLEDGE AND SONS, LIMITED

BROADWAY HOUSE, CARTER LANE, E.C.

NEW YORK: E. P. DUTTON AND CO.

PREFACE

In accepting the invitation of the publishers to write a book of a popular character on modern chemical discovery, I am conscious of undertaking a very serious task. The difficulties to be encountered are twofold; first, there is the complexity and diversity of the subjects to be dealt with, and, secondly, the difficulty of rendering an account of many of them in language at once intelligible to the non-technical reader and free from serious inaccuracy. The author is indebted for assistance from many friends, and in connection especially with the chapters dealing with manufacturing processes such help was indispensable. To all in the following list who have thus kindly given information or supplied illustrations he desires to return his most grateful thanks.

Professor S. Arrhenius, of Stockholm.

Professor A. J. Brown, of Birmingham.

Professor W. A. Bone, of the Imperial College.

The Council of the Chemical Society.

Sir William Crookes, President of the Royal Society.

Lt.-Col. Professor A. W. Crossley.

Sir James J. Dobbie, Chief Government Chemist.

C. S. Garland, Esq., of the Volker Lighting Company.

J. C. Maxwell Garnett, Esq., Principal of the Municipal School of Technology, Manchester.

The Gas Light and Coke Company.

R. S. Giles, Esq., of Rangoon.

Messrs. Hird, Chambers, and Hammond, Chemical Engineers, Huddersfield.

The Governors of the Imperial College of Science and Technology.

Professor A. Liversidge.

William Macnab, Esq., F.I.C.

J. Lewis Major, Esq., Messrs. Major and Co., Ltd., Hull.

Professor W. A. Noyes, University of Illinois, U.S.A.

Professor W. J. Pope, University of Cambridge, England.

Dr. Frederick B. Power, of the United States Department of Agriculture, Washington.

Lt.-Col. Sir David Prain, Director of the Royal Botanic Gardens, Kew.

The Society of Dyers and Colourists, Bradford.

The Society of Chemical Industry.

The Rt. Hon. Lord Rayleigh.

The late Professor Sir William Ramsay.

Sir Boverton Redwood, Bt., Adviser on Petroleum to the British Government.

Professor T. W. Richards, Harvard University, Cambridge, Mass., U.S.A.

Professor Sir J. J. Thomson, Cambridge, England.

J. C. Umney, Esq., of Messrs. Wright, Layman, and Umney, Wholesale Druggists.

Herbert Wright, Esq., F.L.S.

I am also indebted to my friends, Professor Morgan, Professor Philip, and Dr. Martha Annie Whiteley for kindly reading portions of the manuscript.

I am under a special obligation to the Council of the Society of Chemical Industry for permission to adorn the cover of the book with the beautiful obverse of the medal awarded by the Society every two years for discoveries or inventions or other distinguished services rendered to industrial chemistry. The medal was designed by Miss Margaret May Giles (Mrs. Bernard Jenkin). It represents symbolically the four elements of Democritus and Aristotle, namely, Earth, Water, Fire, and Air, together with the seven planets, of which the names and symbols were, by the alchemists, associated with the seven metals known in ancient times. The crescent represents Luna, the moon (silver), and in order from left to right follow Jupiter (tin), Saturn (lead), Mars (iron), Mercury (mercury or quicksilver), Venus (copper). The tortoise above and the fish below belong to earth and water respectively. Sol, the sun (gold), occupies the centre.

W. A. T.

NORTHWOOD, *October*, 1916.

CONTENTS

PART IV

MODERN PROGRESS IN ORGANIC CHEMISTRY

LIST OF ILLUSTRATIONS

PORTRAITS

SUBJECTS

CHEMICAL DISCOVERY AND INVENTION IN THE TWENTIETH CENTURY

INTRODUCTION

In selecting the subjects to be dealt with in the following pages I have been influenced by the reflection that the nature of the operations in which the chemist is engaged, the objects he has in view, the subjects and methods of study, and the uses to which his theories may be applied are still very little understood by the public. I am therefore in hopes that my readers may be assisted in forming new views about all these subjects, and any confusion existing in their minds concerning them may be cleared away. Considerable enlightenment may be hoped for from the fact that in nearly all the universities in the world at least one professor of chemistry is now to be found, while in most of the modern universities it is recognised that the subject extends over too wide a field to be efficiently cultivated by one man, and three main divisions are generally recognised, namely, inorganic, organic, and physical chemistry. To these are sometimes added departments of applied chemistry in which the relations of systematic chemistry to industry or manufacture, such as fuel, metallurgy, dyeing, and bleaching, etc., are studied. But the extension of knowledge from the universities to the mass of the people is still a slow process, and notwithstanding the quickening effect which recent events have produced on the public mind, in England at any rate, it will be long before the practical economic importance of a knowledge of chemistry will be fully recognised by government departments, municipalities, and the public generally.

The ignorance of scientific things which exists among people commonly said to be well educated can only be regarded as scandalous, but evidence as to the gullibility of the public is supplied daily by every newspaper. To confuse Faraday with Fahrenheit may be thought a fallacy which offends only against sentiment, but the grossest practical mistakes would be avoided if a small amount of knowledge of principles and some capacity for independent observation were in the possession of everyone.

But in view of the too common credence accorded to the supposed phenomena of spiritualism, to Christian science and the water dowser it is hopeless to expect to see the world freed from baseless superstitions, the parasites associated with ignorance, for many a generation to come. In the meantime no antidote is likely to be so successful as the cultivation of natural science and the provision of easy and attractive opportunities of learning something of the accumulated stores of knowledge derived from careful observation of nature.

With regard to the branch of Science called Chemistry and to its applications one considerable source of confusion in England is to be found in the fact that the name *Chemist* has long been associated with the calling which is more appropriately styled "pharmacy," that is the dealing with drugs (φάρμακον, a medicine or drug).

In Germany the compounder of medicines is entitled "apotheker," while in France he is called a "pharmacien." In the British Isles, however, the assumption of the title Chemist and Druggist for a century or more by the competitors of the apothecaries has become so completely established in the public mind that to speak of a chemist invariably implies in common parlance the person who compounds and sells medicines. And this custom has been ratified by the Act of Parliament (Pharmacy Act, 1868) which secures to the members of the pharmaceutical body the exclusive right to the title "Chemist and Druggist" or "Pharmaceutical Chemist," according to their qualifications. It is not to be denied that a small number of pharmacists do actually possess such a knowledge of chemistry as to qualify them to undertake analytical work or to manufacture chemicals or pursue chemical research, but this is not the essential part of their calling, and the anomaly lies in the fact that if Sir Humphry Davy himself were now living he could not legally call himself a chemist, his name not being on the pharmaceutical register.

Chemistry has for centuries been associated, perhaps naturally, with medicine, and, confining attention to modern times, it is easy to recall the names of many men who became eminent as chemists having begun by the study of medicine in some form or other. Black, Davy, Berzelius, Wollaston, Wöhler, Wurtz, Andrews, and W. A. Miller began by the study of medicine, while Scheele, Rose, Liebig, Dumas, and Frankland received their earliest notions of chemistry in the druggist's shop.

Chemistry has been gradually emancipated from these associations with enormous advantages to the progress of knowledge. The systematic study of chemistry and provision for teaching it in schools and universities belong to comparatively recent times.

In the middle of last century there were no laboratories for practical work in any of the British universities and chemistry was not a subject which led up to a degree. The Professor of Chemistry at Oxford was also Professor of Botany, while at Cambridge the Professor of Chemistry was a country clergyman who came up once a year to give a course of lectures, and it was even thought very creditable on his part to do so much. In those days no school had a resident science master who gave the whole of his time to teaching elementary physics or chemistry, though a few received a visit about once in a fortnight from a peripatetic teacher who brought with him a box containing portable apparatus.

Happily times are changed in all these respects and, notwithstanding the substantial grounds for hoping that still greater progress will be made, the position of the science master in all the most important schools is now fully assured. But the conflict between literature and science for the possession of endowments, time of teachers and pupils, energy of both, and prominence in the educational field is even now not at an end, and never will cease until both sides are duly influenced by respect for the work and aspirations of the other.

From the circumstances connected with the great war in Europe the importance of a knowledge of chemistry has become particularly prominent. The appropriation of the larger part of the manufacture of dyes and drugs by Germany has attracted particular attention, especially in those countries in which such industries have been less developed, and it has become at last obvious to the public that the imperfect recognition of the value of scientific knowledge is the main cause of their deficiencies. That this has been known to chemists during the last forty years is made evident by the numerous speeches, papers, and presidential addresses which have been issued by some of the most eminent professors of chemistry in the United Kingdom. Now that the dyehouses of Yorkshire and Lancashire are almost brought to a standstill for lack of material, the manufacturers in the north of England have begun to realise that the position is serious, and in response to the call for assistance the Government has assigned a sum approaching a million sterling to an attempt to resuscitate the dye industry, and to provision for a department of research in connection with the making of colours from coal tar hydrocarbons. And further it was announced in the House of Commons on May 13th, 1915, by Mr. Pease, the then Minister of Education, that the Government intended to institute forthwith an Advisory Council on industrial research with the object of bringing the British universities and technical colleges into closer association with industry. He promised that for the current year a sum between £25,000 and £30,000 would be placed on the estimates for this purpose. He added that the demand for money in this direction would increase as time goes on, and that if the scheme is to succeed it must depend on the State for help in the years to come and that State help must steadily progress. The chief advantage to be expected from this scheme is, however, probably not to be found in the subsidy itself and the actual development of industrial research in the universities and colleges. It is rather to be hoped that this tardy though substantial recognition of the importance of training a larger number of young men in scientific method, will operate in stimulating public interest in such matters and rousing it out of the state of indifference which has so long prevailed.

This movement, however valuable as an encouragement to reform, will not afford a permanent guarantee of commercial success unless the spirit of respect for accurate scientific knowledge and its daily application by highly qualified chemists within the works themselves are diffused generally among the manufacturers. It has been alleged that the universities have not provided the sort of instruction required by students who look forward to employment of a technical character, that the subjects taught have been too far removed from possible applications to industrial purposes and that the habits and methods of research have been neglected. So far as research is concerned it cannot be denied that the British universities in times now long past have been to blame, but during recent years such allegations can no longer be sustained.

And when the relative importance of "pure" chemistry and "applied" chemistry is considered it seems, at least to the writer, that the former stands clearly first. For all history of science shows that progress has been accomplished only when research has been released from the restrictions with which it is trammelled when the eye of the worker is on the look out only for immediately useful results.

The Atomic Theory of Dalton and later Kekulé's theory of the benzene ring must have appeared to contemporary manufacturers as mere academic theses of no industrial import. But we have it on the testimony of a very eminent colour-maker that the benzene theory lies at the foundation of the industry. In fact where science has been respected and scientific knowledge cultivated there has been industrial success; where it has been neglected industrial failure has been the consequence.

On the other hand, it seems quite practicable to avoid the reproach which has been so often cast by the practical man at the higher chemical schools, by bringing more prominently and more frequently before the student problems which are connected with industry. This can easily be done by the judicious teacher without loss to the academic value of the training given. For a variety of reasons which are obvious the actual investigation of questions connected with industrial processes cannot be undertaken profitably before an advanced stage is reached in the student's career.

Admitting that there is need for closer attention to the use of science for practical ends, it seems scarcely open to doubt that the greatest benefits to the world have accrued from the pursuit of knowledge for its own sake, and without regard to the possible applications of the knowledge gained to immediate useful purposes. Without such untrammelled enquiry the world would be still in the condition of Europe in the Dark Ages. The relation of the earth to the sun and the rest of the heavenly bodies, the cause of the seasons, the effect of the atmosphere and its several constituents on plant life and the respiration of animals, the composition of common air and water, and the knowledge of the earth's surface facilitated to every traveller by modern means of locomotion would be still unknown. These things are surely of greater political, social, and moral importance to the human race than even the modern valuable discoveries of new dyes, new drugs, or new sources of light. The possession of such knowledge distinguishes the civilised man from the barbarian and the savage, and is the foundation of all the future hopes of mankind so far as life on this earth is concerned.[1]

As already mentioned the number of chairs of Chemistry in the universities is gradually increasing, and each one forms a centre from which many chemical students pass into the outside world and so help in the diffusion of knowledge which may, and frequently does, become very valuable in a practical sense. The university laboratories are also at the present day the source of a good deal of positive knowledge derived directly from researches carried on within their walls. It was not always so, and within the last forty years many reproaches have been directed against the British universities on account of the comparatively small part formerly played by these schools in the production of new chemical knowledge. The departure of so large a proportion of the coal tar colour industry, which originated in this country, to Germany, where the connection between the universities and the chemical industries of that country has been more definite and intimate, has been repeatedly attributed to the neglect of organic chemistry by the universities and the want of mutual respect between academic learning and industrial needs. This want of co-ordination has doubtless something to do with the neglect of science generally by the British Government. But amid the many evils of the European war one good seems likely to arise and that is the awakening of the authorities and the public to the necessity of scientific principles in the work of the Empire, and the utilisation of the knowledge and skill of trained scientific men. Hitherto a large part of the intellect of the country has been attracted into other callings and the career open before a highly trained university or other student of science has offered little temptation either in emoluments or social position.

As already explained there is now some considerable ground for hope that this is going to be changed.

[1] See "Modern Scientific Research," a lecture by Sir W. A. Tilden. *Nature*, Vol. LXXXV, Nov. 3rd, 1910.

The study of chemistry as a subject of intellectual interest or of commercial importance has led to the formation of various associations of persons engaged in the same pursuit. Formerly a discovery was usually communicated to the world through the medium of one of the Academies of which one generally exists in each of the chief countries of Europe, and is represented in our own country, so far as physical and natural science is concerned, by the Royal Society which was founded in the reign of Charles the Second. The British Academy founded in 1902 is an institution having somewhat similar aims, but is composed of men distinguished in literature, history, and philosophy.

As to chemistry the institutions of greatest importance in Britain are the following. The oldest of these bodies is the Chemical Society of London, founded in 1841. The objects of the Society were then defined to be chiefly—" The promotion of Chemistry and of those branches of Science immediately connected with it by the reading, discussion, and subsequent publication of original communications." This object has been carried into full effect, and the Journal of the Chemical Society is now the recognised repository of practically all the purely scientific researches carried out in the British Empire. It also contains abstracts of all papers on chemical subjects published in foreign journals. The Society numbers upwards of three thousand Fellows, and occupies apartments provided by the Government in Burlington House.

The Société Chimique de Paris, founded in 1858, the Berlin Chemical Society, founded in 1867, and the American Chemical Society, founded in 1876, are engaged in similar work in the respective countries.

The Society of Chemical Industry (which includes two American sections with one Canadian and one Australian section) was started in 1881, and, as its name implies, has for its object the study and publication of papers relating to the application of chemistry to manufacturing or other practical purposes. The membership of all these societies implies no professional qualification, but some forty years ago a proposal was brought forward among the Fellows of the Chemical Society to restrict the Fellowship to persons who could produce evidence of scientific training and qualifications for practice as analytical and advising chemists. This after prolonged discussion was found to be impracticable and a new body was formed, namely, the Institute of Chemistry of Great Britain and Ireland, which in 1885 received a Royal Charter. The Institute holds periodical examinations, and in other ways tests the qualifications of candidates for its Associateship and Fellowship. The Institute stands therefore toward the profession of chemistry somewhat in the same relation as the Institutions of Civil and Mechanical Engineers to the calling of Engineers. It does not possess the exclusive powers of the Royal Colleges of Physicians and Surgeons, but the value of the Fellowship is now recognised by many Government Departments.[1]

Other voluntary associations of chemists interested in particular applications of chemistry exist, but sufficient has been said to indicate the nature and extent of the organisation now existing in the British Isles.

As it often happens that parents are uncertain how to gratify the aspirations of a boy to become a chemist a few remarks may be made here as to the course most advisable to pursue. It may be added that there is nothing in the nature of things to prevent a woman following the same course of study. A few women students have successfully taken up chemistry, and they now occupy in most cases important positions as teachers. It is advisable at the outset to point out that success can only be looked for by those who have had a sound general education and that the course of study to be pursued after school age extends over at least four years and may in many cases be usefully prolonged to five or even six years before entry on professional or industrial life.

If the student enters one of the universities with a view to the pursuit of a science course three or four years from matriculation will be occupied according to the quality of the degree which he wishes to take. His studies up to this point will extend into several branches of science, especially mathematics, physics, and chemistry. Having secured his degree he should then devote one or two years to special chemical work or research under the direction of his teachers. Should he not be in a position to enter one of the universities he may attend the courses of instruction given in one or other of the great technical institutions such as the Imperial College of Science and Technology at South Kensington. Students, however, are not admitted to

[1] *The History of the Institute of Chemistry*, compiled by R. B. Pilcher, Registrar of the Institute, contains full details of progress from 1877 to 1914.

these institutions without giving evidence of good general education, and four years will generally be occupied in securing the diploma. In all such cases it is necessary to bear in mind that a thorough grounding in general principles is essential to the successful study of the technical applications of the different branches of science concerned. Mathematics are becoming more and more indispensable to the student of chemistry.

It is sufficient here to point out that an ordinary schoolboy cannot be made into a chemist in a day or in a year or two, and the process is expensive.

The relation of our educational systems to practical ends in manufactures and in trade and commerce has been the subject of much discussion in recent times. So also has the difficult problem as to the arrangements by which manufacturers can make use of scientific assistance and the chemists engaged for industrial employment can gain that knowledge of the business into which they are introduced which alone can enable them to apply their scientific training and skill to the problems which confront them in the works.

There can be no doubt that in the universities and colleges, and perhaps also in secondary and even elementary schools, the view to practical applications of knowledge will in future be kept more distinctly before both teachers and taught. Incentives to industry and concentrated attention will probably be found for the majority of both boys and girls when they realise that what they are expected to learn at school will have a direct influence on their material progress when they go out into the world. And in this connection it will be an advantage if the principles of economics and the sources of the wealth of nations are not overlooked by the teacher in laying the foundation of a sound knowledge of the physical and biological branches of science. The relation of man to the universe in which he finds himself will continue to exercise a fascination sufficient for the exceptional few, but it is probable, and doubtless for the best, that "bread-and-butter" studies will continue to be most attractive to the many. The most serious charge which can be preferred against the time-honoured classical system of education is not so much that a knowledge of two dead languages and their literature has but little to do with the conditions of modern life, but that in consequence of the defective methods of teaching hitherto prevalent, school teachers have failed to communicate to the great majority of the pupils any real knowledge of these subjects to which they have been forced to give the greater part of their years of school life. This fact should serve as a warning to the teachers who are occupied with mathematics and physical science in schools, and lead them to the use of methods and practices which will encourage their pupils to apply their knowledge daily to the affairs of common life and business.

The importance of research into the yet unknown is acknowledged on all hands. We have, for example, the National Physical Laboratory, where a very competent, if very small, staff of scientific men is employed not only in testing instruments and in other routine work, but in research on the application of scientific principles to objects of national importance. The design of ships, the study of the principles of aeronautics, the testing of metals and alloys are among the subjects which occupy them.

Agriculture has received direct assistance of a most valuable kind, in the first instance from the researches in connection with the composition of soils and manures and the treatment of various crops, which were initiated by Sir John Lawes so long ago as 1837. In 1843 Lawes secured the assistance of Dr. J. H. Gilbert, and together the two men continued and developed this remarkable work during a period of fifty-seven years. Sir John Lawes died in August, 1900, Sir Henry Gilbert in December, 1901. The work continues on the same ground, with the aid of the fund left by Sir John Lawes, under the direction of the Lawes Agricultural Trust, and the Director appointed by them, Dr. Edward John Russell. In more recent times several Agricultural Colleges have been established, and research is carried on to a greater or less extent in all of them. To these must be added the Experimental Farm belonging to the Royal Agricultural Society and the Fruit Farm belonging to the Duke of Bedford at Woburn.

Research has been aided for many years past by the distribution of £4000 per annum, entrusted by the Government to the Royal Society, but the amount thus administered is quite inadequate to the requirements of the many applicants from the whole circle of the sciences, the amount available for any one branch, chemistry for example, being almost insignificant. The Chemical Society has also a small research fund, but the grants made seldom exceed £10, and serve only to lighten the burden of the investigator in the purchase of necessary materials.

It is difficult to estimate the amount of research work which has been undertaken by manufacturers for their own purposes. Obviously the results, if any, are reserved for private utilisation, but whatever has been accomplished the necessity for work of this kind is as yet far from being recognised by manufacturers generally. The introduction of new principles and processes founded on them must have been preceded by enquiry and experiment by competent workers. The wide application of electricity for the production of high temperatures and in electrolytic operations, the use of catalytic agents, as in the contact process for sulphuric acid, the hydrogenation of fats and oils, and several other chemical manufacturing operations which have actually been established in England during the last few years, give evidence of advances which have been accomplished, and which give proof of the application of physics and chemistry. There are, however, so many other directions in which this country is dependent for supplies from other countries that much more remains to be done.

The question how the science of chemistry can be brought into the service of industry most advantageously has still to be answered. We may suppose that it is agreed that the young chemist, equipped with a full knowledge of theoretical chemistry and well practised in all the analytical and other operations of the scientific laboratory, requires an elementary knowledge of engineering, and of the properties of materials for construction used in the works, before he is qualified to take charge of operations on a manufacturing scale. This is a kind of knowledge which can be acquired to a certain extent at college, but experience of operations on a large scale is still desirable and the question arises how he is to get it.

In the German and, to some extent, the American works a system has prevailed for many years which seems to have the double merit of being reasonable and practically successful. Over each department an experienced scientific chemist presides. When an assistant is required a graduate of one of the universities, recommended by the professor under whom he has worked, is engaged under a contract to serve for a term of years at a salary which is modest, but which is sufficient to enable him to live till promotion comes. The first year or two is devoted to learning, under the direction of the chief, the business of that part of the works into which he has been admitted, and the assistant is not expected at first to do more than make himself thoroughly competent in routine work, at the same time the degree of diligence and ability shown determines the rate at which he may look for advancement in pay and position. There is usually a clause in the agreement which forbids the employé to leave this employment and enter the service of another firm in the same business within a certain distance. Men who enter the chemical works under these conditions, however, are not prone to wander, and the prospects for a man of real ability are satisfactory or even brilliant. Unfortunately no system of this kind has become general or even frequent in this country. Too often the manufacturer looks for practical results impatiently, and does not realise the fact that a chemist may have a full knowledge of his subject from one point of view but has to apply that knowledge to problems with which he has no previous acquaintance, also that discoveries cannot be made to order. It is only necessary to know a little of the history of chemistry and its applications to be aware that experiments carried on for many years by clever men do not always lead to a successful result.

Another method by which manufacturers who do not choose to establish on their own premises a scientific laboratory and staff may obtain assistance in the endeavour to improve processes, to overcome difficulties or irregularities in existing processes, or to test new ones, may be found in the proposal to institute industrial fellowships. This idea was inaugurated a few years ago by the late Professor Kennedy Duncan in connection with the universities of Kansas and Pittsburg.

The plan is as follows: any manufacturer desiring skilled assistance may apply to the university for a chemist qualified to prosecute research. The chosen person, nominated by the Chancellor and the Director of Industrial Research, is provided with a separate laboratory and with necessary materials in the chemistry department of the university, together with facilities for large scale experiments provided by the manufacturer. The latter also provides the remuneration payable to the Fellow for one or more years. The Fellow works under the general supervision of the Director of Industrial Research, and through him periodical reports on the progress of the work are forwarded to the employer. Any discoveries made by the Fellow during the tenure of his Fellowship become the property of the employer, subject to the payment of royalty or other consideration, the amount of which is determined by the Board of Arbitration provided for in the Scheme.

Many other details are considered and arranged in the programme, which appears from a report by Professor Duncan to have met with remarkable success during the four years it had been in operation. Eighteen Fellowships had been established in the University of Kansas, and twenty were about to be instituted in the University of Pittsburg. This appears to show that the industrial employer had been satisfied with the results. There appears to be no reason why this plan should not be adopted in many other universities, as it could manifestly be put into operation in all those cases in which direct daily observation of processes going on in the works is not essential. That is the qualification which seems to indicate a greater convenience in the other system, previously described, in which the chemist in the laboratory has immediate access to the manufacturing operations, for the advantage of which he is supposed to be at work.

Sufficient has now been said to convey a general idea of the position of chemistry as the basis of a calling or profession. In the chapters which follow a description will first be given of some chemical laboratories, which may be regarded as typical, and of the more important operations which are carried on in them, in order that the reader who is not a chemist may gain some notion of the work carried on in the scientific laboratory and in the chemical manufactory.

Chemistry is that department of natural knowledge which is concerned primarily in determining composition, or in other words finding out what things of all kinds are made of. The chemist has to deal with gases like common air, with water and other liquids, and with solid matters of all kinds whether mineral or organic. He is therefore not confined to the study of composition only, but, with the aid of methods and instruments drawn from other departments of science, he examines the properties of bodies and the conditions under which compounds are formed or are decomposed. In every operation of nature chemical change is incessant, in the material of the earth's crust, in the sea and in the air, in life, death, and decay. The chemist has all nature for his province.

By careful study by many generations of men, more particularly during the last century and a half, knowledge has been substituted for ignorance, system for chaos, and a body of theory has been established which enables the chemist to classify the multitudinous facts of nature and so render them more or less intelligible. The application of specialised portions of this knowledge to manufacture provides mankind at the present day with many of the conveniences of modern life which could never have been dreamed of by our ancestors.

The later portion of the book will contain an account of the most important discoveries which not only find practical applications, but give entirely new views of the constitution of the world in which we live.

PART I

CHEMICAL LABORATORIES AND THE WORK DONE IN THEM

CHAPTER I

LABORATORIES FOR GENERAL TEACHING

THE word *laboratory*, which merely signifies a workshop, has by long custom been applied chiefly to the room or building in which chemical experiments are carried on, or at any rate experiments in natural science in which operations more or less chemical in character are practised.

The chemists of the past were content with very modest accommodation, provided a sufficient amount of light was secured together with a supply of water and the means of obtaining heat. Berzelius, the famous Swedish chemist, who lived till 1848, carried out the greater part of his accurate estimations of atomic weights as well as other researches in a room communicating with the kitchen of his house, where Anna his servant maid acted as his only assistant.

At this time the teaching of chemistry in the universities was everywhere conducted solely by the method of lectures which were rarely enlivened by experimental demonstrations. The student desirous of learning something of chemical analysis or other practical chemical work had to seek the privilege of admission into the private laboratory of some professor of chemistry. Liebig tells us that he had to leave his own country, Germany, where in his youth there were no chemists of any importance, in order to apply to Gay-Lussac in Paris for permission to work under his direction. With this experience in his mind it is not surprising that on his return home two years later he should have determined to found in his own country an institution in which students could be instructed in the art and practice of chemistry, in the use of chemical apparatus, and the methods of chemical analysis. Such was the origin of the famous laboratory at Giessen which, from 1824 onwards, for many years attracted students from every civilised country. It was but a modest place with none of the appliances with which we are now familiar.

It was twenty years later before a laboratory for instruction in chemistry was opened in this country, and even then it was not either Oxford or Cambridge which took the lead in this important reform.

The first laboratory in this country opened for the use of students of chemistry was provided by the Pharmaceutical Society of Great Britain at their premises in Bloomsbury Square. In 1844 places were furnished for twenty-one students. The laboratory was a single apartment, the ventilation of which was very imperfect, and as many of the operations required the use of coke furnaces the place was full of smoke and fumes. Almost immediately after this the Royal College of Chemistry was founded, and for the first year or so carried on operations in George Street, Hanover Square. It was then transferred to its permanent home in Oxford Street, where a building had been provided which still exists, and, with an additional upper storey, contains the offices of the General Medical Council. The building had a frontage of only 34 feet with a depth of 53 feet.

The whole of the first floor was occupied by the Students' laboratory, while on the ground floor were a private laboratory for the Professor, a balance room, and a lecture room at the back. The basement contained furnaces and a steam-boiler and stores.

It is unnecessary to pursue this retrospect any further, for the example set at Giessen, when once the movement had begun, was followed in all the great centres of instruction. But even the new laboratories were very inferior in size and equipment to those which have been erected in more recent times.

The rate of progress during the last thirty or forty years has been very rapid, and stimulated by the rivalry between nations and by the rapidly increasing numbers of students seeking instruction, the buildings provided for the accommodation of the chemical departments in the universities and modern colleges as well as the numerous technical schools, have gradually assumed more and more palatial features. The first important step in this direction was taken by the German Government, when, after the Franco-Prussian war in 1870–71, Strassburg became a German town.

Possibly animated by the desire to placate the Alsatian

population, splendid separate institutes were erected in Strassburg to provide for the several branches of science, chemistry, botany, geology, etc. Each of these institutes contained accommodation, on a scale previously unknown, for laboratories, lecture rooms, and museums, as well as residence for the chief professor. Even the Strassburg chemical institute is now surpassed in dimensions and outfit by some of the establishments more recently erected in various parts of the world.

Before proceeding further it will be convenient to review the purposes for which the very numerous chemical laboratories have been erected in all the civilised countries of the world. In the first place it must be remembered that they are not all devoted to the purposes of instruction. Many are occupied with purely practical objects in connection with analysis of products for control of quality, or for fiscal purposes, or in association with manufacturing operations. And in these later times the importance of research is becoming so generally recognised that institutions have been founded and endowed with the sole object of providing facilities for carrying on such work independent of teaching, on the one hand, and of industrial or practical purposes on the other. The following classification of laboratories must be understood to be only illustrative, and with a few exceptions applies only to the British Isles. The total number of universities and of technical schools in Britain alone is very large, and any attempt to enumerate completely even these would require a volume to itself. The reader must be informed therefore that if the universities of other countries and such famous technical schools as the Massachusetts Institute of Technology at Boston, are not included in the analysis it is from no want of sense of their importance.

LABORATORIES FOR INSTRUCTION

1. Universities (15 British).

2. University Colleges.

   Special Departments for Agriculture; Brewing; Dyeing; Leather; Metallurgy.

3. Technical Schools.

   Among the most important are: The Imperial College at South Kensington; The Royal College of Science, Dublin;

The City and Guilds of London, Finsbury; The Pharmaceutical Society; The Royal Technical College, Glasgow; The Municipal School of Technology, Manchester.

4. Agricultural Colleges.
5. Public Secondary and Elementary Schools.

The laboratories in some of the great public schools of England are now as well equipped as those of the universities.

LABORATORIES FOR ANALYSIS (Chiefly)

1. Government Laboratories.

Central; Admiralty; Woolwich Arsenal; Imperial Institute.

2. Public Analysts.
3. London County Council.
4. Many private analytical.

LABORATORIES CONNECTED WITH MANUFACTURES

These are private establishments connected with individual works for the production of iron and steel and metals generally, also with alkalis and acids, drugs, dyes, and chemical products of all kinds. One which is at the present time attracting much attention is the laboratory for research financed by the Government for the assistance of "British Dyes Limited."

LABORATORIES FOR RESEARCH ONLY

The Royal Institution, established under Royal Charter 1800. With it is associated the Davy-Faraday Laboratory, founded and endowed by the late Dr. Ludwig Mond.

The Lister Institute, corresponding in aims with the Pasteur Institute in Paris.

The National Physical Laboratory, dealing with metallurgical research among other subjects, chiefly physical and mechanical.

The Lawes Agricultural Trust. Experimental Station and Laboratories, Harpenden, Herts.

Fig. 1.—IMPERIAL COLLEGE OF SCIENCE AND TECHNOLOGY, LONDON.
DEPARTMENTS OF CHEMISTRY AND PHYSICS.

To face page 20

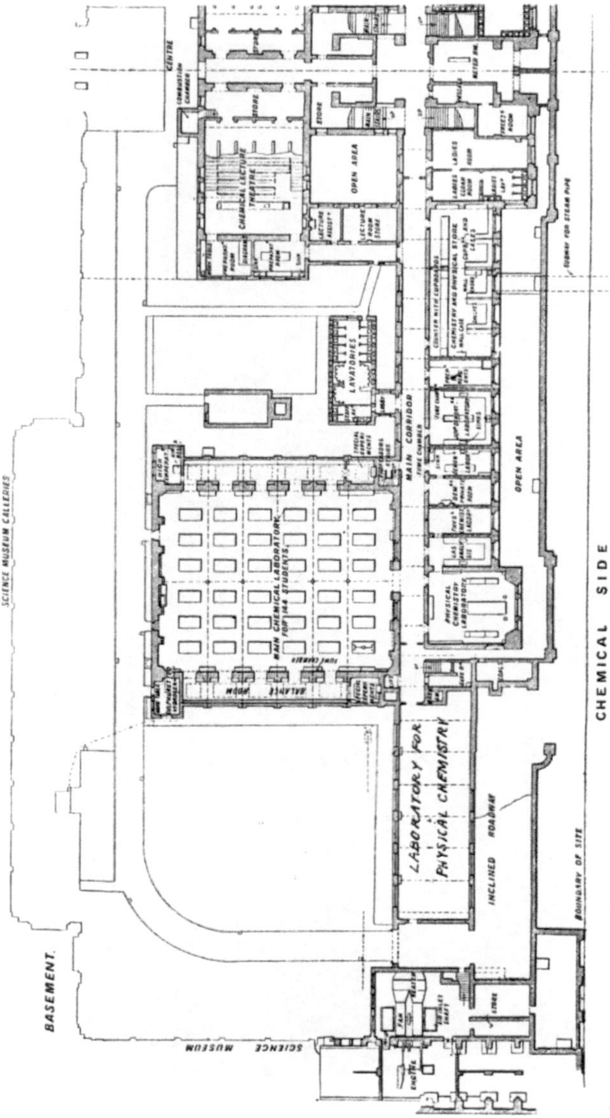

FIG. 2.

To face page 21.

The Kaiser Wilhelm Institute for Chemistry, opened October 23, 1912, at Dahlem, near Berlin.

A few of the more important of these will now be described.

I.—THE IMPERIAL COLLEGE OF SCIENCE AND TECHNOLOGY, LONDON

One of the largest and most completely equipped chemical departments was erected by the British Government for the accommodation of the Royal College of Science and Royal School of Mines at South Kensington, London, and was occupied for the first time in 1906. The architect of the building was Sir Aston Webb, R.A., but the arrangement and fittings of the interior were designed by the then professor of chemistry, the present writer.

A general view of the exterior of the principal building is shown in Fig. 1, from which it will be seen that it consists of a central block with two wings. The eastern half of the building is devoted wholly to chemistry, while the western half is occupied wholly by physics. The central mass contains the entrance and stairs leading to upper stories occupied by the Science Library which forms a part of the South Kensington Science Museum. This provision for pure chemistry is supplemented by the Department of Fuel and Chemical Technology, of which the separate building has been more recently erected and occupied for the first time in 1915. The College and the School of Mines are now united, together with the City and Guilds of London Institute, into one chartered body, the Imperial College of Science and Technology.

The buildings contain complete suites of laboratories, lecture rooms, and accessory apartments, with accommodation for the teaching staff in the four divisions of

1. General and Analytical Chemistry ;
2. Physical Chemistry ;
3. Organic Chemistry ;
4. Fuel and Chemical Technology.

There is a professor at the head of each division, with a number of assistants.

As the internal arrangements are typical of what is aimed at in many chemical institutions it will be worth while to glance at the plans of the several floors which are shown in Figs. 2, 3 and 4, which show respectively the basement, the ground floor, and the first floor.

From the main entrance stairs descend on the left to the lower ground floor, upon which level are found the chief lecture theatre, the large chemical laboratory, with separate places for 144 students, and the series of laboratories for physical chemistry. The balance rooms for the big laboratory extend along each side of the building, access being provided at five points on each wall. The lecture theatre provides comfortable sitting space for 150 students, but there is a large floor at the back and considerable room in front, so that about twice that number of auditors can be provided for when occasion requires. At the back of the table are blackboards, means of hanging diagrams, and two screens for projected pictures or lantern views of experiments. There are several wide pipes leading downwards from the surface of the table by which even copious fumes can be sucked away and prevented from reaching the audience. There are also numerous connections, visible in the picture, by which water, gas, electric current, and vacuum can be at once utilised for experiments to be shown on the table. The room can be rendered completely dark, when necessary, by the provision of black blinds to all the windows.

In addition to these there is a spacious store for physical and chemical apparatus, of which a large quantity in the form chiefly of glass flasks, beakers, and other necessary vessels is always kept in stock. Close at hand is the freezing room, in which there is a machine, electrically driven, for the production of liquid air.

Ascending to the floor above there is a series of apartments which provide a lecture room with seats for about fifty students, chiefly occupied by the professor of physical chemistry, a library of reference furnished with the principal chemical periodicals, dictionaries, and other large special treatises. Adjoining this is the private room for the professor of general chemistry, who is also director of the laboratories, and this leads to his research laboratory, where there is room for about eight or ten workers. The floor above this, called the first floor, is occupied almost entirely by the professor of organic chemistry who has a separate research laboratory. A room, also close by, is occupied by the technical artist who prepares diagrams.

The laboratory devoted to organic chemistry is over the large

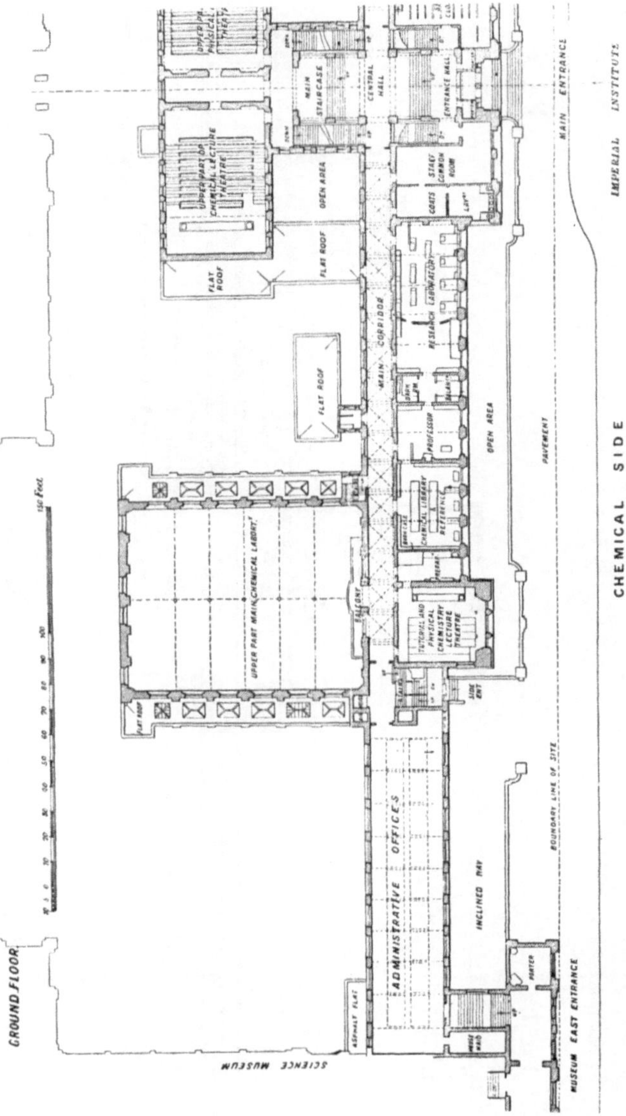

Fig. 3.

To face page 22.

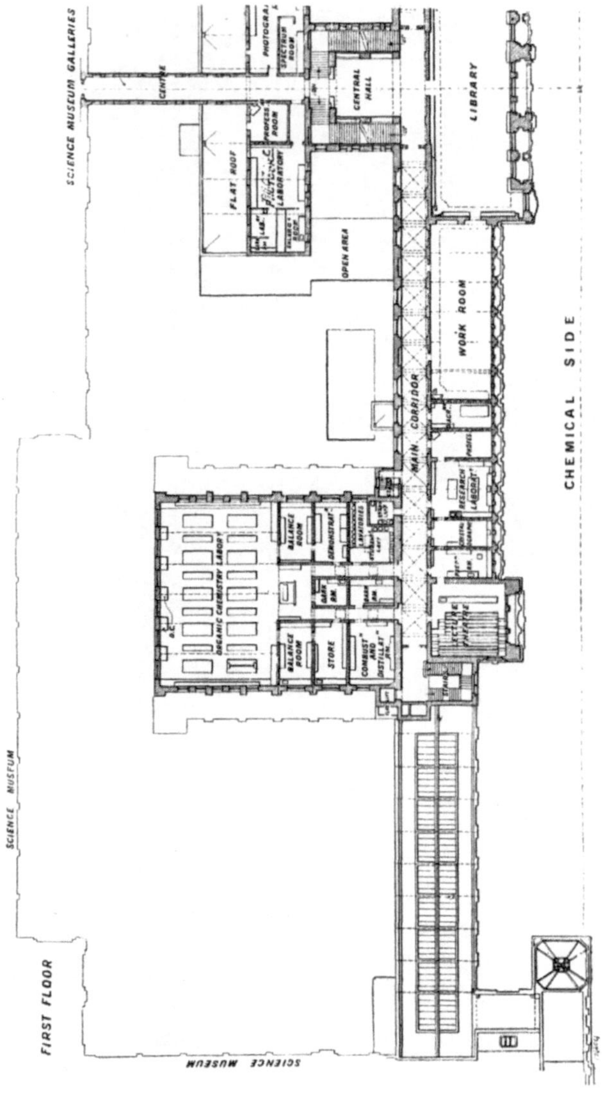

Fig. 4.

To face page 23.

laboratory depicted (Fig. 6), and occupies about half its area. The fittings here differ somewhat from those below, as it is necessary to give each student a larger share of space and to provide for certain operations not usually practised in the general laboratory. The organic chemistry laboratory therefore provides for only forty workers. The other half of the space contains rooms for the demonstrator, for balances, for stores, and for special operations, such as combustions and distillation of very volatile liquids such as ether.

The top floor is occupied by the advanced laboratory, with places for one hundred students, half of whom face each way toward the desk of the presiding demonstrator. A supply of balances is found at each end of this laboratory, and there are several rooms for special operations, such as water analysis.

The three illustrations (Figs. 5, 6, 7) afford views of the interior of the largest of the laboratories and of the chief lecture theatre.

At the back of the building there is a complete suite of rooms fitted with black blinds and special sinks, and a copious supply of cameras and other apparatus for a class in photography.

The laboratories just described were designed for instruction and research in pure scientific chemistry. A very large amount of work of this kind has been done in the new laboratories with results which have been communicated chiefly to the Chemical Society of London, but a good deal has also been done in connection with problems of a more or less directly technical character. To add to the facilities already provided for extension in this direction the new building in the neighbouring Prince Consort Road now provides for two sub-departments, namely, the study of fuel—solid, liquid, and gaseous—and chemical engineering, that is, the design and working of chemical plant for industrial purposes. It is hoped to add a third sub-department for electro-chemistry as soon as funds are available.

The new buildings at present provided form part of a plan which would ultimately include a building with a front and two wings, leaving a space between the wings large enough to allow of the erection of temporary buildings for special technical researches. The wing so far constructed contains the departments for fuel and chemical engineering. At the present time it consists of two stories only, but the walls have been built in a very substantial manner, with a view to the addition of two floors above.

Entering on the ground floor the corridor leads past a lecture room, workshop, and a storeroom, to the refractory materials laboratory (36 feet by 33 feet) and the analytical laboratory (55 feet by 35 feet), with accommodation for twelve students. The floor above contains private laboratories for the professor and staff, and a photometric room, beside the large research laboratory and the furnace room, which are respectively situated over the refractory materials and analytical laboratories, already mentioned, and are therefore of the same dimensions.

A valuable addition to the department of fuel is the experimental gas producer plant presented to the college by Mr. Robert Mond. The plant consists of a producer with the type of grate introduced by the late Dr. Ludwig Mond, and from this the gas is directed through washing towers and scrubbers, so arranged with valve control that any degree of scrubbing can be given that is required. The gas then passes through a tar extractor, and finally through a sawdust scrubber. The gas is now clear of fog and cooled to about the temperature of the air, and after passing through a meter is collected in the gas-holder, which has a capacity of 3000 cubic feet. The pipes leading to the gas-holder and the connections are so arranged that it may be filled with producer gas alone, or coal gas from the supply alone, or a mixture of the two in any desired proportions. Town gas being liable to vary in composition and from day to day in pressure, the experimental gas-holder provides the means of storing, at the beginning of a week, a gas which can be delivered into the research laboratory at constant pressure and of constant composition for experimental purposes during the days following.

In addition to the large holder just described there is a smaller holder of 100 cubic feet capacity. This can be used for experiments on the heating value of fuel gas or of mixtures of gases. In this holder also may be stored single pure gases, such as hydrogen and carbon monoxide. This holder is connected by a $\frac{3}{4}$-inch pipe with the laboratories on the upper floor.

The experimental work going on under these conditions will doubtless prove of immense value in connection with the applications of coal gas or producer gas to industrial purposes.

The department of fuel, etc., has been designed and arranged by Professor W. A. Bone, and contains much gas apparatus and

FIG. 5.—IMPERIAL COLLEGE OF SCIENCE AND TECHNOLOGY, LONDON.
CHEMICAL LECTURE ROOM.

To face page 24.

FIG. 6.—IMPERIAL COLLEGE OF SCIENCE AND TECHNOLOGY, LONDON.
MAIN CHEMICAL LABORATORY.

To face page 25.

other devices connected with his well-known researches on combustion.

The reader who is not a chemist will require some further information as to what is to be found in such establishments as those just described, and what sort of work goes on therein.

Returning therefore to the main laboratory in the pure chemistry department, it will be seen from the picture that the working benches are arranged in groups of four over the entire area. Each student is provided with a share of the table top, five feet from side to side, a sink on one side, and a suite of drawers and cupboards below, which provide for all ordinary requirements. On the top of the table and in front of the worker are shelves on which stand bottles containing the liquid and solid reagents most commonly in use. There are two gas taps to which the Bunsen burners employed can be attached by flexible rubber tubing. The water supply is also duplicated, the one tap serving for the washing of glass or other apparatus, while the other is permanently connected with a high pressure water pump, by which a reduced pressure or "partial vacuum" may be established in any vessel connected with it. This is especially useful in hastening filtration through the paper filter commonly used.

Another agent indispensable in analytical work is hydrogen sulphide or sulphuretted hydrogen gas. This gas has a disgusting smell and is very poisonous; hence precautions are necessary to prevent the escape of any appreciable amount into the atmosphere of the laboratory. In the South Kensington laboratory it is generated by the action of hydrochloric acid on sulphide of iron, and is collected in a gas-holder standing in oil, the whole process being conducted in a chamber devoted exclusively to this purpose, and situated in one corner of the building with door and window opening outwards. From the gas-holder the sulphuretted hydrogen is conveyed in a system of distribution pipes to each working place, where a little glass cupboard, to be seen in the illustration, contains a tap from which a stream of the gas may be obtained when required. This glass cupboard is connected with a wide pipe leading into the system of ventilation conduits situated beneath the floor, into which, by the operation of an electrically driven fan in the basement, the whole of the laboratories and other rooms are cleared of noxious gases and supplied with fresh air.

So much for the special accommodation provided for the student individually. But everyone has access also to the further arrangements for common use.

In so large a laboratory it is necessary to consider the distance to be traversed in reaching the balances and larger fume chambers. In order to reduce this as much as possible these are distributed along the two sides so that no student requires to walk further than half the width of the laboratory for such purposes. Spacious fume chambers and doors, leading directly into the two long balance rooms, are to be found alternately the whole length of the room. The fume chambers are glazed on all sides except the back wall against which they are placed. Each is fitted internally with gas and steam pipes, having cocks at suitable intervals, so that all sorts of operations in which fumes are evolved, or evaporation is required, can be conducted. Water taps are also to be found in them, so that condensers used in distillation can be kept cool, and at the back of each chamber is a shallow groove or trough cut in the slate floor of the chamber so as to carry off the water to the drain.

In each balance room, which is 10 feet wide, on a somewhat narrow slab, supported independently of the floor to avoid vibration, is an array of some twenty-five balances. Each balance therefore serves not more and generally less than three students. A chemical balance will be described later on.

In the laboratory itself are also provided two apparatus for the condensation of steam and production of distilled water, which is another indispensable material necessary in all analytical work and in a great many other chemical operations.

In connection with the distilled water apparatus are several copper ovens, heated by the entering steam, which serve to dry any materials placed within.

There are two small rooms connected with the main laboratory in which processes of electrolysis can be carried on, and in which certain delicate operations can be conducted in a comparatively pure atmosphere, free from contamination by dust or gaseous impurities.

The laboratory is lighted by electricity, one lamp being placed over each working place, while clusters of five are hung from the ceiling for the purpose of general illumination.

The average floor space for each worker is nearly 50 square feet. It may be added that the floor is covered with wood

FIG. 7.—IMPERIAL COLLEGE OF SCIENCE AND TECHNOLOGY, LONDON. ADVANCED CHEMICAL LABORATORY.

To face page 26.

Fig. 8.—IMPERIAL COLLEGE OF SCIENCE AND TECHNOLOGY, LONDON.
REFRACTORY MATERIALS DEPARTMENT.

To face page 27.

blocks set in concrete, and immediately beneath and easily accessible are channels in which gas and water pipes, etc., are laid, as well as open troughs by which the waste liquids from the sinks above are delivered into the drain. Obstructions resulting from the accumulation of solid matters in pipes are thus avoided.

From the main laboratory a few feet away across the corridor is the laboratory or rather series of laboratories devoted to physical chemistry. Here the general arrangements of working tables, fume chambers, light and heat, water and drainage are, in all essential particulars, the same as in the large laboratory. But some of the special arrangements deserve notice. One of the first of these is the room for gas analysis. It would be impossible in this place to describe the several forms of gas apparatus employed for purposes of this kind, and the chemical student must be referred to one of the special technical treatises which deal with the subject, which for successful handling requires experience, and accordingly it is usually assigned to a comparatively late stage of the course of instruction.

It is sufficient to say that as mercury is much used, and is expensive, it is necessary to provide for collecting it in case of its being spilt: the floor therefore is smoothly cemented and slopes very slightly to channels which lead into a small reservoir. In the next room is a table specially fitted against the wall with connections to a battery of storage cells, and resistances by which a current of any desired strength or voltage can be used for the purpose of electrolytic deposition of metals or for other purposes to which the electric current is now conveniently applied, especially to oxidation or reduction in liquid media.

Distillation under greatly reduced pressure is a method of purification often applied to substances which cannot be heated to their boiling-points under ordinary atmospheric pressure without suffering decomposition. Accordingly the means of obtaining a pretty good vacuum is frequently needed, and in the physical laboratory is installed a rotary pump, electrically driven, and capable of reducing the pressure in a vessel of moderate size to about 5 millimetres of mercury or less. The pump is connected with a large vessel placed in the area outside so that slight leakage of air through taps or joints may not appreciably disturb the vacuum, which is laid on by means of a narrow lead pipe to several places where it is required in other rooms.

This is an appliance of a different character from the mercury pumps, which are capable of giving a really high vacuum, such as would be necessary in experiments on the electric discharge through gases and for other purposes to be described later in the book.

In the laboratory for organic chemistry the chief modification in the arrangements of the working tables is in the direction of giving more space to each student. Instead of 5 feet from side to side as in the analytical laboratory, 7 feet are allowed, and between each block of work tables there is a narrow operation table, covered with lead, and supplied with gas, water, and steam taps. Here distillations may be conducted as well as any other operation which requires an extended train of apparatus. The operation table has a raised edge so that in the event of accident any liquid spilt will not run to the ground, but may be swept down a central channel to the drain. This is necessary in view of the fact that volatile and inflammable liquids are so frequently used in this class of work.

In this laboratory, the floor of which is about 35 feet from the ground, reduced atmospheric pressure is obtained (for distillations or filtration) by the use of Sprengel water-pumps; the fall pipes consisting of narrow composition gas-pipe, and necessarily more than 30 feet long, are carried down outside the building to the ground, where they discharge into the drain.

The fume chambers in this room are similar to those in the other laboratories, with the addition of cupboards below closed by sliding iron doors. In these spaces are placed the iron boxes in which sealed glass tubes can be safely heated. If an explosion occurs from the bursting of one of these tubes no damage will be done.

Another everyday requirement in the organic chemistry laboratory is a supply of ice and an ice chest. For the former a means of crushing it will be necessary where there are many workers, and one of the machines designed somewhat like a large coffee mill serves the purpose.

The staff of the Chemical Department at South Kensington is as follows :—

1 Professor of General Chemistry.
1 Professor of Organic Chemistry.

FIG. 9.—IMPERIAL COLLEGE OF SCIENCE AND TECHNOLOGY, LONDON. FUEL DEPARTMENT. RESEARCH LABORATORY, SHOWING GAS AND AIR COMPRESSING PLANT AND KNIGHT'S EXPERIMENTAL FURNACE.

To face page 28.

FIG. 10.—IMPERIAL COLLEGE OF SCIENCE AND TECHNOLOGY, LONDON. FUEL DEPARTMENT. EXPERIMENTAL MOND GAS PRODUCER PLANT.

To face page 29.

1 Professor of Physical Chemistry.
1 Assistant Professor.
1 Lecturer in Organic Chemistry.
5 Demonstrators.
4 Assistant Demonstrators.
1 Research Assistant.

There is accommodation for about 300 students, and in normal times the places are full.

CHEMICAL TECHNOLOGY

1 Professor.
1 Associate Professor.
1 Lecturer in Chemical Engineering.
1 Lecturer on Refractory Materials.
2 Demonstrators.

Chemistry is represented in other departments of the college by 1 Assistant Professor in Biochemistry, and by the staff in the department of metallurgy.

II.—THE ROYAL COLLEGE OF SCIENCE FOR IRELAND
FACULTY OF APPLIED CHEMISTRY

The new buildings of the college though quite magnificent are probably less known than many other public edifices in Dublin, owing to the fact that they are at present concealed by some old houses, which, though now utilised as Government offices, are ultimately to be removed and replaced by a new building designed for the same purpose. The college buildings occupy three sides of a quadrangle, which will be completed by the new Government offices. The old college was situated in Stephen's Green, and housed in the building occupied formerly by its forerunner, the Museum of Irish Industry, founded in 1845. The College of Science came into existence in 1867. Under the late Professor Sir Walter Hartley's regime, the chemical division achieved fame as an active centre of spectroscopic investigation. Within more recent years organic chemistry has been prominent under Professor Morgan.

The new buildings are from designs by Sir Aston Webb, the foundation-stone was laid in 1904 by the late King Edward VII, and the college was opened in 1911 by His Majesty King George V. It is therefore one of the newest chemical institutions in existence, and in extent of accommodation approaches closely the laboratories of the Imperial College at South Kensington. The chemical division occupies the upper ground floor of the main building and includes a large lecture theatre, with seating accommodation for about 200 persons, a smaller lecture room, three general laboratories, a laboratory for chemical technology, and a number of smaller laboratories for special purposes and for research.

The general chemical laboratory is a lofty room 30 feet high, and about 70 feet square, with bench accommodation for 120 students. There is another general laboratory for mineral analysis and one for organic chemistry which provides for sixteen students.

As becomes a school of applied chemistry the technological laboratory is fitted with appliances adapted to operations on a larger scale than those which are carried on in the general laboratories. Thus there is a reverberatory furnace with movable experimental hearth, a large drying oven, vacuum stills, filter presses, evaporating pans, wooden vats with mechanical stirrers, and crushing and grinding mills.

The production of the highest temperatures is provided for by the installation of electric furnaces of different types, while low temperature operations are made possible by the presence of a liquid air machine.

There is also provided for the more advanced students a combined lecture and laboratory course in electro-technology, in view of the practical importance to the industrial chemist of a knowledge of the methods of utilising electric energy.

A course of instruction in practical bacteriology also forms a useful addition to the ordinary curriculum. Needless to say research in every direction is practised by the staff and encouraged among the students of the Royal College in Dublin.

III.—THE UNIVERSITY OF HARVARD, U.S.A.

The University of Harvard, at Cambridge, near Boston in Massachusetts, U.S.A., has long boasted a famous chemical school. The present director and Erving Professor of Chemistry, Theodore W. Richards, having in 1915 received the award of the Nobel Prize for Chemistry at the hands of the administrators, the Swedish Academy of Sciences, it may be assumed that his remarkable record of researches in physical chemistry is admired and accepted by the world of science. For some years past plans for the provision of more extensive and more convenient buildings for the accommodation of the division of chemistry have been under contemplation, and a new area, shown in the reduced copy of the plan (Fig. 11), has been appropriated and laid out for this purpose.

The first of a group of eight buildings as shown in the plan

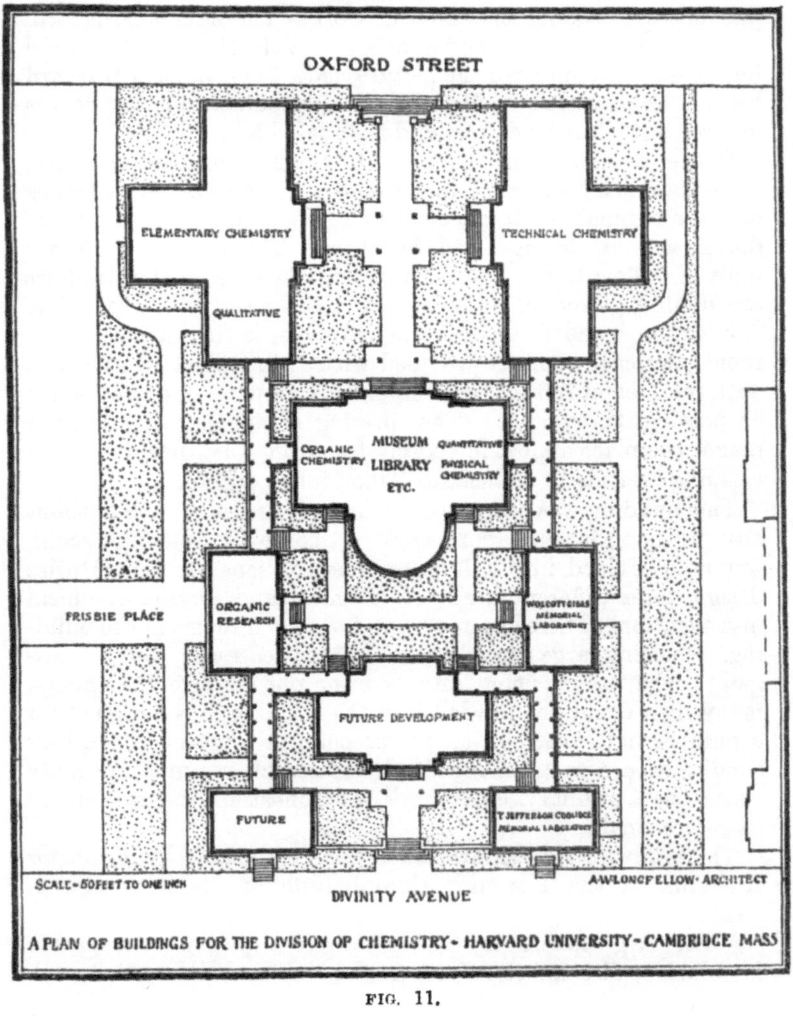

FIG. 11.

was completed in 1913. This is the Wolcott Gibbs Memorial Laboratory, erected in memory of the former Rumford professor of that name, and endowed by a body of subscribers, including more than one name famous in connection with scientific research. Professor Richards' work on the atomic weights of the elements and other of their properties had been carried out under conditions which greatly added to the natural difficulties attending that kind of work. The new building will remove this reproach. Externally the Wolcott Gibbs Memorial building shown in one of the pictures (Fig. 12) is built of Harvard brick with limestone facings and granite foundations. It covers an area of 71 feet by 41 feet and is 48 feet high.

Inside, the laboratory is built of brick and reinforced concrete. Great care has been taken in the construction to avoid materials of an inflammable nature, and there is no wood except for the doors, window frames, and furniture. The outside walls are built of hollow bricks and terra cotta, and most of the windows are doubly glazed in order to save heat as much as possible. Air, warmed and filtered through canvas, is driven into every room, and each room is provided with an auxiliary heating coil with thermostat so that the temperature may be kept as constant as possible throughout. The filtering of all the air admitted makes the place almost incredibly free from dust, but a vacuum cleaning plant is to be installed when funds permit.

This building was constructed wholly for research and without any provision for classroom work. It contains no lecture room, but it is divided into a large number of separate laboratories designed for different kinds of chemical and physico-chemical investigation. There are in all over forty such rooms in the building, one being large enough to contain, if necessary, four investigators, the others being intended for one or two. Although many men could be crowded into the building it is designed for a number not exceeding twelve or perhaps fourteen if the best conditions are to be maintained. Research requires far more space than ordinary class work, and most of the rooms are therefore small.

The following details are taken from an article by Professor Richards in the Harvard Alumni Bulletin for 26th March, 1913:—

"The large laboratory has within it a pier on a separate foundation, disconnected from the rest of the building, for work

Fig. 12.—GIBB'S MEMORIAL LABORATORY. HARVARD.

Fig. 13.—COOLIDGE LABORATORY, HARVARD.

To face page 32.

Fig. 14.—ELEMENTARY CHEMISTRY. HARVARD UNIVERSITY.

To face page 33.

with sensitive instruments; and two similar piers have been placed in the apparatus room in the north-east corner. The latter room is intended not only for the storage of apparatus, but also for accurate measurements; and it is proposed to keep here, ready mounted but suitably protected from dust, various frequently used assemblages of apparatus such as that employed for the measurement of the conductivity of electrolytes.

"The western half of the first floor contains a compact, flexible, and highly convenient arrangement of laboratories in connection with a balance room and dark room. Both balance room and dark room may be used freely by men working either on the southern or the northern side of the building, and the balance room is protected on all sides by other rooms and a passage with glass walls in such a way as to be as free as possible from air currents or changes in temperature. The dark room contains a hood and a solid pier on separate foundations, two rather unusual accessories to such a place, but highly desirable for accurate work in spectrometry.

"On the second floor, at the top of the stairs leading from the floor below, is the library, with an alcove for the librarian and stenographer. This room is connected with the office and study of the director, and also with an adjoining chemical preparation room. The latter leads into a small analytical room for very precise chemical work, and this connects with a physical laboratory and dark room to the east, arranged in connection with a balance room and another laboratory to the north in the convenient manner adopted in the western end of the first floor. Thus more or less space may be placed at the disposal of an assistant. Flexibility in arrangement of this kind has been sought throughout the building; it is especially important in a laboratory designed for a great variety of investigations, where some students may need much space and others little. To provide for yet wider flexibility, some of the partitions between the smaller rooms are arranged in such a way that they could be entirely removed without weakening the building if more large rooms were ever needed.

"The third floor contains a double repetition of the convenient arrangement of the western half of the first floor, and in addition a small common laboratory provided with a still and other general apparatus needed by all the workers.

"All the rooms are supplied with many pipes for various purposes. Steam, illuminating gas, and hot and cold water are everywhere available; distilled water in block tin pipes, compressed air, and vacuum for experimental purpose are to be provided in almost every laboratory. Large pipes for a vacuum cleaning plant (not yet installed for lack of funds) have outlets at convenient places. Yet another pipe for oxygen or any other gas which may be needed throughout the building has been placed in most of the rooms. In addition to these conveniences the walls between the adjoining rooms are pierced in numerous places by porcelain tubes, which provide a convenient means of leading material in pipes or electricity with wires from one room to another. In one corner of the building similar openings through the floors permit the installation of continuous apparatus from the basement to the open air above the roof. This is designed to permit the measurement of pressures by means of a high mercury column or for any other work needing considerable vertical height. Electricity of four kinds is available through many outlets; namely, direct current of 500, 110, and 20 volts, and alternating of 110 volts. An automatic electric lift (provided by a second generous gift of the younger of the original donors, after the building was almost finished) will greatly facilitate the transportation of light apparatus from floor to floor. Space has been left for the installation of a larger passenger elevator, when funds permit.

"The numerous hoods have straight flues of tile pipe, running each one individually straight to the roof without connecting with any other outlet. The chimney-pots on the roof which provide for the emerging air are so arranged by automatic devices that a wind will increase rather than diminish their action. Large porcelain thimbles finish the lower outlets of these flues, which form the only exits for impure air in those rooms thus fitted. The wall spaces back of the hoods are covered with impervious encaustic tile, of a pale warm grey tint, and the same tile is used to form a high dado, reaching five feet above the ground, surrounding all the chemical laboratories. At the top of this dado is fastened everywhere a horizontal strip of wood which gives convenient opportunity for attaching shelves or securing apparatus.

"Some of the desks are covered with tiles like those used for the dado; others are finished in modern lava tops, or thick glass, soapstone, or wood, according to the purpose for which they are intended. The floors are made of various materials; some are of wood, some of painted concrete, and one of rubber tile, while most of them are covered with 'battleship linoleum.' All the laboratory rooms have curved hospital bases where the walls meet the floor, to facilitate cleaning. Especial attention was given throughout the building to the exclusion of dust; weather strips were put on all the windows, and the maintenance of a slight excess of air pressure within the building causes the ordinary leakage to take place outward rather than inward.

"As a place for the prosecution of exact physico-chemical work it will probably offer better conditions than are to be found at any other institution of learning. Its solidity and fire-proof character give promise that it may endure for many years. But even such a building is not an end in itself; it is a means, and its value (other than that due to its memorial character) must depend upon the sort of work done in it. With incompetent workers or inadequate apparatus it would remain insignificant as far as important service to humanity is concerned. Hence the generous subsidies made by the Carnegie Institution of Washington for the provision of apparatus and the securing of assistants are peculiarly helpful, for without these subsidies the work would be greatly hampered. The income from the remainder of the original gift used as endowment provides only for heating and janitor service.

"It may not be amiss to say a word in conclusion concerning the ultimate value to the world of physico-chemical investigation,—a province of research which to some people may appear to be very remote from practical usefulness. Inorganic and organic chemistry are concerned with the study of material substances, analytical chemistry identifies and weighs these substances, and industrial chemistry applies this knowledge to their practical production; but physical chemistry seeks to discover the mechanism of chemical change and the laws which underlie all the other aspects of the subject. Thus physical chemistry is the most fundamental of all the branches of chemistry. It is profoundly interesting and significant considered as a pure science; and its bearings upon the problems of medicine, agriculture, and manufacture are incalculably important. Because the animal and the plant upon which it feeds are both chemical mechanisms, the thorough understanding of the fundamental laws of chemistry is essential for their adequate physiological treatment. Much is already known about the relation of matter to the Protean energy which quickens the universe, but far more remains undiscovered. To a part in this discovery, which gives high hope for the future, the Wolcott Gibbs Memorial Laboratory is dedicated."

The building of which an account has just been given represents a departure from the usual type of university buildings inasmuch as it is dedicated exclusively to research, which can of course be undertaken only by those who have passed through the introductory studies leading to graduation.

In such establishments as those at South Kensington, London, and at Urbana, Illinois, research is carried on, but it is at these places associated more closely with the teaching which goes on under the same roof. A great deal of discussion on this point has taken place in the past. Undoubtedly it is important to instil into the minds of students that what they are then learning is not the last word that can be said on the subject. They should as early as possible be led to realise that as more knowledge is accumulated new views will arise, and they should be encouraged to enquire into the methods by which this new knowledge is acquired. Accordingly in many institutions researches are carried on not only in the presence of students, but with the active participation of some of them in the work.

The fertility of many of the German laboratories in the production of new results, especially in the department of organic chemistry, is largely due to the system of requiring every student who has got past the preliminary stages of practice, to undertake the manufacture of materials and examination of their properties in connection with a subject which is a mere fragment of a large question in which the professor is himself interested. This is perfectly legitimate and, as far as it goes, is for the benefit of the student, but it does not usually carry him very far as he is chiefly interested in satisfying his teacher and so securing his degree. The best way to arouse a real interest in such work is to let the student undertake a piece of independent research selected, if possible, by himself.

The Gibbs laboratory, however, provides for a very different kind of work which can with difficulty be carried on in the presence or by the aid of ordinary students. Here we may expect to find a body of mature specialists engaged in extending the boundaries of knowledge by methods which require not only experience and knowledge in planning, but great skill in the use of the most refined methods and apparatus requiring suitably adapted conditions.

These workers will chiefly be drawn from the ranks of the alumni of Harvard, but there will probably be no great difficulty in the way for competent workers from other universities who desire to make use of the special facilities provided in the Gibbs building.

A moment's reflection will show, however, that other classes of students must be provided for. The body of students at Harvard needing laboratory accommodation has numbered as many as 783 at one time. On reference to the plan it will be seen that the Gibbs Laboratory is only part of a comprehensive plan of connected but individual buildings, each to be devoted to some special branch of chemistry. The library and museum and other parts needed in common are to be placed in the central building, while the others are connected together by colonnades or cloisters.

The new laboratories will supply places for 950 students, and views are given of two of the buildings. One of these devoted to instruction in elementary chemistry is not yet erected, and is shown in the form of the architect's design. The other, harmonious, though not identical, in appearance with the Gibbs Memorial Laboratory, is the Coolidge Memorial Laboratory (Fig. 13, facing p. 32), given by the Hon. T. Jefferson Coolidge, to provide instruction in physical and electro-chemistry. It is unnecessary to describe these buildings in detail. For the present and until the new buildings contemplated in the scheme are erected, the Coolidge Laboratory is occupied chiefly by students of quantitative analysis, under the direction of Professor Baxter.

It is perhaps appropriate to mention here that the University of Harvard is closely associated with the famous Massachusetts Institute of Technology at Boston, though up to the present the Institute is not an organic part of the University. The departments in which the two institutions specially co-operate are engineering and mining, and the President of Harvard stands at the head of the joint faculty.

IV.—THE UNIVERSITY OF ILLINOIS, U.S.A.

When the chemical department of the Imperial College at South Kensington was originally designed it was thought that the laboratories about to be erected were larger than any others then existent, certainly in the British Empire, and probably in the world. But the demand for instruction in chemistry in view of its applications in so many directions and especially in industrial and manufacturing occupations has increased rapidly during the last twenty years. The consequence is that the chemical departments in most of the European universities are crowded with students, and admission of foreigners to some of the most famous laboratories has become increasingly difficult. The pressure of this demand has been recognised in the United States, and new and extensive buildings are being added to several of the universities in that country. On account of its great dimensions a description will now be given of the new chemical laboratory at the University of Illinois, where accommodation is about to be provided for no fewer than 1500 students, or probably four times the number to be found in the University of Berlin. An official paper gives us the following information:—

"When the chemical laboratory of the University of Illinois was built in 1901 there were 238 students in the department of chemistry, and the instruction was cared for by ten persons,—two professors, one associate professor, three instructors, and four assistants.

"During the first week of the year 1914–15 nearly 1500 students have registered in the department. There are now 54 persons in the instructional staff. . . .

"The rapid growth of the department has rendered the chemical laboratory, built in 1901, and already one of the few very large laboratories in America, wholly inadequate for the needs of this large body of students. The large appropriation made for the university by the State during the session of 1913 has made it possible to set aside $250,000 for an addition to the laboratory. . . .

"The cost of the addition will be more than twice the cost of the original building, and a considerable additional appropriation will be required to furnish suitable equipment. When finished the laboratory will be one of the largest and best-equipped laboratories in the world.

FIG. 15.—CHEMICAL LABORATORIES, 1901. ILLINOIS UNIVERSITY.

To face page 38.

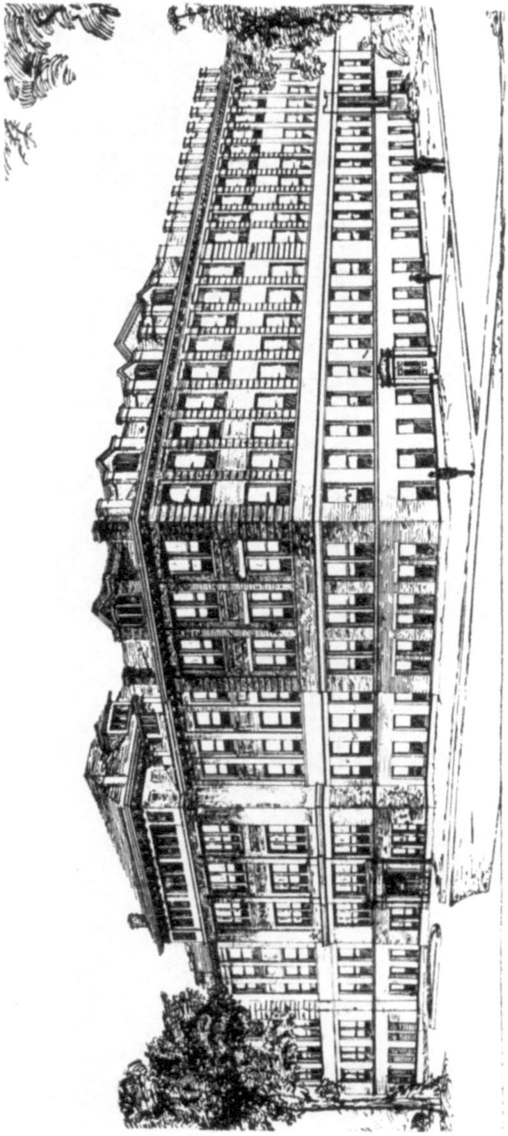

FIG. 16.—NEW CHEMICAL LABORATORIES. UNIVERSITY OF ILLINOIS.

To face page 39.

" The war in Europe has called the attention of the public to our dependence on Germany for many kinds of chemical products. This will undoubtedly prove a great stimulus to many lines of manufacture in which we are now deficient, and this, in turn, will create an increased demand for chemists. The increased facilities which the addition will afford will make it possible for the University of Illinois to do its full share in supplying this demand."

The following description of the chemical laboratory at the University of Illinois is from the pen of Dr. B. S. Hopkins, a member of the staff of the university.

An addition to the chemical laboratory is being built which makes a completed building in the form of a hollow square 231 feet by 202 feet. The centre is occupied by the main lecture room, which is lighted by a skylight ; two large ventilation fans are housed in the court, which arrangement prevents annoyance from noise and vibration. The building is of red brick with sandstone trimmings. The old part is not fire-proof, but is divided into three sections by fire walls, while the new portion of the building is built of fire-proof material. The floors are a combination of reinforced concrete joists and hollow tile, the concrete covering the tile to a depth of 2 inches, thus giving a T effect. Upon the concrete the electrical conducts are laid, and these are covered with a top layer of concrete, rubbed to a smooth surface. The top surface consists of a layer of " rezilite mastic " about $\frac{1}{8}$ of an inch in thickness. This is a preparation of elaterite containing some asbestos fibre, prepared by the Wearcrete Engineering Company of Chicago. It gives elasticity to the floor, and is superior to asphalt because it does not yield under the pressure of heavy furniture. The floors in the halls have the Terrazzo finish.

The roof is constructed of concrete slabs which are covered with wood sheeting, building paper, and slate. The purpose of the wood sheeting is to give an air space for insulation purposes and to furnish a better means of laying the slate. Being entirely covered on all sides by fire-proof material the sheeting does not increase the fire risk.

The minor partitions are made of " pyrobar," a hollow tile made of gypsum. A brick wall separates the old and the new portions of the building, making it possible to completely shut off either side. There is almost no wood used in the new part, the largest amount being for doors and window frames. Some rooms also contain a chain rail and a picture rail.

Abundant hood space[1] is provided in all parts of the building. The hood construction used throughout the building consists of the following: the top is a single frame of plate glass with wire reinforcement, so set as to avoid as far as possible the gathering of dust; the doors are counterpoised, the weights being attached by means of a creasoted hemp rope. The pulleys and axles are made of wood. Each hood is connected with a flue which runs to the top of the building independently of other hoods. Ventilation is by forced draft. In the larger laboratories the air is brought in at the corners of the room, and distributed at various openings along the ceiling. The foul air is forced out through the hood flues. In this way there is an even distribution of the fresh air in all parts of the room and the air is changed six times per hour. In the laboratories for elementary chemistry there are upon the student desks special ventilation conduits which are connected with exhaust fans capable of changing the air in these laboratories eleven times per hour. The toilet rooms are provided with special exhaust fans giving very thorough ventilation.

Alberene stone is used for window-sills, shelves, and table-tops. All tables are supplied with gas, water, waste, and suction; some also have air blast, high-pressure steam, distilled water, and hydrogen sulphide. The building is completely wired with five electric systems: 10, 110, and 220 volt direct current, and 110 and 220 volt alternate current. Many laboratories are also connected with the storage battery system. In the attic a large water distillation apparatus is placed as well as a hydrogen sulphide generator. The hydrogen sulphide is stored in a 500-gallon gas tank, which permits the supply of the gas to various parts of the building under constant pressure.

Besides the main lecture room, which seats 390, there is one smaller lecture room; 7 recitation rooms and 4 seminar rooms are also provided.

Fire-proof vaults and an elevator are available from all floors.

The general plan of space distribution is to have most of the offices, the library, and the research laboratories in the new part, with the routine laboratories in the old; the office of the director and the general executive offices of the department

[1] This is understood to refer to chambers for carrying off fumes.

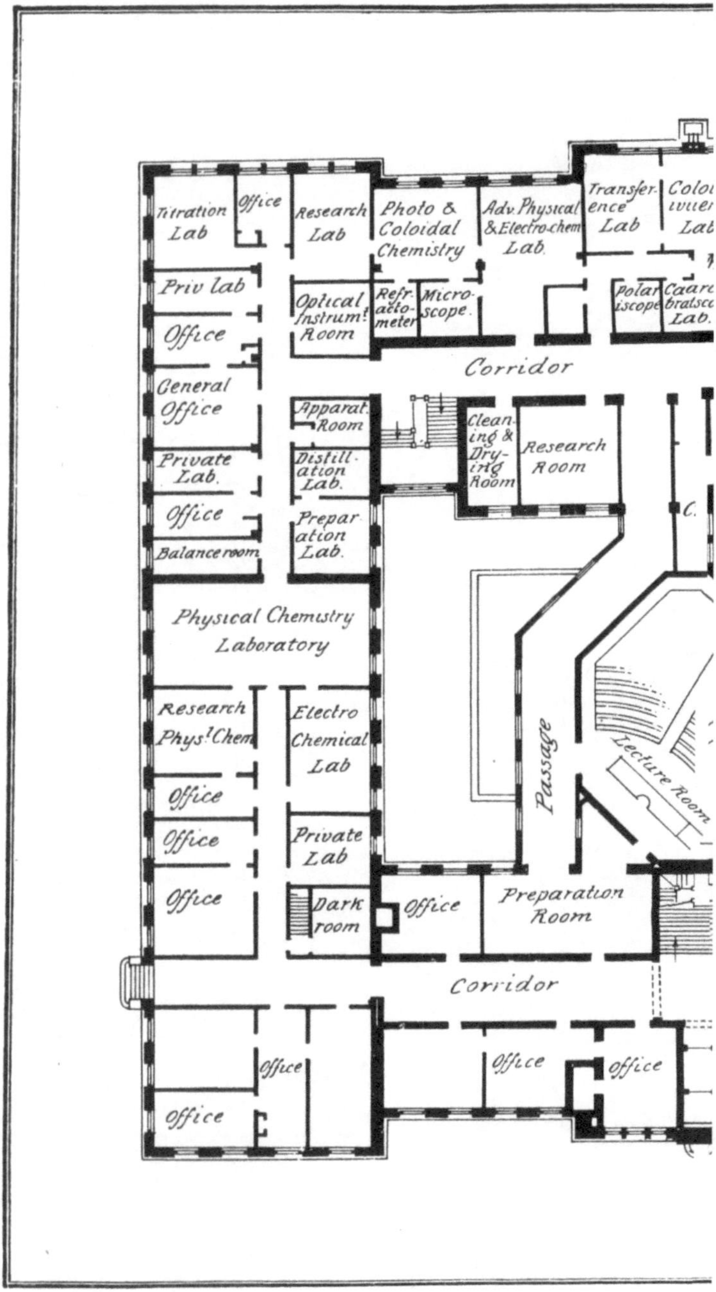

Fig. 17.—ILLINC

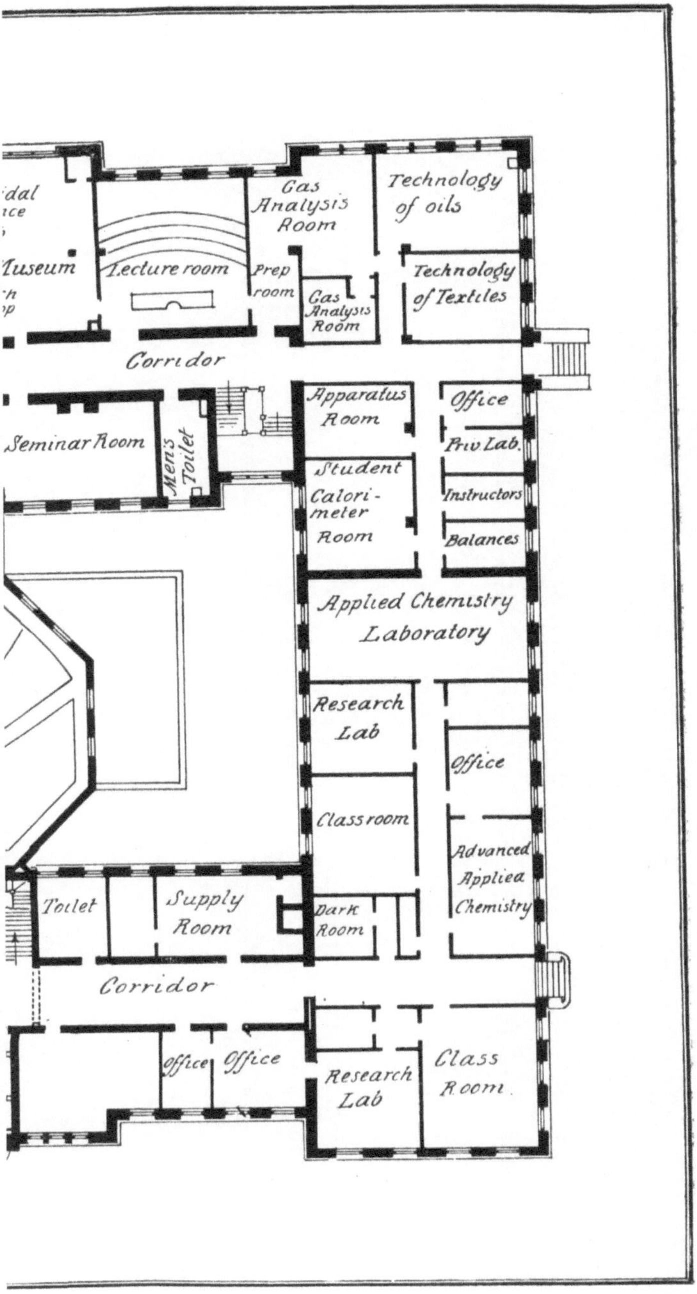

)IS UNIVERSITY.

Between pages 40 and 41.

remain in the old portion of the building at the main entrance. The large stock room and the ventilation equipment are in the old part. There are five entire floors available for laboratory purposes in the new addition.

THE GROUND FLOOR

The ground floor is divided between the department of *industrial* chemistry and the State Water Survey. The south side is occupied by *industrial* chemistry. The metallurgraphic laboratory is supplied with solid piers giving a firm support for the microscopes, which permits high magnification. The assay laboratory has eight furnaces, burning oil, gas, and coal. The oil is stored outside the building and forced in by compressed air through a constant pressure tank; this arrangement prevents fluctuation of temperature. There are also grinding and sample rooms, a laboratory for wet assay, and a thoroughly equipped dark room.

The State Water Survey has the north end of the ground floor. Besides the offices and routine laboratories there are rooms devoted to research, mineral analysis, sanitary analysis, and bacterial examination. Glass partitions make these rooms especially attractive. There are also two incubator rooms, one for 20° and the other for $37\frac{1}{2}°$.

THE FIRST FLOOR

The division of *physical* chemistry occupies the entire north end of the building, both in the new and old parts. The equipment includes the following: a constant temperature room, insulated with rubber; a titration laboratory with white tile finish; a conductivity laboratory with a sound-proof booth;[1] a polariscope room; a room for ultra-microscopic work; an optical room; and a precision calorimeter room. There are in use a 10-kw. 12-volt motor generator set and a 30-kw. transformer with variable voltage for electric furnace work.

The south half of the first floor is occupied by the division of *industrial* chemistry. There is an insulated room for calorimetric work, where both gas and coal calorimeters are used. Stirring is done by a central motor. A laboratory for advanced courses in gas and fuel analysis; a laboratory for the analysis

[1] This is required in the use of the telephone.

of oils, tars, asphalts, and paint; a paper and textile laboratory and a room set aside for the use of the Chemical Club are at this end of the building.

THE SECOND FLOOR

Quantitative analysis and *food chemistry* occupy the north end of the second floor. There is a total capacity of 400 students, the desks being so arranged that the space occupied by each student covers three lockers. In this way each worker has an abundance of room, and by dividing the students into three sets, each one has a private locker. The Kjeldahl room is equipped with special ventilation[1] and provides for 150 digestions and 50 distillations. The steam-bath is arranged in a terrace with three steps in order to make the back rows more accessible. There is a polariscope room and an electrolytic laboratory.

The south end on the second floor is devoted to *organic* chemistry. The student desks, besides the usual equipment, are supplied with individual steam cones, making steam distillations exceedingly simple. This also saves much gas and avoids much of the risk of fire. A measurement room permits accurate conductivity measurements in a pure atmosphere; the instruments are hung from a solid wall. A fire shower is also provided for use in case a student's clothing catches fire.

The library and reading-room occupy the centre of the east front on the second floor.

THE THIRD FLOOR

The entire third floor, with the exception of a few rooms in the north-east corner, is used by the division of *inorganic* chemistry and *qualitative analysis*. There are 980 lockers for individual students, two lockers under the working space of each student. Special ventilation in the new laboratories is provided by conduits which rise a few inches above the level of the table-top. Gases are drawn downward, then through main conduits to the attic, where they are discharged by suction fans into a large flue. Besides gas, water, and waste, the students have easy access to compressed air, suction, distilled water, steam, hydrogen sulphide, and electrical circuits.

A few rooms on this floor are at present used by the depart-

[1] The Kjeldahl process involves the use of fuming sulphuric acid.

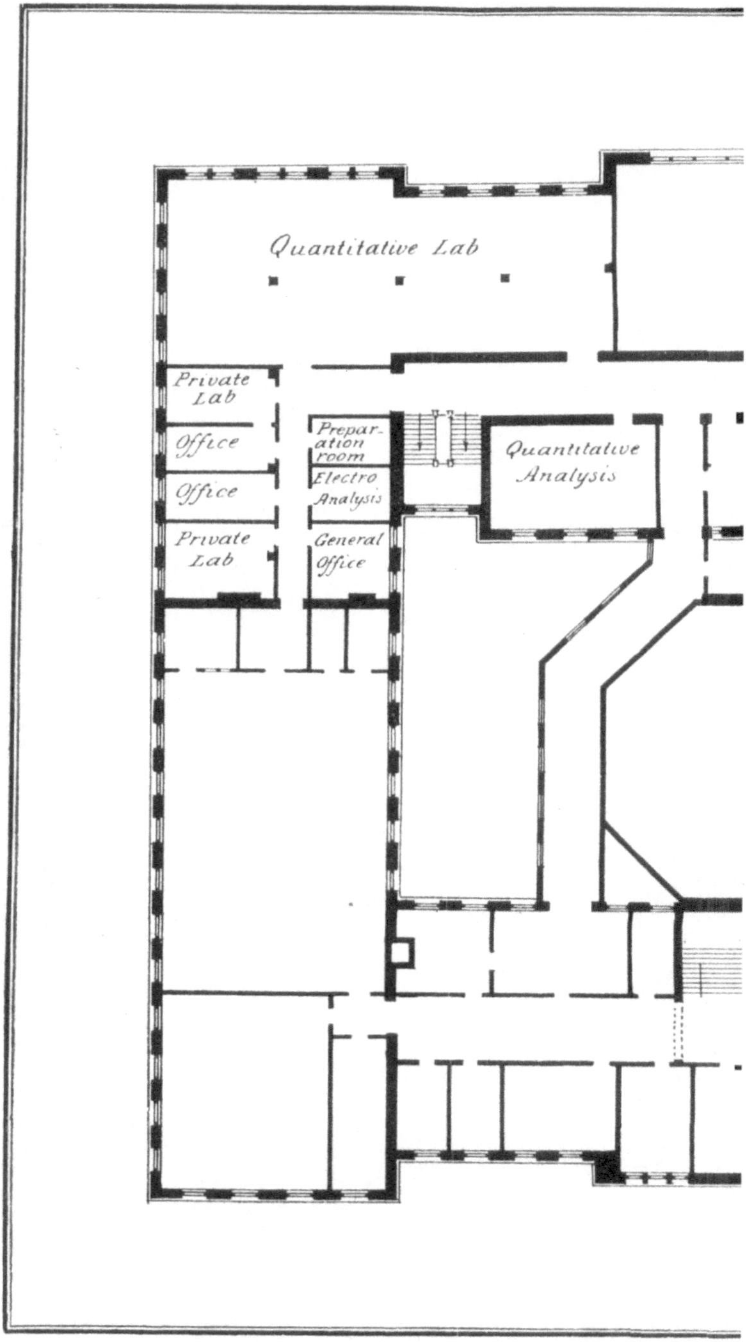
Quantitative Lab
Private Lab
Office
Office
Private Lab
Prepar-ation room
Electro Analysis
General Office
Quantitative Analysis

FIG. 18.—ILLIN(

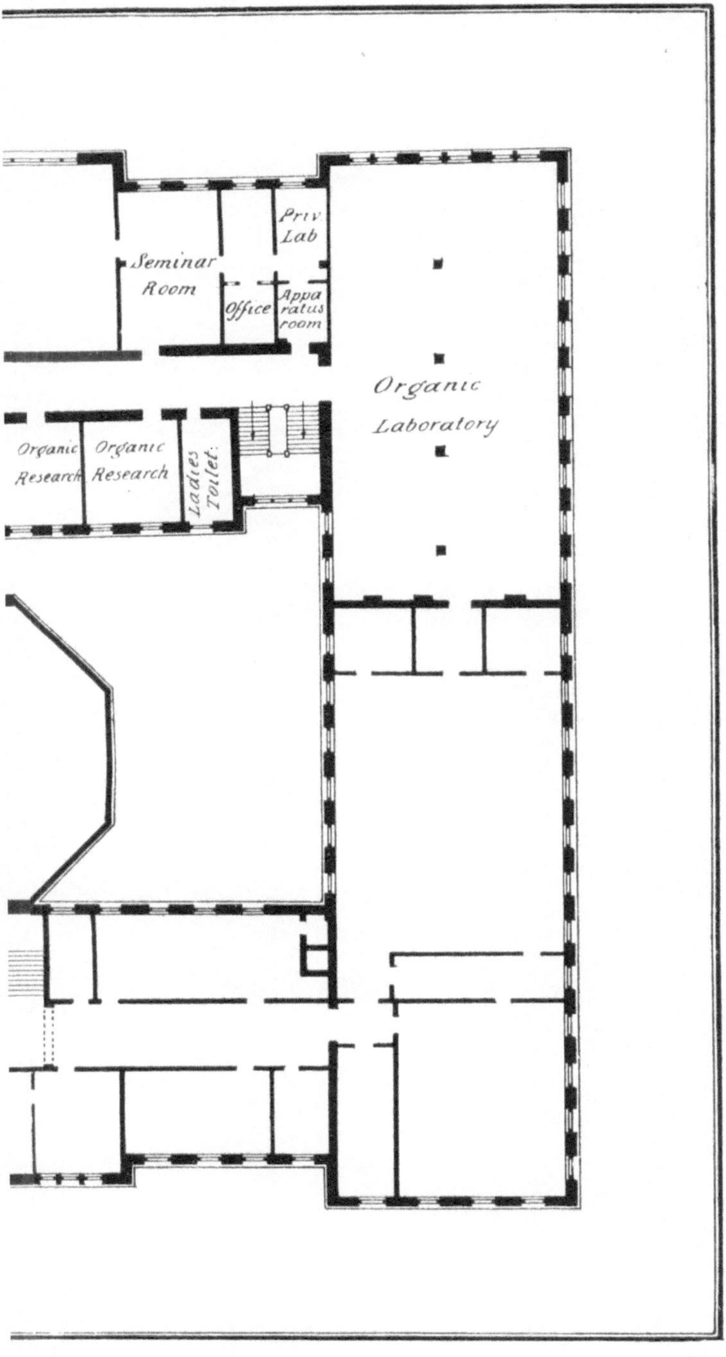

)IS UNIVERSITY.

Between pages 42 and 43.

ment of bacteriology and the State Drug Inspector. These rooms will eventually be used by the classes in elementary chemistry.

THE FOURTH FLOOR

Only the new portion of the building is used upon the fourth floor. By the use of skylights and dormer windows very pleasant and satisfactory rooms are made available. One end of this floor is equipped for use by classes in *inorganic* chemistry and *qualitative analysis*. The equipment in this end is similar to that used on the third floor, accommodation being provided for about 300 additional students.

The main portion of the fourth floor is occupied by the division of *physiological* chemistry. The large laboratory has desks for 108 students, with equipment similar to that used in the organic laboratory. There is a research laboratory for 16 students, and two smaller private research laboratories. Offices, operating room, metabolism room, and Kjeldahl room complete this suite.

The work of the chemical department at the University of Illinois is organised under seven divisions with the following staff:—

GENERAL CHEMISTRY AND QUALITATIVE ANALYSIS

1 Professor	3 Instructors
1 Assistant Professor	8 Assistants
1 Associate	19 Graduate Assistants

QUALITATIVE ANALYSIS AND FOOD CHEMISTRY

1 Assistant Professor	1 Instructor
1 Associate	7 Assistants

ORGANIC CHEMISTRY

1 Professor	3 Assistants
1 Assistant Professor	1 Graduate Assistant
1 Instructor	

PHYSICAL CHEMISTRY AND ELECTRO-CHEMISTRY

1 Professor	1 Instructor
1 Associate	1 Assistant

PHYSIOLOGICAL CHEMISTRY

1 Associate	1 Assistant

INDUSTRIAL CHEMISTRY

1 Professor	1 Instructor
1 Assistant Professor	1 Assistant

WATER ANALYSIS AND SANITARY CHEMISTRY

1 Professor	1 Instructor

In addition to the above there are 2 research assistants, 6 fellows, 1 graduate scholar, 1 glass-blower, 1 mechanician, 1 clerk, 2 stenographers, 1 lecture assistant, 4 storekeepers and laboratory helpers.

V.—UNIVERSITY OF SYDNEY, AUSTRALIA

The laboratories of the University of Sydney have been chosen for illustration partly to show what has been done in a distant part of the British Empire, but also on account of the close association of chemistry with mineralogy and metallurgy, subjects in which Professor Liversidge who designed them has always been specially interested. The need for proper accommodation had been felt for many years, but the use of temporary arrangements had led to procrastination, and it was not till 1889 that the first buildings were erected. Since that date it has been necessary to add to them, to accommodate the greatly increased number of students in metallurgy and assaying.

The accompanying plan shows the general arrangement of the rooms, which are contained in a building of one storey only above ground, part of the metallurgical stores and other rooms occupying the basement beneath the chemical lecture rooms and laboratories. The slope of the ground allows this without inconvenience, as may be seen by reference to the view of the exterior. The buildings have no great architectural pretensions, but are conveniently situated near to the physical, engineering, and biological laboratories, a grouping which is convenient and prevents loss of time to students in passing from one department to another.

The buildings are constructed of brick and cement, external appearance being to some extent sacrificed with a view to economy.

The main chemical laboratory is a room 72 feet by 36 feet, with height of 22 feet in the central part. In this room all, except mere beginners (principally medical) and research

Fig. 19.—CHEMICAL AND METALLURGICAL LABORATORIES. SYDNEY UNIVERSITY.

To face page 44.

FIG. 20.

To face page 45.

students, are accommodated. At first it was intended to provide separate rooms for qualitative and quantitative work, but the advantage to the students of being associated with various kinds of work going on simultaneously, and the greater facility in superintendence and instruction led to the decision to throw the two rooms into one.

The view of the main laboratory (Fig. 21, facing p. 46) shows that it is a cheerful and well-lighted apartment, the working benches being provided with the usual supplies of gas, water, and reagent bottles, together with a very convenient small glass chamber for carrying off surplus sulphuretted hydrogen which is brought to each bench from a gas-holder out of doors. These glass chambers are connected with a draught flue, and are sufficiently large to allow sufficient space for boiling or evaporation of liquids which give off noxious vapours. Each student has a table space of 5 feet from side to side, and the tables provide for four students to whom the glass fume chamber in the middle is common.

The principal lecture room is 34 feet by 47 feet and is 22 feet high in the central part. The seating is arranged for 180 students, but if necessary a larger number can be accommodated. The principal entrance is from the corridor, but in case it should be necessary to empty the room quickly, a door on the opposite side leads directly into the open air. The lecture table is furnished with the usual appliances for demonstration, and the windows are fitted with convenient black blinds by which the room can be rendered dark when necessary.

The plan shows that in arranging the space the requirements of research have not been forgotten and special rooms are provided for balances, spectroscopic and polariscopic work, as well as for microscopes, which of late years have come more and more into use in connection with the examination of metals.

The assay laboratory, of which a view is given (Fig. 22, facing p. 46), is a lofty room with 40 feet by 54 feet floor space, containing twenty fusion furnaces and twelve muffles arranged down the middle of the room. The flues are carried beneath the floor to a central stack which appears in the view given of the exterior. There are twelve fusion furnaces and four muffles, in addition, in another room in the main building. The working benches are fitted with drawers, draught cupboards, gas, water, and exhaust pumps nearly like those in the chemical laboratory.

A feature in the assay laboratory, as in the chemical laboratory, is the connection of the water baths and drying ovens with supply cisterns fitted with ball-taps by which they are kept full of water, and there is no danger of running dry or unnecessary waste of water.

It will be gathered from this brief account that the joint chemical and metallurgical department of the University of Sydney is a compact and efficient establishment not comparable as to size with some of the laboratories in other countries, but serving for the present the needs of the colony, though in all probability large additions will be necessary before long. About 200 students work in the laboratories; apart from those who attend lectures only.

VI.—CHEMICAL LABORATORIES OF THE FEDERAL POLYTECHNIC, ZÜRICH

The Zürich Polytechnic owes its fame as a school of chemistry to a succession of distinguished teachers, among whom may be mentioned Johannes Wislicenus, Victor Meyer, Arthur Hantzsch, Richard Willstätter, and Georg Lunge. Their work in pure and applied chemistry, of which a large part was done in the laboratories in Zürich, and before their removal to other universities, is among the most brilliant to be found in the literature of chemistry.

The chemical laboratories and lecture rooms occupy the whole of one large building divided into two wings, which are devoted respectively to general and analytical chemistry on the one side, and technical or applied chemistry on the other. The former division need not be described in any detail inasmuch as the arrangements and fittings do not differ in essential particulars from those of the other large laboratories elsewhere, some of which have already been illustrated in previous pages. The laboratories for applied chemistry are under the direction of a separate professor, and are arranged with the idea of providing not only for purely scientific and analytical operations, but for manufacture, though of course on a reduced scale, of many products of industrial chemistry. The students, as a rule, are required to devote their first year chiefly to analytical work, but in the second year they are occupied exclusively in the production of chemical compounds by processes of technical interest. Similar studies engage them during the third year, while in the fourth they undertake research on some subject

Fig. 21.—MAIN LABORATORY. SYDNEY UNIVERSITY.

Fig. 22.—ASSAY ROOM. SYDNEY UNIVERSITY.

To face page 46.

Fig. 23.—LARGER LABORATORY FOR APPLIED CHEMISTRY. ZÜRICH.

To face page 47.

usually selected for them by the teacher. This is practically the system prevalent in the German universities.

The laboratory shown in the illustration (Fig. 23) contains about sixty working benches. There is another laboratory of about the same capacity on the same floor, and below it a large room, fitted with stone benches, where operations on a larger scale can be carried on which are unsuitable for the laboratories proper. Here the worker finds close at hand the various mills for breaking up and grinding hard substances, such as minerals or the products of furnace operations, with the machinery for driving the mills. There are also compression pumps for gases, shaking machines, hydraulic presses, centrifugal machines for separating solids from liquids and drying the separated solid, beside filter presses and drying ovens. There is also close by a furnace room where operations may be carried on at various temperatures up to the highest.

The two laboratories on the floor above are associated with the usual arrangement of balance rooms, dark room for work with the spectroscope, polariscope, and photometer, and with a library of chemical books. There is also a long gallery of communication open to the sky, but supplied with water, gas, and stone tables which provides conveniently for many operations in which stinking or poisonous gases are evolved.

Large lecture rooms, each with seats for 160 persons, are on the top floor.

CHAPTER II

LABORATORIES FOR SPECIAL PURPOSES

I.—THE BRITISH SCHOOL OF BREWING, UNIVERSITY OF BIRMINGHAM

The British School of Malting and Brewing, now a department of the University of Birmingham, was founded in the year 1899 in connection with the Mason University College, from which sprang the existing university.

Possibly some readers may be disposed to ask why such a business as making beer should form part of a university, or why a laboratory and a professor are called for in connection with such a long-established industry. The answer to such a question can only be supplied by a review of the history of the matter during these later times.

It is not necessary in this place to peer into the mists of antiquity with the idea of discovering the origin of the production of an intoxicating drink from barley, in those countries in which the vine was unknown. But it is within the memory of many persons now living, that home-brewed ale was to be found in many country houses, and the practice of producing the beverage at home was one of very long standing, which history tells us may be found recorded in the annals of many centuries. One of the characteristics of the domestic product fifty or sixty years ago was the uncertainty of its quality, and this arose from the state of total ignorance, in which even the experienced brewer then carried on his operations, as to the nature of the process in which he was engaged. It was known that a solution of sugar mixed with yeast undergoes a change, in the course of which it loses its sweetness and, if strong enough, yields a liquid which has an alcoholic flavour and intoxicating properties, but the true nature of yeast was not understood, and it was thought to be merely a form of very unstable albuminous matter. And this idea was supported to some extent by the known fact that a solution of sugar containing vegetable albuminous matter, as in the case of fruit juices of all kinds, enters into fermentation apparently spontaneously and without the recognised addition of yeast. The state of error was also encouraged by the introduction of a peculiar theory of fermentation on the authority of Liebig, at the time referred to at the height of his fame. To cut a long story short it may be at once pointed out that by the researches of Pasteur, published from about 1857 to 1861, the facts were established, and the vitalistic doctrine of fermentation placed on a secure foundation. That the yeast which develops spontaneously in a fermenting fruit juice, or which is added to the infusion of malt or wort in the process of brewing beer, is a mass of living matter and that its growth and development are in some way directly connected with the destruction of the sugar in the wort, and the production of alcohol, was the outcome of Pasteur's experimental studies, of which an account is given in his famous book *Études sur la Bière*. But since Pasteur's time many steps have been taken which, while superseding some of his conclusions, detract in no way from the merit of the important pioneering work which he accomplished.

Ordinary yeast (*Saccharomyces cerevisiæ*) when seen under

the microscope presents the appearance of nearly spherical cells containing partially granular substances. Immersed in a suitable fluid at the temperature of 60° to 70° Fah., some of these cells will quickly proceed to multiply by budding, which is the common mode of reproduction. The result is the formation of long strings of cells resembling the parent cell. Under other circumstances yeast is formed by generating within itself minute granules or spores which, escaping from the parent cell when the latter bursts, grow into individual yeast cells of the ordinary form.

It has been found by modern researches that the cell wall of the yeast organism is not necessary to the process, for the liquid contents of the cells when added to a fermentable liquid are capable of setting up true alcoholic fermentation without any process of growth or multiplication. In fact the decomposition of the sugar has been traced to the presence in the liquid of a soluble unorganised nitrogenous substance of the class of compounds called " enzymes " (q.v.), to which the name *zymase* has been given. It appears, however, that this zymase *alone* is incapable of producing fermentation of sugar, but is dependent on the presence of another substance, the exact nature and composition of which is unknown, but which differs from zymase in not being rendered inactive by the temperature of boiling water. Fermentation is also dependent, as was found out by Pasteur, on the presence in the liquid of a small quantity of a phosphate, which has more recently been found to associate itself in a remarkable way with the sugar to form a compound, from which both the sugar and the phosphate are again regenerated. The process of fermentation is therefore essentially a chemical process resulting from a succession of changes in which the complex organic materials are supplied by the yeast cell, and are not as yet producible by purely chemical means.

Pasteur's classical researches, however, led to another very important practical conclusion. The uninstructed reader may perhaps be led to think of yeast as though it consisted of one kind of substance or organism, and in well-conducted operations in skilled hands the examination of a sample of yeast under the microscope might substantially confirm that idea. But part of Pasteur's work consisted in showing that there are many varieties of yeast, and that each one has properties of its own, and further that many other organisms, present in the air or water or brewing materials, are capable of seriously modifying the process of fermentation.

The study of the various forms of yeast and their associated organisms led to the modern views as to the cause, progress, and means of communication of many forms of disease, and laid the foundation of the modern science of bacteriology. Moreover, great practical advantages have arisen in enabling the brewer to control the operations carried on in the brewery, which are no longer empirical as in the time of little more than a generation ago, but completely under the direction of well-known scientific principles.

As evidence of the importance attached by the practical man to the results of Pasteur's work, mention may be made of the establishment of the laboratory at Carlsberg, founded and endowed in 1875, by the late J. C. Jacobsen, a well-known Danish brewer. In this institution the investigations carried out by Dr. E. C. Hansen led to the adoption of a system of pure yeast culture, which has found its way into many continental breweries. After trial in each brewery to find out which variety of yeast suits best the local conditions and furnishes most satisfactorily the quality of beer required, the process begins in the laboratory by isolating a single cell, under proper conditions to avoid contamination, and from this by allowing it to produce other cells by budding, a larger number of parent cells are generated, and these again in successive generations produce a sufficient crop for practical use in the brewery, where further cultivation can be carried on and a sufficient quantity of yeast obtained for any quantity of wort.

The use of a brewing school therefore is to supply instruction first of all in the scientific principles of physics, chemistry, and biology sufficient to enable the student to understand the application of this kind of knowledge to the problems connected with fermentation. It has to be remembered that alcoholic fermentation is not the only kind of chemical change to which saccharine solutions are liable in the presence of micro-organisms. The student must learn to recognise and distinguish those which produce acetic, lactic, and other acids, and those which give rise to other morbid conditions of beer, such as turbidity or disagreeable flavours. He must also make himself familiar with all the different varieties of carbohydrate, the sugars, starches, dextrins, which occur in brewing materials. He must know the influence exerted on the product by the composition of the water supplied to the brewery. He ought to be familiar with all the operations of water analysis, and to understand how to modify the quality of the water when found to be unsuitable. The use of antiseptics for destroying moulds and bacteria and so preventing infection from lurking impurities in brewing plant or barrels has long been recognised. The sulphites are the most commonly used agents, but many others such as salicylic acid have come into use in more recent times. In the brewing school the student will have opportunity of comparing them as to efficiency, and of learning when their employment may be regarded as useful and legitimate.

The brewing school in the University of Birmingham is the only one of its kind in Great Britain, and it has been endowed almost entirely by the brewers of Birmingham and its neighbourhood. The buildings are in the city itself and are independent of the new chief buildings of the University which is situated nearly two miles away. The accommodation provided includes a main laboratory with places for more than twenty students, and well equipped with special apparatus suitable for the study of general brewing and bacteriological subjects; an analytical laboratory adapted more especially to the examination of brewing materials; a microscope room and library; a balance room; a dark room for polariscopic and photographic purposes; a research laboratory and a lecture room. The courses of study in the School are to a large extent, but not entirely, technical in character, and on the purely scientific side include biochemistry in its relation to fermentation. A Diploma in Brewing is granted by the University to students who pass successfully through studies covering a period of three years, while a Degree course in the Biochemistry of Fermentation is provided for students who contemplate taking up biological or chemical work connected with the fermentation industries, agriculture, water supply, sewage treatment, etc.

Professor Adrian J. Brown, F.R.S., has been Director of the School since its foundation, and his practical experience for many years in the brewery of Messrs. T. Salt and Co., at Burton-on-Trent, added to his previous scientific training and reputation based on his published researches, are ample demonstration of his eminent fitness for the post. Much of Professor Brown's work is rather too technical to be described here, but two important results may be made intelligible. The first published in 1892 related to the question long under discussion as to the influence of the presence of air on fermentation by yeast. At that time Pasteur's view was predominant, namely, that yeast, as an organism capable of living either in contact with air (aerobic) or without it (anaerobic), behaves actively as an alcoholic ferment only in the absence of free oxygen. Brown showed, however, that Pasteur's view was untenable and that the presence of oxygen stimulates rather than retards the fermentive activity of yeast.

In 1912 he discovered that the seeds of the barley and other plants of the *Graminaceæ* are covered with a membrane which has the remarkable power, when immersed in various solutions, of permitting the entry into the seed of some substances, while excluding others.

Thus the membrane completely excludes sulphuric acid, and caustic soda so long as the membrane is unaltered, while it allows iodine to pass in slowly, as indicated by the deep blue colour assumed by the starch grains within. This property of selective permeability does not appear to be a function of living protoplasm, as it is exhibited after the seeds have been immersed in boiling water, and it is evidently not a case of ordinary liquid diffusion by which crystalloids are separated from colloids.

These results bear upon not only the question of what happens when barley is steeped in water as in the process precedent to germination, but raise further questions in connection with the already much-debated theories of solution.

II.—THE MUNICIPAL SCHOOL OF TECHNOLOGY, MANCHESTER

The Municipal School of Technology, like so many other educational institutions which within the last generation have risen into a position of prominence, arose out of a comparatively humble origin.

The earliest of the Mechanics' Institutes, which owed their existence to the enlightened and far-seeing philanthropy of such men as Dr. Birkbeck and Lord Brougham, were established in London about 1820. During the greater part of the nineteenth century these Institutes provided, by means chiefly of evening classes, almost the only opportunity for those who felt the disadvantages of ignorance, to make up for the deficiencies in their early training. Nowhere in this country did the establishment of Institutes of this kind proceed more rapidly and successfully

FIG. 24.—MICROSCOPE ROOM. BREWING DEPARTMENT. UNIVERSITY OF BIRMINGHAM.

FIG. 25.—BREWING DEPARTMENT. UNIVERSITY OF BIRMINGHAM.

To face page 52.

FIG. 26.—MUNICIPAL SCHOOL OF TECHNOLOGY, MANCHESTER.

To face page 53

than in the manufacturing districts of Lancashire and Yorkshire. In 1824 the Manchester Mechanics' Institute was founded, and proved the pioneer of many important movements. The Owens College, opened in Manchester in 1851, has grown into the Victoria University of Manchester.

During the period, approaching a century, which has elapsed since its foundation, the Mechanics' Institute has developed into an establishment which, in its influence on the industry of the district, in the number and successes of its students, and the high range and character of the instruction given within its walls, has risen into a position comparable with that of the University itself. This is a phenomenon not peculiar to this country, but happily in Manchester any question of rivalry, such as is said to prevail, for instance, between the Technical High School at Charlottenburg and the Friedrich-Wilhelm University of Berlin, has been avoided. With the aid largely of local benefactions, but also with assistance of Government funds, new buildings were erected and opened in 1902, the City Council having already decided to change the name of the Technical School to that of "The Municipal School of Technology," and to place at the head of each of the more important departments a highly qualified professor or director. A Faculty of Technology in the University was established in 1905, and the Principal of the School of Technology was appointed Dean of the Faculty, while the heads of several departments in the School became Professors of the University with seats on the Senate.

The courses of instruction provided by the School of Technology lead to the degrees of Bachelor and Master of Technical Science (B.Sc., Tech. and M.Sc., Tech.). Such courses are necessarily controlled by the Senate of the University, but the Education Committee of the City Council has also a voice in the matter. The remaining work relates to the part-time classes for evening students and others whose ordinary avocations occupy the greater part of their time. These are controlled by the City Council alone, through the agency of the appropriate Boards of Studies.

These general statements are made here because the Manchester School of Technology, in virtue of its association with the University, occupies a unique position, at any rate in England. For though in some cases one department of a technical school is incorporated into the University, while in others the

University has itself initiated and established departments of applied science, no other technical school has at present been incorporated as a whole into the University. At Charlottenburg the degrees,—Dip. Ing. and Dr. Ing., and the corresponding degrees of the Munich Technische Hochschule,—Dip. Ing. and Dr. Tech.,—are given by the Technical Schools themselves independently of the University.

The relations subsisting between the Massachusetts Institute of Technology and the University of Harvard seem to be intermediate between the complete incorporation at Manchester and the independence at Berlin. Students in Engineering and Mining are eligible for degrees both from Harvard and the Institute simultaneously, provided they have satisfied the conditions prescribed by the respective institutions. In brief both professors and students in these departments belong to both institutions at once. Probably in time a closer union will be entered into, though at present the Institute maintains its corporate individuality.

The buildings of the Municipal School of Technology at Manchester are six stories in height, and cover a plot of land 6400 square yards in area. The value of the site, buildings, and equipment is upwards of £370,000. Many departments are provided for, but we can in this place only describe, and that but briefly, the department of Applied Chemistry. The chemical laboratories comprise a laboratory for inorganic chemistry with bench space for 90 students working simultaneously; a laboratory for organic chemistry with bench space for 36 students; a laboratory for physical chemistry and special laboratories for practical work in gas analysis, water analysis, and others for research and technical work in metallurgy, brewing, rubber working, dyeing, bleaching, calico printing, paper-making, photography and its applications to photo-mechanical reproduction and colour.

A view of the dyeing laboratory is shown in the illustration (Fig. 28, facing p. 55). It is fitted with experimental dye baths, jacketed colour pans, and drying cupboards heated by steam, hand printing machines, and other necessary appliances for carrying out experimental and comparative dye and print trials on a small scale. With this laboratory is associated a pattern room and a laboratory for the analytical and research work which relates to colouring matters, to dyeing, and the allied industries.

Chemical technology necessarily requires a knowledge, not

Fig. 27.—MUNICIPAL SCHOOL OF TECHNOLOGY, MANCHESTER.
INORGANIC CHEMISTRY LABORATORY.

To face page 54.

FIG. 28.—MUNICIPAL SCHOOL OF TECHNOLOGY, MANCHESTER.
DYEING LABORATORY.

To face page 55.

only of chemistry, but of those parts of mechanics and physics which are applied in industrial operations. Such subjects as fuel and its uses in the solid, liquid and gaseous forms, steam, gas and oil engines, the dynamo and electric transmission of power, mechanical transmission of power, pumps, especially those which deal with acids and other corrosive liquids, the composition and uses of cements and concrete, the testing mechanically of all kinds of materials such as iron, steel, brass, and bronze, cement, brick, stone, fireclay, and fuel, all these and others must receive careful attention from the chemical student at some point in his career, if he aims at applying his chemistry usefully to industrial purposes. But if he hopes to attain to any position of importance in this direction he will also need some knowledge of various kinds of manufacturing plant, and the best way of laying it out to advantage in the establishment of new works or extension of those already in existence. And if he becomes a manager he cannot neglect the various Factory and Workshops and Public Health Acts, protection against fire and accidents, to say nothing of the regulation of workpeople.

Information on these and many other topics must be supplied in the courses of instruction carried on in any well-directed School of Technology, and all these are therefore to be found in due proportion at Manchester.

III.—THE BERLIN TECHNICAL HIGH SCHOOL, CHARLOTTENBURG

This great institution stands in Charlottenburg facing the main road from Berlin. Until quite recent times the range of education associated with the term " high school " represented in Germany a standard far higher than anything existing in this country. But as already indicated in previous pages we now possess in the Imperial College at South Kensington, the Royal College of Science for Ireland, the Royal Technical College at Glasgow, and the Municipal School of Technology at Manchester, and perhaps one or two other Colleges, institutions which are comparable with Charlottenburg in variety of subjects taught, in the standard aimed at in the teaching, and in the reputation of the professors who are responsible for maintaining that standard. None of these are now places of mere elementary instruction administered to part-time students in the evenings. Those who frequent their classrooms are required to give evidence of the necessary preliminary education, their studies are spread over three, four, or even five years before they can receive their certificates or be admitted to degrees. In these institutions, equally with Charlottenburg, the activities of professors and students are not confined to teaching and learning, the aim is to foster every kind of knowledge which can be turned to practical account in manufactures and industries, and to extend the boundaries of existing knowledge by research. The importance of cultivating this spirit of enquiry into the unknown, the study of new problems, the encouragement of invention, and the exercise of the imagination are now being generally recognised by governing bodies, and receive favour, perhaps not yet liberally enough, from authorities generally.

The technical high school at Charlottenburg is entirely independent of the University, appointing its own professors and teachers, of whom there are about four hundred, arranging its own curricula, and granting its own degrees. The object in view is not education in the general sense, but the cultivation and extension of all kinds of knowledge likely to be of service in manufactures and commerce.

The main building at Charlottenburg is 750 feet long by 300 feet deep, and four stories high. It was inaugurated in November, 1884, by the Kaiser. The chemical laboratories stand in a separate building to the east (Fig. 29, facing p. 60). This building is 215 feet long by rather less than 200 feet deep. It encloses two courts separated from each other by a building which contains two of the principal lecture rooms. In the original description it was stated to be the largest building of the kind in the whole of Germany.

IV.—THE LABORATORY IN THE WORKS

The business of a works laboratory is, naturally, not general, but is concerned exclusively with the operations actually carried on or prospective. Hence it is divisible under two distinct heads, namely, routine testing or analysis merely for the purposes of control, and, on the other hand, research with a view to improvements or developments. As to the routine work again this relates generally to raw materials bought for manufacture, or products of the factory for the market, or to the testing necessary at several stages of a manufacturing process. Another kind of work is that which has to be done in order to comply with some legal requirement, such as the Alkali Act, which applies to the escape of corrosive gases from alkali works, vitriol and glass works. The gases allowed to pass up the chimney with the smoke, and thus to escape into the atmosphere, are required to contain in one cubic foot not more than $\frac{1}{5}$ grain of muriatic acid, and from vitriol works the sulphurous and nitrous gases are not permitted to produce an acidity corresponding to more than 4 grains of anhydrous sulphuric acid per cubic foot before admixture with the air or smoke which passes into the atmosphere. Tests of this kind require to be made either continuously or at frequent intervals.

Another kind of chimney test which is adopted in many works involves the analysis of the gases passing up a flue connected with furnaces, with the object of regulating the combustion. Here samples of the gas are drawn from the flue, and a measured volume is passed through an apparatus where, in successive vessels, the constituents of the gas are absorbed and measured. The carbon dioxide is absorbed by caustic potash, the oxygen by phosphorus, the carbonic oxide by an ammoniacal solution of cuprous chloride, while the nitrogen remains. Other gases would have to be provided for by appropriate reagents.

As an example of the tests which have to be applied in order to check the progress of a manufacturing operation the case of phosphorus in steel may be quoted. When phosphoric ores are used the pig-iron made contains the phosphorus, and to eliminate this in the production of steel from iron of this kind the basic process is employed. When the operation in the Bessemer or other furnace is supposed to be complete a sample of the fluid metal is taken, plunged into water to cool it, and some borings are immediately obtained, these are dissolved in acid, the phosphorus is precipitated in the form of phospho-molybdate, which is collected, dried, and weighed. From the weight of this precipitate, by reference to a table, the amount of phosphorus present in the sample is known. As the furnace is waiting with the charge of molten steel in it, these operations from first to last must be accomplished in a very short time, usually less than half an hour.

Raw materials must be analysed not only because they are commonly bought according to a specification as to quality, but their treatment by the manufacturer depends in many cases on the amount present of some one constituent or the percentage of one or other of several impurities. Thus the soap-maker must know exactly the percentage of soda in the soda-ash or other alkali he buys; the chemical manufacturer who makes ammonia or ammoniacal salts must know the percentage of ammonia in the gas-liquor he obtains from the gasworks; the manufacturer of bleaching powder requires to know the percentage of lime in the material he buys from the lime burner; the dye maker, who starts from benzene, toluene, etc., the hydrocarbons present in coal-tar naphtha, requires to know exactly the percentage of each present in the naphtha obtained from the tar distiller, according as the hydrocarbon is to be converted into aniline, toluidine, etc., the immediate parents of the dye stuff.

The research laboratory, where it is found in connection with chemical works, differs in no important respect from laboratories elsewhere. It contains the usual appliances for weighing, heating, drying, and fittings for reagents, removing noxious fumes and so forth. It must necessarily be provided with the special instruments appropriate to the subjects to be investigated, and should contain or be immediately connected with a good library of works of reference, including some of the principal technical and scientific journals. But in too many cases in the past there has been either no provision for systematic research in the works, or where some attempt has been made it has too often been assumed in English works that any building not required for other purposes is good enough to accommodate the chemist. The respect with which this part of the business is treated in Germany is indicated by the style of building, specially erected in many works, in which industrial research is carried on. An example is shown in the illustration which gives a view of the scientific laboratory, erected in 1901, on the works of Schimmel and Comp., at Miltitz, five miles from Leipzig. This firm manufactures essential oils and perfumes, both natural and artificial.

"A special building contains the research and analytical laboratories, seven large, light, and airy work-rooms, each for two or three chemists; further additional rooms for weighing and combustion, cellars for storing chemicals and glass-ware. This building also contains the collection of drugs and chemical preparations, which includes many objects of ethnological interest, and finally the library with more than 3000 volumes, and some 1000 pamphlets, reprints, and dissertations. Here are found the most important chemical journals, some right from their first issue, and a complete collection of the pharmacopœias of all countries; chemical, botanical, pharmacognostical, medical, and technical cyclopædias," etc.

This firm has issued for many years a semi-annual report which contains not only information of commercial importance, but the results of scientific work which in any way bears on the subject of manufacture, whether proceeding from their own laboratory or from the published works in any part of the world. Such reports doubtless savour of advertisement, and are intended to have that effect. Manufacturers who produce such publications may be trusted to reserve information which they think likely to be of value to rivals in business. Nevertheless the reports give a survey of the whole field which has its value to the world outside.

The laboratories which have been described in the foregoing pages are chosen as representative because it is believed that they represent various types. But it must not be inferred that they are the only large and well-equipped establishments in the world or in the British Empire. During the last thirty years or more a number of new colleges and universities in the British Isles alone have been called into existence, and in 1912 it was estimated that the universities and technical schools together numbered nearly three hundred. If to these are added the very large number of spacious, well-fitted, modern laboratories for instruction in chemistry, independently of physics and other subjects, which are to be found in the great schools of England, there can be no doubt that opportunities are not wanting for those who are in a position to make use of them. Nor in a certain sense is there a lack of support and appreciation of the work, but this comes unfortunately from a comparatively small number of enlightened people who know something of the purposes, aims, and methods of the scientific chemist. There is some reason to hope, however, that the importance of the study of chemistry not only for the sake of its useful applications, but as giving a new view of the natural world, will ultimately be recognised everywhere.

Even in this time of upheaval when the machinery of civilisation is out of gear and the man of science is called away from his peaceful studies to the practice of war, we hear of new laboratories being built for the University of Oxford, for University College, London, and developments elsewhere. As will be shown in the later chapters of the book vast fields are open to research. The colossal pile of knowledge which has been accumulated by the labours of generations of chemists, is still as nothing to the incalculable mass of our ignorance.

V.—THE GOVERNMENT LABORATORY, LONDON

The following account of the history and work of the Government Laboratory is extracted from Reports of the Government Chemist for 1912 and 1914. It contains much interesting information as to the nature of the enquiries which have to be made in such an institution as to the quality of foodstuffs and various beverages supplied to the public, the testing of alcoholic liquids and tobacco for the purpose of levying duty, the investigation of various questions which arise from time to time in relation to public health, such as arsenic in beer and lead in the glazes of pottery and china ware, and the supply of materials for the public services.

The figures given as to the number of samples analysed attest the activity of the Government Chemist and the large staff of skilled workers under him.

A review of the work done in the laboratory also gives a fair idea of the multifarious character of the business undertaken by the Consulting and Analytical Chemist generally, and of the officials known as "Public Analysts" throughout the country.

The origin of the Government Laboratory dates back to 1843, when a laboratory was established at Somerset House in connection with the Inland Revenue Department, mainly for the purpose of checking the adulteration of tobacco. The scope of this laboratory was afterwards extended so as to include the analysis of almost every excisable commodity.

A laboratory was also established at the Custom House in connection with the Customs service for the analysis of articles liable to duty on importation.

In 1894 the two Revenue Laboratories were placed under one Principal, and from that time were known officially as the Government Laboratory. Each branch of the laboratory, however—Excise and Customs—remained, as formerly, subject to the administrative control of the department with which it was immediately connected, and this anomalous state of things continued until the amalgamation of the Customs and Excise Services in 1909.

While the laboratory at Somerset House was established

Fig. 29.—THE CHEMICAL LABORATORIES, CHARLOTTENBURG.

Fig. 30.—SCHIMMEL & CO., MILLITZ. RESEARCH LABORATORY.

To face page 60.

FIG. 31.—GOVERNMENT LABORATORY. CUSTOMS AND EXCISE MAIN LABORATORY.

To face page 61.

primarily to assist the officers of the Excise Department of the Inland Revenue in the performance of their duties, other Government Departments obtained the permission of the Treasury to avail themselves of the services of the " Somerset House " chemists. This larger sphere of usefulness was gradually extended until in course of time the laboratory came to undertake work for nearly every department.

By the Food and Drugs Act of 1875 the chemical officers of the Inland Revenue were made the official referees in disputed cases under the Act, and provision was made for the examination of tea on importation, by persons to be appointed by the Commissioners of Customs. At a later date, the principal of the laboratory was appointed Chief Agricultural Analyst by the Board of Agriculture in connection with the administration of the Fertilisers and Feeding Stuffs Act.

With the growth of the demand on the part of the various Public Departments for advice and assistance in matters involving chemical knowledge, it became apparent that the arrangement under which the staff of the Laboratory was composed entirely of Revenue Officers appointed by and subject to the authority of the Commissioners of Customs and Excise was no longer a satisfactory one. With the view, therefore, of placing all Departments on the same footing as regards the use of the laboratory and of promoting the centralisation, as far as practicable, of government chemical work, the Treasury intimated, towards the end of 1910, its intention of asking Parliament to provide for the expenses of the laboratory under a separate vote as from 1st April, 1911.

The new department thus constituted is known under the official title of " The Department of the Government Chemist."

The change in administrative arrangements was accompanied by important changes in the method of staffing the laboratory.

As has already been stated, all posts on the staff of the laboratory were formerly filled by Revenue Officers, and there was no avenue by which admission could be obtained to the establishment of the laboratory except through the Customs and Excise Service. While it was generally recognised that there were many advantages to be gained by continuing to employ Revenue Officers on Revenue work, no good reason seemed to exist for confining laboratory appointments to Revenue Officers in so far as the non-revenue work was concerned. It was decided, therefore, on reorganising the staff, to continue to utilise the services of the Revenue Officers for purely Revenue work, but to employ chemists recruited from the open market for non-revenue work.

The staff now (1915) consists of: The Government Chemist, the Deputy-Government Chemist, 4 Superintending Analysts, 9 First-Class Analysts, 20 Second-Class Analysts, 35 Temporary Chemical Assistants, 60 Revenue Assistants.

The duties performed by the staff of the Government Laboratory are of a very varied character. They include the analysis of samples in connection with the assessment of Revenue and drawbacks; of stores supplied to Government Departments on tender and on contracts; of dairy produce imported into this country; and of samples referred by Magistrates under the Food and Drugs Act, and by the Board of Agriculture and Fisheries under the Fertilisers and Feeding Stuffs Act. The laboratory staff also deals every year with a large number of questions referred by Government Departments for advice, and conducts investigations in connection with such references and with the enquiries of Royal Commissions and Parliamentary and Departmental Committees.

The chemical work of the following departments and other public bodies is now carried out wholly or in part in the Government Laboratory:—Board of Customs and Excise; Admiralty; Board of Agriculture and Fisheries; Department of Agriculture and Technical Instruction for Ireland; Colonial Office; Crown Agents for the Colonies; Foreign Office; Home Office; India Office; Board of Inland Revenue; Local Government Board; Post Office; Public Record Office; Stationery Office; Board of Trade; Trinity House; War Office; Office of Works, London and Dublin; Geological Survey.

The work of the Excise branch of the laboratory, which was originally conducted at Somerset House, having outgrown the accommodation provided for it there, was transferred in 1897 to the present building in Clement's Inn Passage. The greater part of the analytical work is now carried out in the laboratory in Clement's Inn and in the branch laboratory at the Custom House. There are also twenty-nine provincial stations in different parts of the United Kingdom for testing work for Revenue purposes. Up to the present time there have been separate Customs and Excise stations in some of the principal ports, but

Fig. 32.—GOVERNMENT LABORATORY. ELECTRIC HEATERS FOR WINE AND SPIRIT DISTILLATION.

To face page 62.

FIG. 33.—GOVERNMENT LABORATORY. MUFFLE FURNACES FOR INCINERATION OF TOBACCO

To face page 63.

in consequence of the fusion of the Customs and Excise Departments the separate stations are about to be amalgamated. This concentration of the work will permit of the number of these provincial laboratories being considerably reduced.

The total number of analyses and examinations made in the two branches of the Government Laboratory during the years ended 31st March, 1913, and 1914 were—

——	1914.	1913.
Government Laboratory, Clement's Inn	148,419	126,493
" " Custom House Branch	86,335	83,009
Provincial Chemical Stations	157,613	145,257
Total	392,367	354,759

The work carried out in the Government Laboratory for the Department of Customs and Excise consists mainly in the examination of samples in connection with the assessment of duty and drawback or with the regulations and licences relating to the manufacture and sale of dutiable articles.

The duty on beer brewed in this country is charged on the wort or unfermented saccharine liquid from which beer is produced by fermentation. The specific gravity of 1055 at a temperature of 60° Fah., distilled water at the same temperature being taken as 1000, is adopted as the standard specific gravity of wort. Technically such wort is said to have a gravity of 55°, and the duty is increased or diminished in proportion to the degrees of gravity over or under 55. In practice the beer duty is levied on the basis of entries made by the brewers in official books supplied to them by the Revenue authorities. These entries contain a statement of the quantity of materials—malt, sugar, etc.—to be used, and of the quantity and gravity of the wort produced.

For the purpose of assisting the officers in assessing the duty on beer, and as a check on the operations of the brewers, samples of the malt and other materials used in brewing are sent to this laboratory in order that their wort-producing value may be determined, for purposes of comparison with the quantity actually brought to charge. These brewing materials are also examined for arsenic and other deleterious substances in the interest of the public health.

For the purpose of checking the brewers' declarations as to the gravity of their worts, a large number of samples in various stages of fermentation are sent to the Government Laboratory or its branches in order that the original gravity may be determined.

With a view to the suppression of the practice of diluting beer, which is prohibited by Statute, numerous samples of finished beer taken from the premises of publicans and other licensed retailers are examined at the laboratory.

During the year 1911–12 7464 samples were taken from the premises of 2081 publicans and other retailers. Of these, 452 or 6·1 per cent were found to have been diluted, this being 1 per cent less than last year. In addition, 683 samples of beer as it left the breweries were taken for comparison with the publicans' samples, 114 samples were examined for the presence of saccharine, and six for the suspected addition of sugar and other substances. In no case was any illegal ingredient discovered.

The percentage of spirit was determined in 444 samples of herb beers and other imitation beers, beer substitutes, and temperance beverages. In 300 of these the percentage did not exceed the legal limit of 2 per cent of proof spirit, 107 contained less than 3 per cent, 30 between 3 per cent and 5 per cent, and 6 between 5 per cent and 8 per cent, while 1 sample contained 10·6 per cent of proof spirit.

Samples of beer and of brewing materials are regularly examined at the laboratory for the presence of arsenic. The total number of samples tested in the course of the year, including beer, wort, malt, sugar, and other materials used in brewing, was 1046. It is satisfactory to have to record that only eighteen of these were found to contain arsenic in excess of the limits laid down by the Royal Commission on Arsenical Poisoning, namely, the equivalent of one-hundredth of a grain of arsenious oxide per pound in the case of solids, or per gallon in the case of liquids.

Of 122 samples of malt examined five exceeded the limit, the highest proportion found being one-fortieth of a grain per pound. Eight samples of worts and beers out of 547 examined contained arsenic in excess of the limit, the highest proportion being one-twentieth of a grain per gallon, while of 377 samples of other materials five samples exceeded the limit, the highest being one-fiftieth of a grain per pound.

In all cases in which the proportion of arsenic was above the

Fig. 34.—GOVERNMENT LABORATORY. ELECTRIC FURNACE ROOM FOR STEEL ANALYSIS.

To face page 64.

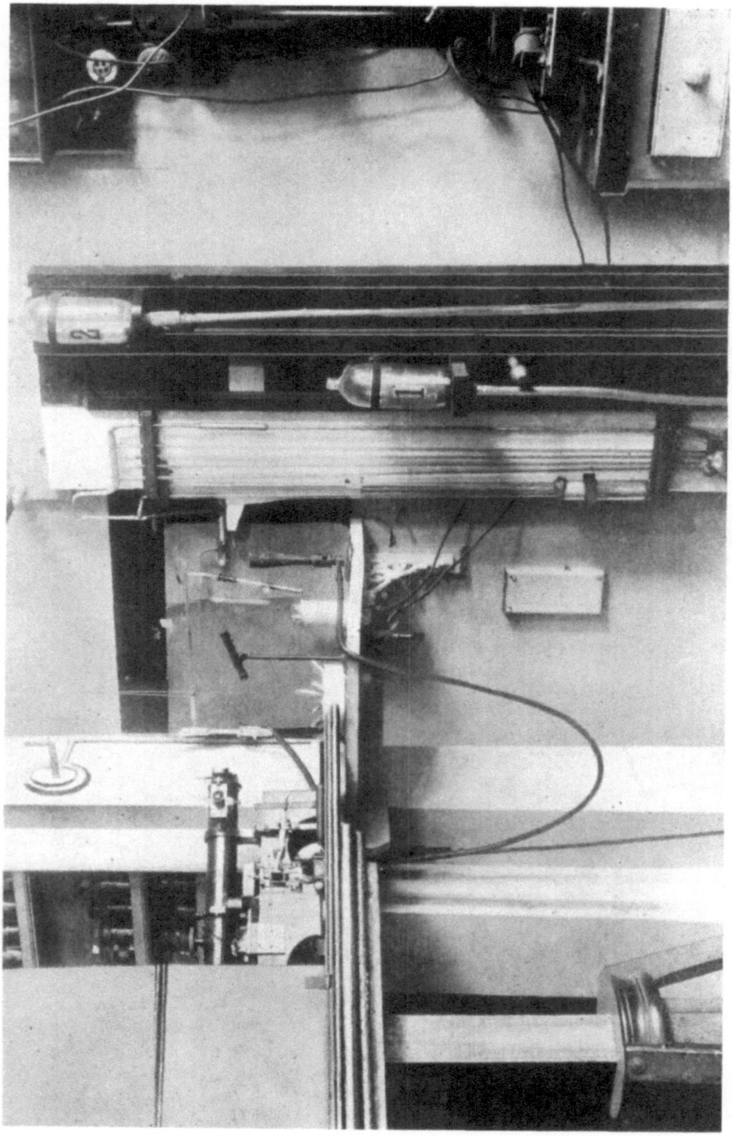

FIG. 35.—SPRENGEL PUMP IN SIR WILLIAM CROOK'S LABORATORY.

To face page 65.

limit, efforts were made to trace its origin and to prevent contamination in future. As in previous years, the arsenic was generally traced to the fuel used in drying the malt on the kiln.

Cider is obtained by fermentation of apple juice. Beverages of a non-alcoholic character are frequently sold as cider or under names such as "Sparkling Cider" and "Champagne Cider," the words "non-alcoholic" or "non-excisable" being occasionally added. Some of these beverages are mixtures of real cider with solutions of sugar and contain less than 2 per cent of proof spirit. But the great majority of the non-alcoholic ciders, so called, are entirely free from fermented apple juice, and are simply solutions of sugar which have been aerated, flavoured, and coloured. The use of any name for such beverages, which suggests that they consist wholly of fermented apple juice or cider, is an infringement of the Merchandise Marks Act, and samples are examined in the laboratory for the Board of Agriculture and Fisheries in the interests of the makers of genuine cider.

Beverages of this class are frequently prepared from liquids or essences supplied by manufacturers who also furnish a recipe for making "cider" from them. During the year, nineteen samples sold as cider were found to be artificially prepared liquids. In one instance the sellers stated that they had prepared the beverage from a liquid supplied to them by a continental firm as concentrated apple juice. A sample of this liquid was found to be a strong solution of sugar flavoured with fruit essence and coloured with aniline dye and to be quite free from apple juice.

The earlier stages of the manufacture of spirits and beer are practically the same, and the methods adopted for checking the distillers' operations during the preparation of the fermented liquor or "wash" from which the spirits are distilled are very similar to those employed in the case of beer.

The duty is, however, charged on the spirits actually produced, and the main object of the preliminary checks on the brewing entries is to ensure that the proper quantity of spirits has been brought to charge, having regard to the quantity of materials used and the quantity of wash distilled.

The duty is not usually paid on spirits immediately on completion of the manufacture or immediately after importation, spirits as a rule being stored for shorter or longer periods in duty free warehouses. During storage soluble matter is abstracted from the wood of the casks, and in the subsequent treatment which is required to convert the spirits into gin and other beverages, small quantities of sugar, and of colouring and flavouring matters are added. In these cases and also when dealing with medicinal spirits, essences, and other preparations distillation or other treatment is necessary since the true alcoholic strength cannot be determined directly by Sikes' Hydrometer, the legal instrument used by the Customs and Excise officers for this purpose. A large number of samples of spirituous preparations must therefore be subjected to a more or less detailed chemical analysis before their true strength can be ascertained. Brandy, rum, and other imported spirits and spirit mixtures require similar treatment before the extent to which their true alcoholic strength is "obscured" by the colouring, flavouring, and other matters present in the spirit, can be determined.

Alcohol of high strength is allowed duty free for scientific and manufacturing purposes, and for domestic uses such as burning in spirit lamps. In most cases the alcohol for these purposes is required to be "denatured" or mixed with nauseous or tell-tale ingredients to prevent its consumption as a beverage or its use in medicines or preparations capable of being taken internally. "Mineralised methylated spirit" which is allowed to be sold retail is ordinary alcohol denatured with a mixture of 10 per cent of wood spirit and three-eighths of one per cent of mineral naphtha. "Industrial methylated spirit" is ordinary alcohol denatured with half this quantity of wood spirit without mineral naphtha. The latter spirit can only be used for manufacturing purposes under regulations approved by the Commissioners of Customs and Excise. Other denaturants are allowed in connection with particular manufactures. The denaturants in all cases must be submitted for approval to the Government Chemist, and samples are taken frequently during the manufacturing processes in which the spirit is employed, as well as samples of the finished articles manufactured, in order to ascertain whether the conditions imposed are being complied with. Beverages and medicinal preparations in which the presence of methylated spirit is suspected are also examined with a view to the prevention of any illegal use of the denatured spirit.

Fusel Oil, a by-product of the manufacture of spirits, consists mainly of alcohols of higher boiling-point than ordinary alcohol. It usually contains some dutiable spirit, but may be delivered by distillers duty-free provided that the amount of proof spirit does not exceed 15 per cent; which proportion of spirit is also allowed duty-free in imported fusel oil.

Wine imported into this country is subject to Customs duty. In connection with the assessment of these duties, it is necessary to determine the alcoholic strength of the wines imported, and for this purpose many samples are annually examined in the laboratory.

During the year ended 31st March, 1914, 89,727 samples of foreign wine were tested as to their alcoholic strength as compared with 102,862 in the previous year. The great majority of the samples consisted of the lighter varieties containing less than 30 per cent of proof spirit. In no case did the strength exceed the 42 per cent limit. Of 110 samples examined as to character, 59 were syrups or fruit juices containing less than 2 per cent of proof spirit, 43 had the character of British wines or sweets, and 8 were foreign wines.

In consequence of regulations made by the Commissioners of Customs and Excise under Section 10 of the Finance Act of 1911 with a view to the restriction of the practice of mixing foreign wines with British wines, there has been an increase for the year in the number of samples of British wines submitted to the laboratory for examination as to character.

Foreign or British wines so medicated or mixed with drugs as to entitle them to be classed as medicines rather than beverages may be sold by registered chemists and druggists without a licence. In order to ascertain whether they were entitled to this exemption forty British and nine foreign wines were examined during the year.

Wine which has become unsound may be delivered free of duty after the addition of sufficient acetic acid or commercial vinegar to preclude its consumption as wine. Seventy-four samples of acid to be used for this purpose were submitted to the laboratory in the course of the year for determination of the quantity necessary for each operation.

Unfermented grape juice is sampled and tested for spirit on importation. Of seventy samples, a chargeable proportion of spirit was found to be present in thirty-five, but these were all from the same consignment.

The work of the Government Laboratory under the head of tobacco consists chiefly in the examination of manufactured tobacco for home consumption and of tobacco, commercial snuff, and offal tobacco for drawback.

Samples of manufactured tobacco are examined for the purpose of controlling the amount of moisture and oil, under regulations based on the provisions of the Customs and Inland Revenue Act 1887, as amended by the Oil in Tobacco Act 1900, and the Finance Act 1904, whereby the amount of moisture which may be present in manufactured tobacco is restricted to 32 per cent, and the amount of oil to 4 per cent. Two kinds of samples are taken, distinguished as " General " and " Special." The various officers, to whom the duty of inspecting the manufacture of tobacco is entrusted, are required to take from each tobacco factory in the United Kingdom one or more samples every week in order to ascertain whether the tobacco conforms to the legal limits as regards moisture and oil. These samples are termed " General " samples. When there is reason to suspect that the limit of moisture or of oil is being exceeded, samples are taken at the factory under such conditions that their identity can be established if necessary in a Court of Justice. The samples so taken are called " Special " samples, and in this category are included also all samples taken from the stocks of retailers of tobacco.

Samples of tobacco are also examined as occasion arises for the presence of ingredients, for example, sugar, liquorice, and colouring matters, the use of which is prohibited by Statute.

In the operations involved in the manufacture of tobacco from the original leaf a considerable quantity of the material is discarded as unsuitable for the preparation of commercial tobacco or commercial snuff. By regulations based on the provisions of the Finance Act of 1904, manufacturers of tobacco are permitted to deposit this refuse or offal tobacco on drawback for abandonment, denaturing, or exportation. For Revenue purposes offal tobacco is classified as follows :—

1. Tobacco Stalks—the midrib of tobacco leaves.

2. Shorts or Smalls—small pieces of tobacco broken off in the process of manufacture.

3. Offal Snuff—consisting either of ground tobacco or of siftings from tobacco sufficiently fine to pass through the meshes of the official standard sieve.

FIG. 36.—GOVERNMENT LABORATORY.
APPARATUS FOR ELECTRO-ANALYSIS.

To face page 68.

FIG. 37.—GOVERNMENT LABORATORY. SPECTROSCOPE ROOM.

To face page 69.

Some of this offal tobacco is employed in the manufacture of sheep dips, fumigating powders, and nicotine for agricultural and horticultural purposes.

The articles examined under this head included "joggery"—a mixture of tobacco and opium with sugar and molasses—used by Asiatics.

The samples examined for assessment of duty on sugar on importation include, besides sugar, articles made with sugar and also those containing glucose, molasses, saccharin, and other sweetening agents. Glucose is largely used for brewing purposes and in confectionery; molasses in the preparation of foods for cattle and in the manufacture of spirit; and saccharin in the manufacture of mineral waters and as a substitute for sugar in foods intended for diabetic subjects.

The number and variety of the preparations containing sugar are so great that it has been necessary to adopt fixed rates of duty in the case of those which are imported frequently or in large quantities. The articles so dealt with comprise biscuits, cakes, catsup, chutney, confectionery, condensed milk, crystallised fruit, desiccated cocoanut, drugs, fruit pulp, infants' and invalids' foods, lozenges, invert sugar, jam, milk powder, pickles, and soy. There are, however, many articles for which it has not been found practicable to fix a special rate of duty, and which have therefore to be tested on each importation and assessed with duty according to the percentage of the dutiable ingredients present. Amongst these may be mentioned egg yolk, gelatine, glue, honey, manna, meat extracts, parchment paper, printers' roller composition, and tanning extracts.

Owing to the heavy duty on saccharin, which has approximately five hundred times the sweetening power of sugar, with a rate of duty in proportion, the inducement to smuggle this article into the country is very great, and numerous ingenious methods devised for this purpose have been detected by the Customs Department. The presence of saccharin has, therefore, to be searched for in all preparations in which there is any probability of its occurrence. Saccharin was discovered in 83 samples specially examined with this object in the year 1911–12.

In order to ensure that only genuine tea shall pass into the country all consignments are examined at the ports by tea inspectors appointed by the Commissioners of Customs and Excise under the provisions of the Sale of Food and Drugs Act,

1875. Doubtful samples requiring a more complete examination than is possible by the inspectors are submitted to the laboratory.

Of the total number of samples submitted, 1104, representing 256,603 pounds, were condemned as containing sand or other foreign matter or as being on other grounds unfit for human consumption. There was, however, no evidence in any case of intentional adulteration. It is to be noted also that the quantity rejected, while absolutely large, is quite insignificant in relation to the total amount of tea imported, namely, 347 millions of pounds.

The rejected tea is allowed delivery duty-free for use in the manufacture of caffeine or theine, the alkaloid which imparts to tea and coffee their stimulating properties, and which is extracted for use as a drug. In such cases the tea has first to be denatured under the supervision of Customs Officers to prevent its possible use for human consumption, and samples both of the denatured tea and of the denaturants used are submitted to the laboratory for test to ensure that the process has been effectively carried out.

Coffee is liable to duty on importation and when exported or supplied for use as ships' stores is entitled to an equivalent drawback when not mixed with chicory or other substances.

The coffee substitutes examined in the course of the year were of the usual character, consisting apart from chicory, of caramel, roasted cereals, dandelion root, and figs.

Formerly the number of samples of cocoa and of preparations of cocoa examined in the laboratory was comparatively small owing to the fact that these articles were subject to a fixed rate of duty of 2d. per pound and no drawback was allowed.

By the Finance Act of 1911 this fixed rate was replaced by duties calculated upon the actual proportion of dutiable ingredients (cocoa, sugar, and cocoa butter) contained in preparations of cocoa.

Under the Act known as the "White Phosphorus Matches Prohibition Act, 1908," which came into force on the 1st of January, 1910, 1665 samples were examined.

In only three instances was white phosphorus found to be present, and in these cases the consignments were refused admission into the United Kingdom.

Seventy-eight samples of medicines, or articles offered for sale as medicines, were examined during the year in connection with the Medicine Stamp Acts. These Acts impose duties, payable by means of special stamps, upon preparations advertised for the cure or relief of human ailments. There are, however, certain exemptions, and it is chiefly in connection with these exemptions that analyses of the samples are required. Thus a single medicinal drug, sold unmixed with any other substance, is not charged with stamp duty, and many of the samples received were analysed in order to ascertain whether they were, in fact, simple drugs or mixtures. Another provision of the Acts exempts, in certain circumstances, medicines of which the composition is known, and analyses are required to establish the identity of samples which it is claimed come within this exemption.

Among points of interest arising may be mentioned an instance of pills which were advertised as a remedy for obesity. The magistrate before whom legal proceedings were brought held that obesity was an " ailment " within the meaning of the Acts, and the seller was convicted of an offence. Specimens of water, and of mud compresses, alleged to have radio-active properties and to exert medicinal effects, were among the samples examined, which included also pills, powders, plasters, ointments, medicinal snuff, herbs, corn cures, embrocations, and various liquid and solid " remedies."

Hydrometers for ascertaining the strength of spirits and saccharometers for use at breweries and sugar factories are tested as to their accuracy at the Government Laboratory before being issued to the officers of Customs and Excise. In addition, graduated vessels of various descriptions for use in the Surveying Department are calibrated at the laboratory.

Chemical work is performed for the Admiralty in connection with the Contract Department at Whitehall, the Naval Yards, the Engineering Department, the Canteen Inspections, the Hospitals and Schools, and the Medical Branch. The work consists mainly in the examination of food substances, including fresh and condensed milk, butter, margarine, suet, lard, tinned foods, jam, lime and lemon juice, rum, ale and stout, pepper, and baking-powder.

It is satisfactory to note that nearly all the samples of dairy produce examined were of genuine character and good quality. In only two instances was the fresh milk found to have been watered and deprived of a portion of its fat, and of 91 samples of condensed milk all but four were up to the specified standard. The samples of butter and margarine were generally satisfactory; in six cases the proportion of water was excessive, in three the amount of salt exceeded the specified limit, and three tender samples for Hospital supplies contained boron preservative contrary to the Specification. One sample from a Naval Canteen was found to contain 22·4 per cent of water and to be mixed with condensed milk.

The tinned foods examined included peas, kidneys, rabbit, and apricot jam. All were in sound condition, but in tinned peas copper was generally present, and tin was found in some of the other articles.

In connection with the Board of Agriculture and Fisheries, and the Department of Agriculture and Technical Instruction for Ireland, a large number of samples of milk, cream, and butter are annually examined.

The examination of the samples of imported butter included in every case the determination of the quantity of water, the examination of the fat as to its purity, and tests for boric acid and added colouring matter. Salicylic acid and hydrogen peroxide are also occasionally used as preservatives.

Sixteen per cent of water is allowed by law, but this is occasionally, though not often, exceeded. There was no evidence of the presence of fat other than milk fat in any of the samples of imported butter examined.

In consequence of the poisonous effects of lead on pottery workers a large number of samples of pottery glazes were examined for the Home Office and for other Departments, in connection with the Home Office regulations on the subject. Under these regulations manufacturers are divided into four classes, each of which is subject to a different degree of control. (1) Those who use "leadless" glazes only, a "leadless" glaze being defined as one which does not contain more than one per cent of lead calculated as metallic lead; (2) those who use only glazes which, tested in the prescribed manner, show no more than two per cent of soluble lead monoxide; (3) those who use no glazes containing more than five per cent of soluble lead monoxide; and (4) those who enter into no engagement with regard to the quantity or state of solubility of the lead present in the glazes they employ.

Government Departments now stipulate that the china and

earthenware articles which they purchase shall be prepared either with leadless glaze or with glaze in which any lead that may be present is mainly in the insoluble condition. A large number of articles, including glazed bricks and tiles, telegraph insulators, and sanitary and domestic ware of all kinds were submitted by the Admiralty, the India Office, the Office of Works, the Post Office, and the Prisons' Department of the Home Office for examination as to conformity with these conditions. Most of the articles examined were of perfectly satisfactory character, but in some cases it was clear from the quantity of lead found that glazes containing lead must have been employed in their manufacture.

Among the curiosities of investigation the following duty devolves on the Government Chemist.

Before an Old Age Pension can be granted, it is necessary that the age of the applicant should be clearly established. In the absence of the Registrar-General's certificate, reliable evidence as to age is sometimes obtained from entries of the date of birth in old Bibles or Prayer Books, from names and dates written in books received as gifts in childhood, and from marriage certificates and other documents. Sometimes there is reason to suspect that such entries have been made recently, or that the original writing has been altered for the purpose of deceiving the authorities.

In the course of the year thirteen documents were submitted for examination, on account of their suspicious appearance. In six of these cases it was found that the writing was of recent date, or that it had been recently tampered with.

In 1913 at the request of the Deputy Keeper of the Records the composition of a series of mediæval wax impressions of seals of various dates, from the thirteenth to the sixteenth century, was examined with a view to obtaining information for his guidance in devising means for the better preservation of the seals under his charge. Most of the seals were found to consist of mixtures of beeswax and resin, the resin in some cases being ordinary colophony. The wax in the case of two of the seals, dated respectively 1399 and 1423, possessed the character of East Indian rather than European beeswax. An impression of the Great Seal of 1350 was found to consist of pure beeswax, and it is remarkable that the wax, although nearly six centuries old, corresponded exactly in properties with wax of recent origin.

The colouring matter of the seals was generally vermilion or verdigris, mixed in some cases with dark organic matter.

In conclusion the work of the laboratory may be roughly classified as Revenue and non-Revenue. The former is carried on partly at Clement's Inn and partly at the Custom House and eighteen provincial stations. The nature of the Revenue work has been sufficiently indicated in preceding pages.

The non-Revenue work consists largely in the examination of samples in connection with Crown contracts for the supply of material required by the various Government Departments. Besides this the whole of the chemical work of the Board of Agriculture and Fisheries (analytical, advisory, and research), and of the Geological Survey is carried out by the Government Laboratory. While the work for such departments as the Post Office, Home Office, and War Department is largely of a routine character, questions involving more or less elaborate research are constantly referred to the laboratory by Public Departments, by Royal Commissions, and by Departmental Committees. The following list contains a *few* examples of the researches which have been carried out in recent years and will serve as an indication of the variety and extent of this aspect of the work of the laboratory.

Nature of Investigation.	Department for which Conducted.
New tables for determining the original gravity of Beer. (Published officially, and by The Institute of Brewing.)	Board of Customs and Excise.
New Sikes Hydrometer Tables.	do.
The amount of Sugar remaining in the refining vessels after the lapse of periods of 7, 12, and 21 days: in connection with the Sugar Bounties. (Confidential; not published.)	Foreign Office.
Growing of Sugar Beet. (Published by Board.)	Board of Agriculture and Fisheries.
Absorption of Tar Acids by Sheeps' Wool.	do.
Action of Alkalis upon Straw.	do.
Deterioration of Liver of Sulphur.	do.
Relation between citric solubility and the availability of the phosphates in slags.	do.

Nature of Investigation.	Department for which Conducted.
Alleged emanation of lead vapours from painted surfaces. (Published in Report of Departmental Committee on Lead Paints.)	Home Office.
Composition and properties of Celluloid. (Published in Report of Departmental Committee on Celluloid.)	do.
Preservation of Roof Timbers of Westminster Hall. (Published in Report of H.M. Office of Works on Westminster Hall.)	Office of Works.
Prevention of Corrosion of Iron by means of varnish painting. (Not published.)	do.
Elimination of ultraviolet rays from the light of electric lamps. (Not published.)	do.
The permanency of colours used in printing of postage and fiscal stamps. (Confidential; not published.)	Inland Revenue and Post Office.
Composition and character of Milk Powders. (Not published.)	Local Government Board.
Preservatives in British-Made Wines. (Not published.)	do.
Composition of the Wax of Mediæval Seals. (Subsequently published in Transactions of Chemical Society, 1914, p. 795.)	Public Record Office.
Authenticity of writing in a Book of Revels, 1604–5 A.D. (Results published in "Supposed Shakespeare Forgeries," by Ernest Law, B.A. G. Bell and Sons.)	do.
Preservation of Old Records.	Meteorological Office.

CHAPTER III

APPARATUS

ON entering a chemical laboratory the first impression on the mind of a visitor not conversant with the science of chemistry and its practice is produced by the apparently innumerable ranks of bottles. This is particularly noticeable in those laboratories in which provision is made for a large number of students, inasmuch as each worker requires for his individual use a considerable number (about thirty) of "reagents," liquid or solid, which are so frequently in use that it would be a source of great inconvenience if they were shared with a neighbour. Beside these there are usually stacks of shelves placed in a position of easy access, so that the bottles of solutions or other materials which are in less frequent demand may be ready for common use by any of the occupants of the laboratory.

Laboratories which are devoted to other purposes than the teaching of bodies of students do not generally display so prominently any large array of bottles. By reference to some of the illustrations which show the interior of laboratories employed for special purposes it will be seen that bottles do not always form a conspicuous feature. In such cases apparatus connected with the special business of the laboratory will catch the eye. It may be apparatus for distillation, for the estimation of melting points or freezing points, for electrolytic or other electrical operations, and so forth; in any case glass vessels of all shapes and sizes, glass tubes contorted into a variety of forms, and the appearance of the Bunsen blue cone of burning gas at many places will be prominent features of the chemical laboratory.

Detailed descriptions of apparatus would alone occupy a large volume, and since the purpose of this book is general no attempt will be made to describe apparatus which has long been a familiar part of the indispensable equipment.

Weighing.—Every chemist must possess at least one balance of precision which will carry 100 to 200 grams in each scale and turn with a difference of $\frac{1}{10}$th milligram. With the same instrument, however, and a smaller mass in the pan, and using the method of vibrations, much smaller differences can be recognised. The only change in the construction of chemical balances which has become general within the last thirty years, is the shortening of the beam, in consequence of which its oscillations are quicker, and therefore the business of weighing is abbreviated.

The ordinary balance agrees in principle with the common systems of scales and weights. The accompanying illustration shows an ordinary short beam balance of modern type. It consists of a two-armed lever, the centre point of which is suspended on a knife edge which rests on a plane surface of agate. The pans are also suspended on knife edges resting on agate planes when the balance is in use. When not actually in operation the beam and the pan supports are raised from contact with the planes. This is accomplished by turning the screw head in the middle of the box below. The divided scale above the beam carries a small rider of wire, the position of which can be altered, and the final adjustment of weight effected. The chemist's balance differs from the pillar scales of the grocer or druggist only in superiority of workmanship and materials whereby it is made vastly more accurate and sensitive.

FIG. 38. A SHORT BEAM BALANCE.

The balance about to be described differs in principle from all ordinary balances, and in the skilled hands of the inventors and others exhibits a degree of sensitiveness and delicacy in dealing with small masses which has hitherto been unheard of. It is obviously applicable only to exceptional cases.

The new balance is described by the inventors, Professor B. D. Steele and Mr. Kerr Grant, in the Proceedings of the Royal Society for 1909, and in addition to the interest attaching to the application of the fundamental idea involved, a balance of this construction with some modifications in detail, was used by Professor Sir William Ramsay and Dr. R. Whytlaw Gray in determining the density of the gaseous emanation from radium (q.v.) which Ramsay calls "niton."

The rationale of the method of weighing is as follows: If a bulb filled with air is at the same temperature and pressure as the air surrounding it the weight of the contained air will be nothing. This is in accordance with the principle of Archimedes.

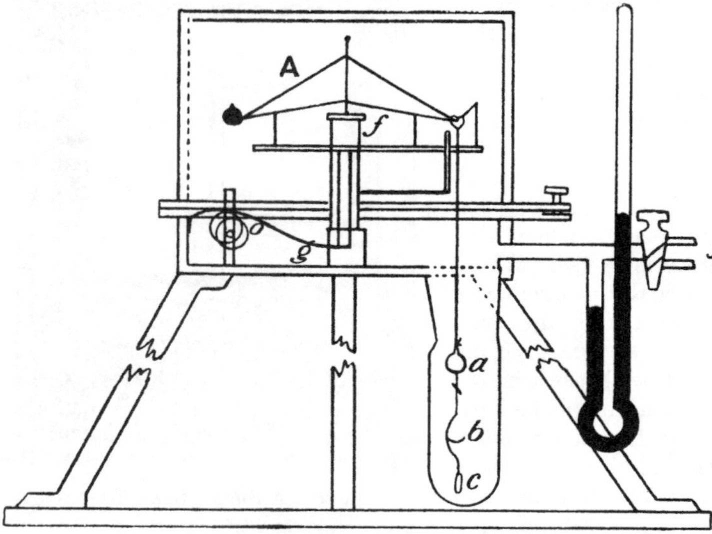

FIG. 39. MICRO-BALANCE OF STEELE AND GRANT.

If, however, the pressure of the air surrounding the bulb is altered, the sealed up air exerts more or less of its full weight. By suspending a bulb containing a known quantity of air at one arm of a balance, and arranging the whole instrument within a case from which the air can be pumped out to any desired extent, the effective weight of the bulb of air can be changed to any amount desired.

Temperature changes are eliminated as far as possible, as well as vibration, by mounting the balance on a stone pillar in a cellar, and placing the brass case of the balance inside a large box of bright tin plate. The above diagram will give an idea of the essential parts of the micro-balance of Steele and Grant.

A is the beam of the balance constructed in the form of two triangles base to base, and made of fused quartz rod, 0·6 millimetre in diameter; the whole weighing less than half a gram. The frame thus formed oscillates about a central knife edge, ground at the end of the vertical rod, and resting on a plane quartz plate *f*. Attached to this beam at its centre (but not shown here) is a tiny mirror, taking the place of the pointer in an ordinary balance.

At one arm of the balance is a quartz counterpoise, while at the opposite end is suspended a small quartz bulb of known internal volume, and sealed up at known temperature and pressure. This is hung by means of a fine quartz fibre within a tube which is connected air-tight to the bottom of the balance case. A fine hook carries the quartz bulb, *a*, and below this a quartz scale pan, *b*, and a quartz counterpoise, *c*. The weight of this suspended system is always adjusted to equilibrate the counterpoise attached to the opposite end of the balance.

The method of weighing is as follows: if the quantity of substance to be weighed does not exceed the total weight of the air contained in the bulb, the pressure inside the balance case and the resting point having been taken with the scale pan empty, the substance to be weighed is placed on the pan and the pressure adjusted till the same resting point is obtained. If w is the total weight of air contained in the bulb, which was filled at pressure P, and P′ represents the difference in pressure required to recover the original resting point, then the weight of the substance is wP'/P. If the quantity of substance to be weighed exceeds the weight of air contained in the bulb it is necessary to prepare one or more counterpoises which must be lighter than the original counterpoise *c*, and must differ from each other by a known amount not exceeding w.

The resting or zero point of the instrument is found by the position taken by the image of a Nernst lamp reflected from the mirror attached to the beam. The case is deprived of air by means of a vacuum pump connected through the two-way stopcock x, and the pressure of the residual air is determined by observing the height of the mercury column in the manometer, which is read by means of a telescope and scale to a tenth of a millimetre.

The attachments for the release of the beam consist of two V-shaped quartz rods which just centre the beam but do not lift it. These can be lowered when required by means of the curved brass wire, *g*, connected as shown in the figure with the upright brass support. The wire is controlled by an excentric cam, *o*, rotated by a handle passing air-tight through a plug in the side of the case.

Such a description as the foregoing is only capable of giving an idea of the way in which the pneumatic principle is applied to the determination of weight. As to the possibilities of such a balance the statement of the authors is as follows: "Weights of the order of one-hundredth of a milligramme may be compared with the standard measures with an accuracy of one five-hundredth of their amounts, i.e. the absolute value of such weights can be determined with certainty to one fifty-thousandth of a milligramme (2×10^{-8} gramme), while changes of weight can be measured of an order as low as one two-hundred and fifty-thousandth of a milligramme."

With such appliances the mote in the sunbeam becomes a ponderable mass!

Heating and Cooling.—The common source of heat for the purpose of ordinary experiment in the chemical laboratory is the combustion of coal-gas, generally in the Bunsen burner, or some modification of that familiar instrument in which the gas is mixed with sufficient air to secure complete and smokeless combustion.

As nearly everyone knows a much hotter flame is obtainable if the air is replaced by unmixed oxygen. The oxyhydrogen or oxy-coal-gas flame has long been used for special purposes, such as the production of the well-known limelight employed in the magic lantern and for theatrical purposes. Such a flame is also capable of piercing a sheet of iron, if not too thick, and may be used even for cutting armour plate.

The temperature of the oxyhydrogen flame is in the neighbourhood of 2000° C., but a still hotter flame is produced when, in place of hydrogen, acetylene is used. The oxyacetylene flame may indeed be used for cutting armour plate six inches in thickness, at a rate more rapid than that of the saw. This, however, is not a laboratory operation.

The use of electric heating for warming houses and cooking has been introduced long ago, but the progress actually made is comparatively slow, owing chiefly to the cost. This method is based on the fact that when an electric current passes through an imperfect conductor more or less of the energy of the current appears as heat. An illustration of this is seen daily in the ordinary incandescent electric lamp, in which a thread of carbon or of an imperfect metal like tungsten is raised to such a temperature that a brilliant light results. The principle is applied in the heaters which for some purposes are used in the laboratory. The current in this apparatus is turned into a box containing carbonaceous material, the resistance of which can be diminished or increased by altering its state of compression, and the temperature resulting can be thus regulated. Apparatus of this kind has several advantages for laboratory purposes; it provides a steady source of heat, spread over a larger surface than a flame, which may be increased at will from a gentle warmth to a low red if required, it is unaffected by draughts of air, and it is unattended by risk of fire when inflammable liquids, such as ether or alcohol, are to be heated. An example of the use of electric heaters is shown (Fig. 32, p. 62) in the distillation of wine, spirits, or beer, for the determination of their alcoholic strength. A measured quantity of the liquid to be tested is placed in the flask, distillation is continued till all the alcohol has passed over and a determinate quantity of liquid is collected. The specific gravity of this liquid, compared with the figures in an alcoholometric table, supplies the percentage of alcohol in the original liquid.

Another application to laboratory purposes is shown again (Fig. 34, p. 64) in the apparatus for steel analysis. The carbon left after dissolution of a weighed quantity of the metal in an appropriate solvent is burnt in a stream of oxygen in a tube heated by the current to redness. The resulting carbon dioxide is collected in weighed tubes containing caustic potash, as shown in the figure, and from the increase of weight the carbon in the sample can be calculated.

The use of the electric arc for the attainments of high temperatures is the result chiefly of the researches of the late Professor Henri Moissan,[1] whose untimely death in February, 1907, deprived the world of a very brilliant and indefatigable worker in science. One form of Moissan's furnace is shown in

[1] Moissan's career is both interesting and instructive. He was born in Paris 28th September, 1852, the son of an employé of the Eastern Railway Company, his mother assisting the slender resources of the family by working as a dressmaker. At the age of twelve he entered the municipal school at Meaux, where he remained till 1870. He then obtained a situation in a pharmacy, which he retained during two years, and then passed into Professor Dehérain's laboratory in the Museum of Natural History. Here he experienced the attractions of

the adjoining illustration. The body of the furnace is formed of blocks of good lime, a material which resists without change the highest temperatures produced by the oxyhydrogen flame. In the arc lime slowly fuses and volatilises. As the diagram shows the carbon poles pass through the sides of the box and, being connected with the source of the current, are brought into contact with each other immediately over the object to be heated

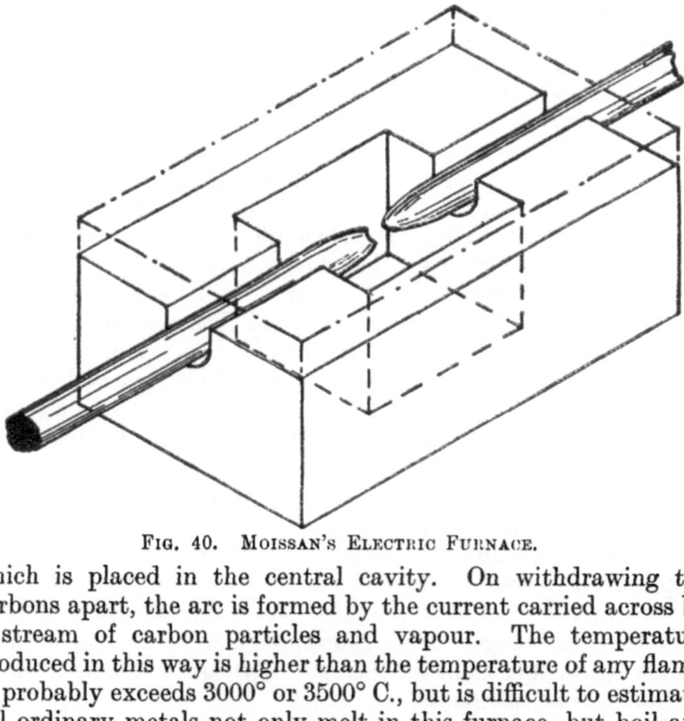

FIG. 40. MOISSAN'S ELECTRIC FURNACE.

which is placed in the central cavity. On withdrawing the carbons apart, the arc is formed by the current carried across by a stream of carbon particles and vapour. The temperature produced in this way is higher than the temperature of any flame. It probably exceeds 3000° or 3500° C., but is difficult to estimate. All ordinary metals not only melt in this furnace, but boil and pass off in vapour. Even lime and magnesia, among the most

research, but, encouraged by Dehérain, he set to work privately to prepare for his degree, which he ultimately obtained. Subsequently he received an appointment at the École Supérieure de Pharmacie, where a few years later he succeeded to the chair of toxicology, and afterwards that of inorganic chemistry. At the time of his death he held the professorship of inorganic chemistry in the Faculty of Sciences of the University of Paris, where he had succeeded Troost in 1900.

To Moissan we owe the isolation (in 1886) of the elusive element fluorine, perhaps his most interesting work from the scientific point of view, and the discovery of the conditions of formation of the diamond. The latter discovery resulted from the experience he had gained in the application of the electric arc to the production of very high temperatures.

refractory oxides known, melt and ordinary charcoal is changed into graphite. In such a furnace Moissan saturated molten iron with carbon and cooled the mass rapidly, so as to produce solidification on the outside. The still fluid portion within had to cool under the great pressure which results from the tendency to expansion during the solidification of such iron. A small portion of the carbon crystallises under these circumstances in the form of the diamond. In order to separate these small crystals the iron has to be dissolved away by means of acids, and the carbonaceous residue is again treated with an oxidising mixture of sulphuric acid and nitre to remove the graphite, while siliceous impurities are afterwards got rid of by use of hydrofluoric acid. In the end the small grains which remain are examined under the microscope. The largest diamond obtained by this process only measured half a millimetre in diameter, but Moissan proved the identity of these tiny crystals with natural diamond not only by reference to their crystalline form, but by their density (about 3·3 to 3·5), their hardness being able to scratch ruby, and by their combustibility in oxygen.

Another product of the electric furnace is the substance carborundum, a compound of carbon with the allied element silicon. This is produced by heating together fine sand (silica) with carbon and a little common salt. Carborundum has become very valuable, on account of its hardness, as a material for grinding and cutting metals in the engineering shop.

The most valuable of all products from the electric furnace is, however, calcium carbide. This again we owe to Moissan, who studied for the first time systematically the carbides of all the chief classes of metals.

The production of calcium carbide, CaC_2, is simple enough. It results from the action of heat, in the electric furnace, on a mixture of ground lime and coke. The carbon divides itself between the oxygen of the lime and the calcium, so that carbonic oxide gas escapes, while the carbide remains at the end of the operation as a grey solid or sometimes in the form of black crystals.

For nearly twenty years calcium carbide has been manufactured as a source of acetylene gas, which is now a familiar illuminant, and, as already mentioned, is employed in conjunction with oxygen for the production of a high temperature blow-pipe flame.

The gas is produced by allowing water to drip on the solid

carbide placed in a suitable generator, and connected with a gas-holder in which the gas is collected over water.

Since 1903, however, calcium carbide has received application in a new direction, having been found to absorb nitrogen when heated to about 1000° C., in the presence of small quantities of calcium chloride or some other salts. The product is the calcium salt of cyanamide, CN.NCa, and is known commercially as "nitrolime." It is a valuable nitrogenous manure.

The application of the electric furnace to such operations as the manufacture of steel is obviously a subject of the highest importance, but is outside the programme of the chemist.

The use of electrical methods for bringing about the combination of atmospheric nitrogen with oxygen, and the fixation of nitrogen and its oxides into solid compounds will be described in a later chapter.

Cooling.—Ice is an indispensable agent for use in the laboratory, especially in connection with the reactions in which carbon compounds are concerned. In very many cases the heat which is produced by chemical change goes on accumulating in the materials which have been mixed together until the temperature of the whole is such as either to cause the volatile ingredients to boil and evaporate away, or to give rise to new changes which may become violent and uncontrollable. The result is that the desired product is not secured and dangerous explosions may result.

The temperature of melting ice is always at 0° C., but lower temperatures are easily obtained by mixing it with due proportions of very soluble salts. With common salt the temperature falls to the zero of the Fahrenheit scale (–17°·7), and a few pounds weight of such a mixture will keep a temperature of –12° to –15° for a long time.

But much of the experiment during the last twenty years has required the employment of temperatures far below such limits, and even mixtures such as solid carbon dioxide in ether or alcohol, which gives a temperature of about –75° C., are not low enough. Moissan in his experiments which resulted in the isolation of fluorine, made use of boiling methyl chloride as a frigorific agent maintaining a fairly steady temperature of –23° C.

The liquefaction of atmospheric air in quantity has placed in the hands of the physicist and chemist an agent which has indirectly furnished the means of reducing the remaining more refractory gases, hydrogen and helium, to the liquid state. This achievement would, however, never have been reached but for the establishment of several important general principles which were only discovered in the course of nearly a century of laborious and sometimes dangerous experimental enquiry.

In all the older text-books of chemistry a distinction was made between the liquefiable or condensable gases and those which were called "permanent gases." No doubt the more philosophical writers on the subject foresaw from the close resemblance between vapours and the more easily condensable gases, such as sulphur dioxide and chlorine, that after all there might be a similar relation between such gases and the so-called permanent ones, and it was expected that had suitable power been available the latter would prove to be also merely the vapours of very volatile liquids.

The difference between a vapour and a gas is now known to be a definite physical difference, as will be explained presently.

The history of the liquefaction of all the gases need not be related in detail, but the following brief record will suffice to show the position of the question in the former half of the nineteenth century.

The list of gases below is accompanied by the name of the experimenter who succeeded in reducing them to the liquid state either by compression alone, or by compression assisted by the lowest temperatures then producible.

Name of Gas.	Observer.
Sulphurous acid (Sulphur dioxide).	Monge and Clouct, about 1800.
Sulphur dioxide and Chlorine.	Northmore, 1805.
Chlorine, hydrochloric acid, sulphur dioxide, sulphuretted hydrogen, carbon dioxide, nitrous oxide, cyanogen, ammonia.	Faraday, before 1825.
Ethylene, hydrogen iodide, hydrogen bromide, phosphoretted hydrogen, silicon fluoride, boron fluoride, arsenetted hydrogen.	Faraday, before 1845.
Hydrogen, oxygen, nitrogen, nitric oxide, carbonic oxide, marsh gas.	Remained unliquefiable.

Experiments carried on by Thomas Andrews, Professor of Chemistry in Queen's College, Belfast, after several years of work gave the clue. In 1861 he described experiments in which he had submitted some of these remaining gases to very great pressures combined with cold, the pressure being limited by the capacity of the thick glass tubes to resist it. The cold was the temperature of the carbonic acid and ether mixture which, under reduction of pressure, would go down to –160° F.

Atmospheric air was compressed to $\frac{1}{675}$th of its volume, in which state its density was little inferior to that of water, oxygen to $\frac{1}{554}$th, hydrogen to $\frac{1}{500}$th, and so on, but in no case was there any appearance of liquefaction. But observations on the compression of carbon dioxide at different temperatures led to the discovery that when the gas contained in a tube is partly liquefied by pressure alone, on gradually raising the temperature to 31° C. (88° F.), "the surface of demarcation between the liquid and gas became fainter, lost its curvature, and at last disappeared. The space was then occupied by a homogeneous fluid which exhibited, when the pressure was suddenly diminished or the temperature slightly lowered, a peculiar appearance of moving or flickering striæ throughout its entire mass. At temperatures above 88° F. no apparent liquefaction of carbonic acid or separation into two distinct forms of matter could be effected, even when a pressure of 300 or 400 atmospheres was applied. Nitrous oxide gave analogous results."

A series of comparisons were then made by Dr. Andrews on the volume of carbon dioxide and air when submitted to increasing pressures at different degrees of temperature. When his results were plotted, the volumes against the pressures, a series

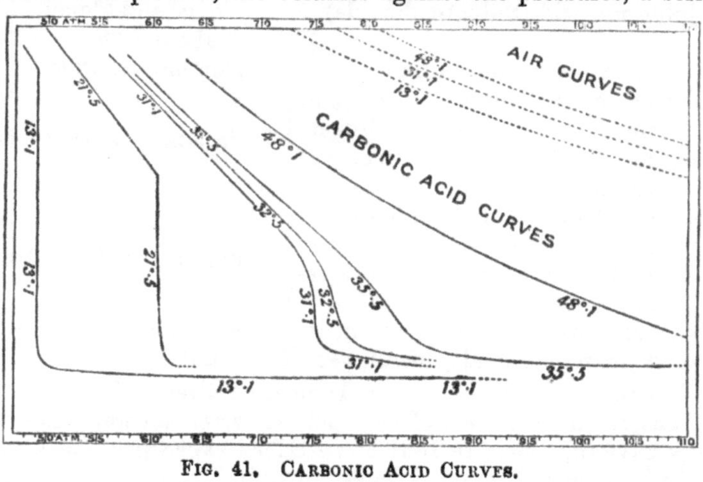

FIG. 41. CARBONIC ACID CURVES.

of curves was obtained, in which it appears that air steadily diminishes in volume as pressure is applied at common temperatures, while carbon dioxide at temperatures considerably above 31° C. imitates it pretty closely. If the temperature, however, approaches within a few degrees of 31° a sudden diminution of volume is observed, which is indicated by a sudden change of direction of the curve, becoming vertical at anything below 31° with a pressure about 75 atmospheres.

There is therefore a critical point of temperature above which carbon dioxide gas cannot be reduced to liquid by pressure. This has been shown to be true of all other gases, and the reason why oxygen, hydrogen, and some others had not been liquefied by pressure up to that time was that the temperature employed in each case had been above the critical point of the gas. The critical temperature for oxygen is about –118° C., while that of nitrogen is –146° C. Cooling the gas below these temperatures is therefore an essential condition for their liquefaction, and in the case of oxygen this result was attained, for the first time at the end of 1877 by the French physicist Cailletet and the Swiss professor Pictet, by two distinct methods.

But for the continuous production of the liquid from the gas yet another principle was necessary. Experiments made nearly seventy years ago by Dr. Joule of Manchester, in conjunction with William Thomson (afterwards Lord Kelvin), resulted in the discovery that if a gas is caused to expand so as to do external work cooling results. Joule and Thomson caused a stream of compressed gas to pass through a long copper spiral, immersed in water, at constant known temperature. The gas then escaped through the pores of a plug of compressed cotton wool, and its temperature was noted. In every case, except hydrogen, a reduction of temperature was observed which, though actually small in amount, was the greater the lower the temperature of the gas before passing through the plug. Thus the effect on carbon dioxide is shown in the following figures :—

Temperature	Degrees Cent.		
Before escape through plug . .	12°·8	19°·1	91°·5
Reduction of temperature . .	1·207	1·144	0·69

The amount of cooling is proportional to the difference of pressure before and after release, and inversely as the absolute temperature.

At low temperatures it has been found that hydrogen behaves like other gases. These results afford information as to the internal constitution of gases, for, on allowing a gas to expand, any change of temperature observed must be due to the performance of work, either in lifting atmospheric pressure or internally in overcoming the mutual attractions of the gaseous molecules. The rise of temperature noticed in the case of hydrogen at ordinary temperatures indicates that the molecules of this gas, under such conditions, repel one another.

By compressing an ordinary gas, such as air, removing the heat produced by immersing the containing vessel in cold water or otherwise, and when cool allowing the gas to escape, a reduction of temperature is obtained. At common temperatures this amounts to about one quarter of a degree centigrade for each atmosphere of pressure taken off. If this cold stream of escaping air is then made to pass over the pipe through which the gas passes previous to release, it may be cooled in a cumulative manner to lower and lower temperatures until it reaches a temperature below the critical point. By continuing the same process with comparatively moderate pressures a portion of the gas becomes liquid. This is the principle of the air liquefying plant to be found in all the greater institutions for chemical and physical research.

Practical results on a large scale were first shown by Linde, an engineer of Munich, in 1895. Patents for the application of the same idea were taken out a little earlier in England by W. Hampson.

The apparatus actually employed consists of a compressor which, usually in two stages, puts the air, previously freed from water and carbon dioxide, under a pressure of 200 atmospheres. In this condition it is delivered into a close coil of copper tube, at the end of which is a small hole, the size of which can be regulated by a screw. Escaping from this orifice the cooled air returns to the atmosphere through a cylinder in which the copper pipe is coiled from which the air has just escaped, or through a coil which encloses the return pipe as shown in the diagram. The cooling effect is thus economised on the regenerative system,

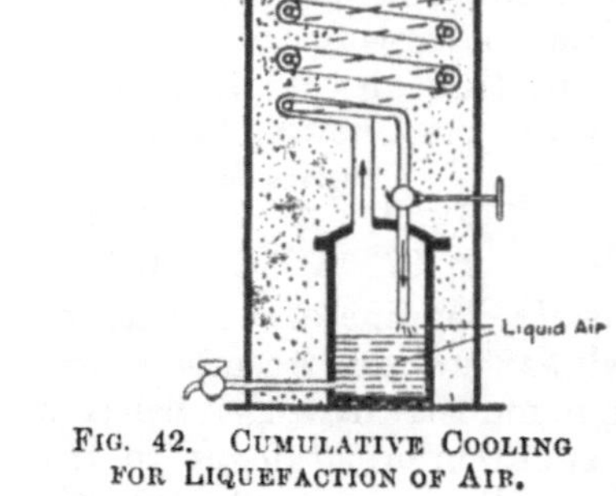

FIG. 42. CUMULATIVE COOLING FOR LIQUEFACTION OF AIR.

just as heat is economised in a regenerative steel furnace. In some machines the compressed air is cooled before escape by means of liquid carbon dioxide, but this is not necessary, as in a well-constructed machine liquid air begins to drip into the receiver half an hour after setting the pumps to work at atmospheric temperatures.

The liquid collected in this way has a very pale blue colour, and it boils at about −181° C. To attempt to collect it in any ordinary bottle would be like trying to catch water in a red-hot vessel. For when the liquid is poured into any vessel in the open air it boils furiously, and continues to do so till the two hundred degrees of difference of temperature has been abolished by the evaporation of a portion of the liquid. Even then heat reaches it from the outside far too rapidly to permit of keeping it for more than a very short time. To meet this difficulty Sir James Dewar's device of a vacuum jacketed glass vessel is everywhere adopted. The device is now familiar in the ordinary "Thermos" flask, which consists of two vessels, one inside the other, with a space between from which the air has been removed as completely as possible. A vessel of this kind is shown in section, page 95.

The vacuum vessel is rendered still more efficient by coating it with a thin but bright deposit of silver.

The low temperature of liquid air is demonstrable by many curious experiments. A tube full of mercury, which freezes at −39°, when plunged into liquid air, sets almost instantly to a solid resembling silver in appearance, and as malleable as lead. As is well known the mercurial thermometer is replaced in some latitudes such as the north of Russia, by a thermometer containing absolute alcohol. This liquid dipped into liquid air at first assumes the consistency of oil, becoming more and more viscous till it sets into a glass-like solid. Such materials as fruit, flesh, and india-rubber cooled in liquid air become as brittle as glass, and when at this temperature they are struck with a hammer they fly to pieces like that substance.

The commercial production of liquid air has not only provided a valuable agent for the investigation of the properties of matter, but has led to some important practical results which will be described later on.

The history of the liquefaction of the gases would not be complete without reference to the two cases which have presented the greatest difficulty, namely, hydrogen and helium. The principles involved were perfectly well understood, the difficulties arise in their application. First of all it is necessary to cool hydrogen gas to a temperature of about –80° C. before the evolution of heat observed in the Joule-Thomson experiments is changed into a cooling effect. Compression and release of hydrogen gas, therefore, bring the experimenter no nearer to the point of liquefaction unless the gas is already below that temperature. Then the critical point for hydrogen is at 238° to 240° below zero, or considerably more than 100° below the critical point of oxygen. There can be no liquefaction of hydrogen, therefore, no matter what pressure is employed, without the intense cooling obtained by causing liquid air to boil rapidly under reduced pressure. The self-intensive apparatus can then be used successfully. Liquid hydrogen is a colourless liquid with a well-defined surface, and having an extraordinarily low density. Bulk for bulk it has only about $\frac{1}{14}$ the weight of water. It boils at about –252° to –253° C., and by rapid evaporation may be cooled till it sets into a wax-like mass resembling solid paraffin. The appearance of this solid at once dissipates the favourite theory advanced by Graham half a century or more ago, that hydrogen gas is the vapour of an extremely volatile metal. It certainly has many of the properties of a metal in the production of salts (acids), and in the readiness with which it exchanges places with metals in ordinary saline reactions. The significance of this exchange is, however, discounted by the fact that it also exchanges with chlorine and the other halogens which are at the opposite end of the electrochemical scale. Hydrogen gas is more nearly the analogue of marsh gas, CH_4, and from one point of view may be regarded as the first term of the series of hydrocarbons called paraffins. Liquid hydrogen in quantity was first produced in an open vessel by Sir James Dewar in the laboratory of the Royal Institution in 1898.

One gas now remained, namely, the inert gas helium, which, discovered in 1895, was obtainable in moderate quantity. But it was another ten years before this exception to the rule, which could now be applied to gases generally, was abolished. Liquid hydrogen was now available as a cooling agent, and the liquefier being supplied with this powerful aid to condensation success was at last achieved.

To Professor H. Kamerlingh-Onnes, in the cryogenic laboratory directed for many years by him in the University of Leiden, science owes this interesting result. More than 60 cubic centimetres (over 2 fluid ounces) of liquid helium were obtained, so that there could be no ambiguity about the result. And it is fortunate that such very definite success was secured, for it is not very likely that so costly and difficult an experiment will be often repeated. The boiling-point of liquid helium is estimated to be 268° to 269° below 0° C. The density of the liquid is 0·15, or less than $\frac{1}{6}$ the density of water, and a little more than twice the density of liquid hydrogen.

By causing liquid helium to boil under reduced pressure a temperature about 2° lower was reached, but the helium did not solidify. This is the lowest temperature known, and is believed to be only about 2·5° C. above the absolute zero.

At these very low temperatures the properties of many solid bodies are considerably modified. Thus the conductivity of metals diminishes with rise of temperature and increases with fall of temperature, and in the case of all the pure metals which have been examined the resistance at successively low temperatures down to about -200° is such that at the absolute zero it would disappear altogether, and all metals would be equally good conductors. The specific heat of solids is also greatly reduced, but not to zero; the atomic heat, that is, the product of the specific heat into the atomic weight, is now found to be periodic, rising and falling at regular intervals with increase of atomic weight. (See *Periodic Law*, p. 126.)

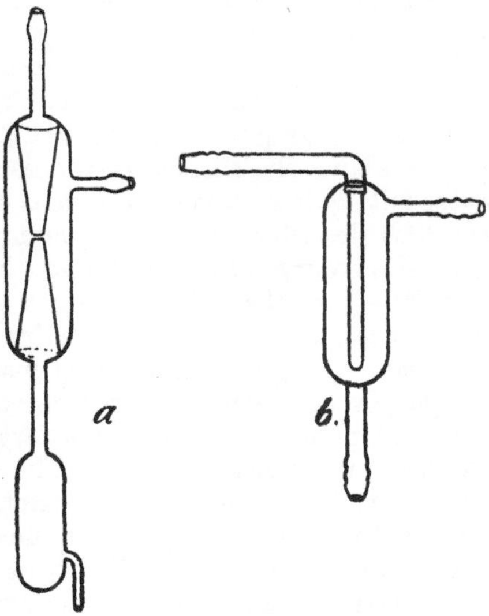

FIG. 43. FILTER-PUMPS.

Reduction of Pressure. Vacuum.—It has already been mentioned in describing the interior of a laboratory (see Imperial College, p. 9) that nearly all the tables or benches for work are provided with water pumps, used chiefly for aiding filtration of liquids. For this particular purpose it is not necessary, nor is it desirable, to arrange for a very great reduction of atmospheric pressure beneath the filtering surface, which is almost always formed of paper, the sides of which are supported by the funnel, while the point of the cone is unprotected, or strengthened only by a small cone of gauze or of toughened paper. The water-pumps used for this purpose are supplied with high pressure water, and their action is very similar to that of the steam injector.

Two forms of these pumps are shown in the preceding figures *a* and *b*. In both a jet of water under pressure escapes from the downward-pointed nozzle, and is discharged into the open end of the tube below, carrying with it air drawn from the space to be exhausted by means of the side tube. The water is usually carried into the sink by means of a piece of flexible tubing attached at the bottom of the apparatus.

The other form of water pump does not depend on the pressure produced by a head of water, but on another principle which requires the pump to be situated at a height a little greater than 30 feet above the surface of the earth or the well into which water falling down a vertical pipe is discharged. The apparatus was first described by Dr. Hermann Sprengel just fifty years ago. If mercury is the liquid used, then the fall pipe requires to be a little over the ordinary height of a barometer, or some 33 or 34 inches, and in this form the Sprengel pump was the parent of most of the more complicated instruments devised later for the purpose of removing air from an enclosed space. These mercurial pumps have played a large part in the modern researches on the gaseous state. The principle is so simple and yet so very important that a description of Sprengel's pump, in the inventor's words,[1] may be introduced here, as it will enable the reader to understand immediately its more recent modifications. The construction and use of the water pump also becomes obvious.

"C *d* is a glass tube longer than a barometer, open at both ends, and in which mercury is allowed to fall down, supplied by the funnel A with which the tube is connected at C. The lower end *d* of this tube dips into a small glass bulb B, into which it is fixed by means of a cork. This glass bulb has a spout at its side, situated a few millimetres higher than the lower end of the tube C *d*. The first portions of mercury which run down will

[1] *Journal Chem. Soc.*, Vol. 18, p. 10 (1865).

consequently close the tube and form a safeguard against the air, which might enter from below if the equilibrium should be disturbed. The upper part of C *d* branches off at *x* into a lateral tube to which the receiver R is affixed. As soon as the stop-cock at C is opened and the mercury allowed to run down, the exhaustion begins, and the whole length of the tube from *x* to *d* is seen to be filled with cylinders of mercury and air having a downward motion. Air and mercury escape through the spout of the bulb B, which is above the basin H, where the mercury is collected. This has to be poured back from time to time into the funnel A, to pass through the tube again and again until the exhaustion is completed. As the exhaustion is progressing it will be noticed that the enclosed air between the mercury cylinders becomes less and less, until the lower part of C *d* presents the aspect of a continuous column of mercury about 30 inches high. Towards this stage of the operation a considerable noise begins to be heard similar to that of a shaken water-hammer, and common to all liquids shaken in a vacuum. The operation may be considered completed when the column of mercury does not enclose any air, and when a drop of mercury falls upon the top of this column without enclosing the slightest air bubble. The height of this column now corresponds exactly with the height of the column of mercury in the barometer; or what is the same, it represents a barometer whose Torricellian vacuum is the receiver R."

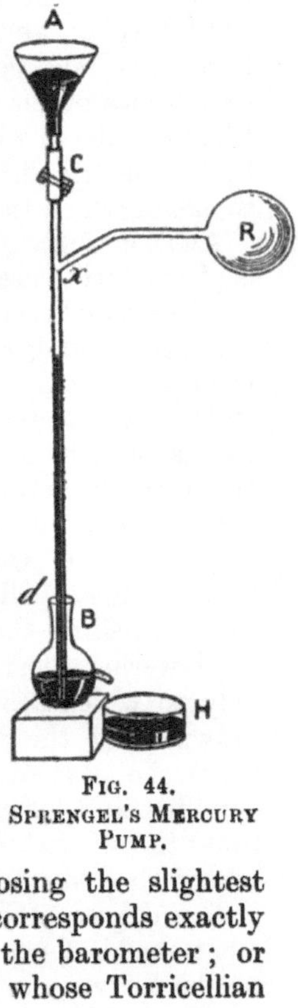

FIG. 44.
SPRENGEL'S MERCURY PUMP.

As a matter of fact the pump in this very simple form does not give a perfect vacuum, for air adheres to the surface of the glass funnel and tube, and india-rubber joints do not exclude the entrance of minute quantities of air. The mercury requires to be admitted to the fall tube with greater precaution, and the whole must be constructed of glass without rubber connections. Figure 35 (p. 65) shows a pump of this kind as mounted in Sir William Crookes's laboratory. It will be seen that there are four fall tubes, down which the mercury drops as it is supplied from the reservoir at the side when the latter is lifted to the proper height. There is a vertical tube placed alongside the fall tubes, and this standing in a vessel of mercury acts as a barometer, and so indicates by the height of the mercury the first stages of the exhaustion. The apparatus shown includes a Plücker tube in which gas under examination can be illuminated by the electric discharge, and the light viewed through a spectroscope, of which the end bearing the slit appears on the left.

Another form of mercury pump was devised by Töpler, and is frequently used for laboratory purposes. A cylindrical vertical vessel is connected at its upper end with an erect straight glass tube, longer than a barometer tube, and dipping into a basin of mercury. The lower end of the cylinder has a branch which is connected with the vessel to be exhausted, and also, by means of a flexible rubber tube with a reservoir of mercury. When the reservoir is raised the mercury rises into the cylinder and cuts off connection with the vessel from which the air is to be withdrawn. On continuing to raise the reservoir the mercury rises and drives all the air out of the cylinder into the barometer tube, expelling the excess of it from the bottom of that tube. On lowering the reservoir again the mercury retreats, but the only air which can take its place in the cylinder is that which is drawn from the vessel to be exhausted. The mercury thus plays the part of a piston which moves up and down in the cylinder, and by repeating the operation a sufficient number of times the air within can be expanded indefinitely, and ultimately a very good vacuum can be obtained.

The degree of exhaustion obtainable by any of the pumps mentioned is dependent not only on the efficiency of construction, but on the vapour pressure of the liquid used at the temperature of the air. Using water there is not only the vapour pressure of water, amounting to between 12 and 15 millimetres of mercury at room temperatures, but the atmospheric gases held in solution by all ordinary water supply. Hence the exhaustion obtainable by a water pump is always far from complete, although it may be amply sufficient for the majority of laboratory operations, such as distillation under diminished pressure.

Even mercury gives an appreciable vapour pressure at room temperatures (at 20° C. it is ·001 mm.), and hence the efficiency of a mercurial pump can never exceed the exhaustion which this represents. Moreover for cases in which the best vacuum is required it has always to be remembered that the surface of glass retains a film of air and moisture with extraordinary tenacity. To get rid of this from the interior of the bulb or tube which it is desired to exhaust, the tube must be heated to a temperature as high as it will bear, while the pump is kept at work.

A very valuable method of obtaining a high vacuum without the use of a pump has been introduced by Sir James Dewar. It has long been known that charcoal absorbs many gases very freely, and this property has been turned to account for many practical purposes in past times. Thus a respirator, consisting of a small metal case filled with coarsely crushed wood charcoal, which could be held over the mouth and nose was actually in use by nurses and dressers before the days of antiseptic surgery. Such an appliance also protects the wearer very completely for a short time if exposed to an atmosphere containing irritant fumes. The most dense varieties of charcoal, obtained for example by heating cocoanut shell, seem to be the most efficient. If a glass vessel filled with charcoal of this kind is first heated, and the moisture and gases removed by means of a mercury pump, and the vessel is then connected with the space to be exhausted and immediately plunged into a vacuum vessel containing liquid air, the gases present are absorbed with great rapidity and completeness. A selection is made by the charcoal when exposed to contact with mixed gases, air, oxygen and nitrogen are absorbed more readily than hydrogen, while helium and neon remain unabsorbed to the last.

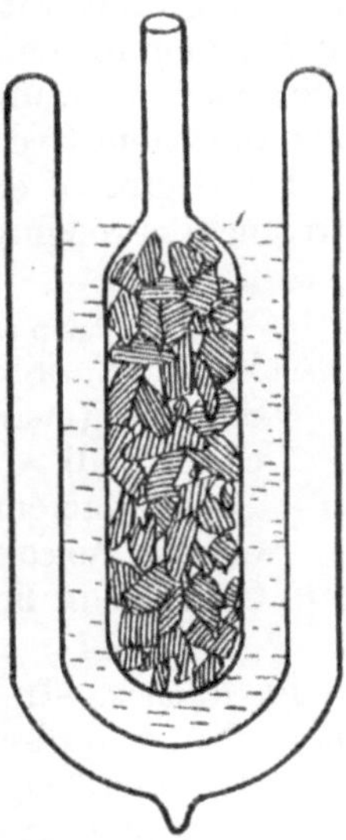

FIG. 45. CHARCOAL VACUUM VESSEL.

Pumps in which mercury is used generally require to be worked by hand, but for many purposes a pump is required which will produce a moderately good vacuum, and which can be made automatic or can be driven by power. It is only possible in this place to mention the names of some of the more important of the pumps of this kind, one or other of which is to be found in most laboratories. The Geryk (Fleuss) pump is a piston pump in which the valves are immersed in oil.

A very remarkable pistonless pump, introduced by Gaede, has been described in *Engineering* for 20th September, 1913. It is claimed for this " molecular " pump that a " vacuum " far beyond that of any other pump is attainable.

In ordinary air pumps a moving piston either solid or, in the case of the mercury pumps liquid, divides the air in the space to be exhausted into two parts, the residue being continually reduced in pressure, but never entirely removed. In some cases by doing away with valves between the space to be exhausted and the pump cylinder, while the working parts in the latter are immersed in oil, imperfections of workmanship are compensated and the attenuated residue of air mechanically assisted to escape.

The degree of exhaustion obtainable by some of these contrivances is roughly indicated by the following figures :—

Pump used.	Pressure reducible to mm. of mercury.
Water injector	7·00
Sprengel (mercury)	0·001
Geryk (oil)	0·000,2
Töpler (mercury)	0·000,01
Gaede (molecular)	0·000,000,2
Charcoal in liquid air	0·000,000,8

Electrolysis.—In the chemical laboratory the electric current is used in two directions, namely, in the production of heat as already described, and for its application to electro-chemical decomposition for analytical and other experimental purposes. In the latter case in all ordinary operations a small current of 10 to 15 ampères is all that is necessary. The current is usually obtained from a battery of accumulators of 12 to 24 cells, which can be charged from the electric lighting circuit. Four cells will usually furnish each working place, which is fitted with suitable resistances, an ammeter for measuring the current, and a voltmeter.

In ordinary gravimetric analysis the weight of some precipitate of known composition formed in the liquid by adding a suitable reagent is the object aimed at. Take, for example, a solution of copper sulphate of which it is desired to know the amount.

If to one-half of the solution an excess of solution of caustic potash is added and the liquid is heated to boiling, a black precipitate of copper oxide is formed, which is then collected on a paper filter, washed by pouring hot distilled water over it, then dried completely, and after heating to redness it is weighed. From the known composition of the copper oxide, CuO, the amount of copper is calculated.

Similarly if a solution of barium chloride is added to the other half of the solution a white precipitate of barium sulphate is produced. If this precipitate is in corresponding fashion collected, washed free from the copper solution, dried, ignited, and weighed, the amount of the sulphion, SO_4, present in the original solution can be calculated from the weight of barium sulphate obtained in the form of the precipitate, and thus the total weight of copper sulphate in the original solution becomes known.

In electro-analysis, which in many cases is more rapid and more exact, the metal in a metallic salt is deposited as such by the action of an electric current on a weighed plate or gauze of metal which can be used as the cathode in the process. In modern processes for rapid electro deposition of metals one or other of the electrodes is made to revolve rapidly in the solution under analysis. The advantage of this is obvious from the consideration that in decomposing, say, a solution of a copper salt, the liquid in the neighbourhood of the cathode surface becomes impoverished as the metal is deposited upon it, and unless the liquid is stirred so as continually to bring the metalliferous solution into contact with the cathode, the operation occupies a considerable length of time, as the extraction of the last portions of metal is then dependent on convection currents bringing those portions of the solution which still contain metal into contact with the cathode. The electrodes should be as close together as possible, and warming the solution is often an advantage. It appears to be of no importance whether the anode or the cathode be made to revolve, or whether the electrodes are stationary, the stirring being accomplished by an independent stirrer. With suitable arrangements the estimation of a metal such as copper, once in solution, can be accomplished in a very short space of time, not exceeding a few minutes.

The arrangement for the use of rotating electrodes is shown in Fig. 36, p. 68.

By suitable gradation of the potential used two metals such

as copper and zinc may be deposited one after the other from the same solution.

The current may also be employed for many experiments in which the oxidising effect at the anode, or the reducing effect at the cathode may be turned to advantage. This method has been employed chiefly in connection with the study of organic compounds. The solution is divided into two parts by a porous partition, the cathode being on one side, the anode on the other, both being immersed in the same solution, the substance to be operated on being dissolved in the one compartment or the other according to the effect, reduction or oxidation, which it is intended to bring about.

The Spectroscope is an instrument now familiar and to be found in some form in every chemical laboratory. The discovery that white light is made up of a great number of rays which give to the eye the sense of colour, was made by Newton at some time previous to the year 1675, when he described many experiments, and gave the explanation of them in his treatise on "Opticks" presented to the Royal Society. But the application of this discovery to the purposes of the chemist and physicist came nearly two hundred years later, when the spectroscope was invented and used by Bunsen and Kirchhoff, professors at the University of Heidelberg. Newton discovered that lights which differ in colour, differ in refrangibility, and when a beam of white light passes through a transparent prism the coloured rays of which it is composed are spread out, and when received on a white screen exhibit a coloured band called the spectrum. In order that the coloured rays may not overlap and confuse one another the light should be made to pass through a narrow slit parallel with the edges of the prism. The spectroscope then is an instrument consisting of one or more prisms, through which the light is made to pass, and by which the coloured rays are separated and dispersed. The light to be examined is admitted through a slit and passes, on its way to the first face of the prism, through a lens fixed in a tube, called the collimator, by which the rays are rendered parallel. The spectrum produced by passage through the prism is observed through a telescope movable through a small arc so as to enable the observer to see either the less refrangible rays at the red end or the more refrangible blue. The eye of the observer is sometimes replaced by a photographic plate on which the spectral lines or bands are recorded. In Fig. 37 (p. 69) is shown a spectroscope with one prism, and a large instrument with several prisms. The latter has a camera attached to the telescope and at the other end, next the slit, is an arrangement for producing electric sparks or a discharge through a tube containing gas or other materials which when thus ignited emit light.

Other Instruments are required for special purposes. The *Refractometer* is an instrument for determining the refractive index of transparent substances, and one form of it is a kind of spectroscope.

The Polarimeter, of which again there are several forms, is used for measuring the angle through which the plane of polarisation of a ray of polarised light is turned, to the right or left, when the light is made to traverse a layer of known thickness of a substance which has the property of circular polarisation. This instrument is much used in connection with the study especially of organic compounds (see Stereo-chemistry), and it is almost always applied to liquids. Solids which have to be examined are dissolved in an appropriate solvent.

The Microscope is occasionally used for examination of precipitates or other fine powders to ascertain if they exhibit crystalline structure, but it is most frequently employed in the study of micro-organisms, especially in connection with fermentation. In the metallurgical laboratory the microscope is much used in the examination of steel and other metals.

The Colorimeter is an instrument by which a comparison can be made between the colour or depth of tint of some liquid and a column of equal length of a standard solution. This comparison is often applied to the case of drinking water, and affords a useful indication of contamination by organic matter derived from the surface drainage of land or from streams which have passed through peaty or other vegetable deposits.

The Goniometer is used for measuring the angles of crystals. The instrument in its original simple form was described by Wollaston in 1809. It is much to be regretted that the goniometer is not to be met with in many chemical laboratories. This is perhaps due to several circumstances. Up to comparatively recent times the measurement of crystals has been applied almost exclusively by the mineralogist, for the mere recognition and characterisation of minerals, and the importance of crystalline form in its relation to chemical constitution has been recognised only within recent years. It is also doubtless in part attributable to insufficient acquaintance with the necessary mathematics, but as the mathematical treatment of chemical problems of all kinds is now common, the study of crystals, their external form and optical properties, will doubtless receive in future a larger share of attention from chemical students.

In conclusion mention must be made of the application of fused silica or quartz glass to the construction of chemical apparatus. This substance melts at about the same temperature as platinum, and is manufactured in two qualities, opaque and transparent. The former is made in an electric furnace from the less pure varieties of massive quartz; the amount of impurity is, however, very small, not usually exceeding $\frac{1}{2}$ per cent, the opacity being due to air bubbles. Muffles, trays, and large evaporating dishes are made of this material, as it resists the action of all acids except hydrofluoric acid, and, at high temperatures to a small extent, phosphoric acid. The transparent silica ware is made from rock crystal, and for some years was made exclusively by fusion in an oxyhydrogen flame. Small flasks, beakers, tubes, crucibles, dishes, and other vessels are now in use in all laboratories with great advantage, owing to the remarkable properties of this substance in relation to heat and chemical agents. Its coefficient of expansion being very small, only about $\frac{1}{17}$ that of common glass, a silica vessel bears sudden changes of temperature without damage. It is an interesting and surprising experiment to see a silica flask heated to redness in a flame plunged at once into cold water without a crack.

At the same time it is not advisable to subject large or thick masses of the material to so severe a test. And of course it must be remembered that although silica resists the action of ordinary acids it is not proof against alkalis. Contact with hot alkaline solutions and especially fused potash or soda and lime or baryta must be avoided.

PART II

MODERN DISCOVERIES AND THEORIES

CHAPTER IV

PRINCIPLES OF CHEMISTRY

CHEMISTRY is a science based on the results of experiment, but its real foundation belongs to quite modern times when experiment began to take the form of exact measurement. For ages all kinds of chemical operations and manufactures had been practised in a crude way, such as the production of soap, glass, dyes, and pigments, the distillation of alcohol from wine, the production of sulphuric acid from green vitriol, and so forth.

It was only in the middle of the eighteenth century when Black, and a little later Lavoisier, began to weigh and measure, as accurately as they could, the materials with which they were working or the products obtained in their experiments, that a body of facts was gradually accumulated on which theories could be safely established.

At different periods in the history of the science various estimates have been formed as to the influence of different men or the importance of different discoveries. Many writers have been accustomed to date the rise of chemistry as a branch of science, from the time of Robert Boyle, " The Father of Chemistry " as he has been called, at the end of the seventeenth century.

Boyle gave the first clear and precise idea of the word *element* so much used in chemistry. For he got rid not only of the Aristotelian four elements, but of the whole brood of fantastic assumptions which for centuries had clouded the brains of the alchemists. Their *tria prima*, the salt, sulphur, and mercury of their occult lore, were henceforth to disappear, and the elements of the chemist were simply those substances which were found to be incapable of further analysis.

An eminent French chemist, Wurtz, about fifty years ago commenced a graphic history of chemical theory with the words " Chemistry is a French science. It was founded by Lavoisier of immortal memory."

For such a statement, if we make allowance for a little exaggeration arising out of not unnatural national pride, justification would be sought in the interpretation of the facts then accumulated about combustion and the overthrow of the then prevalent doctrine of "phlogiston" which science owes to the genius of Lavoisier. The classification of acids, bases, and salts, and the system introduced in his remarkable *Traité élémentaire de Chimie*, as well as the nomenclature which in principle is used, so far as it is applicable, down to the present day were also part of his work.

By some English writers, on the other hand, John Dalton in virtue of his "Atomic Theory" has been regarded as the real founder of the modern science. For this view there is some justification, for the conception introduced by Dalton in 1808 remains to this day the indispensable foundation on which all modern chemistry is built, and without which some departments of our science, notwithstanding the accumulation of facts, would either not exist at all, or would remain a chaotic assemblage of observation and hypothesis.

Whatever may be the verdict about the claims of these older men of science as founders of modern chemistry there have been undoubtedly epochs from which a new departure may be dated. One of these, inaugurated about 1860, arose out of the belated recognition of a principle enunciated clearly enough fifty years before by the Italian physicist Avogadro.

It was a countryman of his, Cannizzaro, for the last forty years of his life professor of chemistry in the Royal University of Rome, who with remarkable insight perceived the importance of Avogadro's hypothesis, and with most praiseworthy insistence persuaded the chemical world of 1858 to listen to his expositions. The principle will be explained a little later.

About this time also began the more systematic study of the physical properties of substances in connection with the enquiry into their composition which had been previously the chief business of the chemist. The melting points, the boiling points, the specific gravities, the optical properties of bodies henceforward occupied attention, and it was by observations of one of these properties, namely, the power possessed by many substances of rotating the plane of polarisation of a ray of polarised light, that one of the most fruitful discoveries of our time was made. The first observations on the fact that for every compound which possesses the power of turning the plane of polarisation to the right, there is another which, while possessing the same composition, rotates equally to the left, was made by Pasteur in 1848 when a very young man. His discovery of the relation between the crystalline forms of the several tartaric acids and their action on polarised light led him to perceive the necessity for some kind of theory to account for the internal structure of the molecules of such compounds. If the atoms composing the molecule in one of such a pair of compounds be conceived as arranged in a particular order, then the atoms in the other must be arranged in the same order but inversely, so that if the atoms could be made visible they would be seen to exhibit the relation of an object to its image in a mirror. Twenty years later the subject again attracted attention, and after the study of the lactic acids by Wislicenus, a theory was put forward, by the Dutch chemist Van't Hoff, and the French chemist Le Bel, which furnished the necessary clue, and provided the basis for that large department of the subject which has since developed so remarkably under the name "Stereo-chemistry," or chemistry in space.

But probably nothing has contributed more to the progress of modern chemistry than the closer study of the relations of chemical to electrical phenomena. The fact of the decomposition of water by the voltaic pile in 1800 was soon followed by the isolation of the metals potassium and sodium by Davy, and later the establishment of the quantitative laws of electrolysis by Faraday. Then came all the wonders of spectral analysis, and so the still greater wonders to be revealed by the phenomena connected with the discharge of electricity through attenuated gases were not discovered till a good many years later.

Another circumstance which has greatly assisted progress in chemical research is the development of many of the improved instruments which are now available for the use of the experimenter. Some of these will be described later on, but the invention of the Sprengel mercury pump in 1864, by which a high vacuum was for the first time easily available, was certainly one of these. So also is the development of the dynamo by which an electric current is now supplied to every laboratory, and made accessible for so many purposes formerly undreamed of. Thus operations in which the current is made to produce chemical decomposition, or electrolysis, may be the object on one occasion, while on another a high tension discharge through a so-called vacuum tube may be wanted or the production of an arc for experiments at high temperatures. Machines for the liquefaction of air and other gases afford, on the other hand, the means of producing great cold, and much has been learned during the last twenty years about the properties of matter at low temperatures, and the influence of temperature on chemical action. The range of temperatures thus attainable in the laboratory stretches from approximately that of interplanetary space to near the central heat of the sun. Many manufacturing operations are now conducted at the temperature of the electric arc, such as the production of calcium carbide, the manufacture of phosphorus and carborundum, the fusion of quartz for making silica vessels, and the reduction of several metals from their oxides.

Among the most remarkable results of the application of modern theoretical ideas should be remembered the success which has attended the production of organic compounds, many of which had been previously known only as naturally occurring constituents of the tissues of animals or plants. The synthesis of alizarin, the red colouring matter of madder, has for many years been conducted on a large scale for industrial purposes, and the synthesis of indigo bids fair to result before long in the almost total extinction of the cultivation of the indigo plant.

Some of the sugars, fats, and proteins or albuminoid constituents of animal matters have in like manner been built up by chemical operations from purely inorganic materials. And it is a matter of common knowledge that many of the drugs employed in modern medical practice,—saccharin, aspirin, phenacetin, antipyrin, sulphonal, etc.—are artificial products of the chemical laboratory.

In theoretical chemistry electrical ideas are predominant. Chemical combination or decomposition is attributed to exchanges of electrical units, and the decomposition of fluid bodies by the electric current is almost universally attributed to the presence of "ions," wandering free fragments of molecules carrying electric charges. From the discoveries which have been made during the last thirty years by Sir William Crookes, and especially more recently by Sir Joseph J. Thomson and his school, coming about the same time as the discovery of radium, we should be perhaps justified in saying that the opening of the twentieth century is a new epoch in chemistry. For in these last few years chemists have had to get accustomed to the idea that the atoms of Dalton, previously supposed to be indestructible, are complex structures all of which can be broken up, some even undergoing spontaneous disintegration.

Leaving generalities we may now give a brief statement of the fundamental principles accepted generally by chemists at the present day, in order that what follows may be intelligible to the general reader.

First of all great principles in which chemistry is concerned is the doctrine of the *Conservation of Mass*. This means that though matter may be transformed in appearance and qualities none of it is lost or destroyed. When a candle is burned it slowly disappears, but when arrangements are made for catching and weighing the gaseous products of its combustion these are found to be made up of the carbon and hydrogen of which the wax is composed, together with oxygen taken from the air. Or when limestone is heated in a limekiln the lime which remains always weighs fifty-six pounds for every hundred pounds of limestone burnt, if the latter is free from impurities. The carbon dioxide, commonly called carbonic acid, which escapes can be shown to represent the missing forty-four pounds, and if combined again with the lime will reproduce the original substance. Experiments of this kind were originally published by Black in 1777.

It is true that elaborate experiments have been made in recent times to test the validity of this principle, but none of the results observed so far have shaken confidence in its soundness.

The practice of weighing carefully led to the discovery, early in the nineteenth century, of the *Law of Constant Proportions*. Here again is a fundamental principle. The law states that any given chemical compound is always composed of the same elements united in the same proportions. This proposition was finally established by a French chemist, Proust, early in the nineteenth century, notwithstanding much controversy and criticism. It is of course a principle on which rests the whole of quantitative analytical chemistry, and from the practice of which daily evidence of its truth is supplied. In plain language it means that, for example, water is always composed of one part of hydrogen combined with eight parts of oxygen, and that no matter from what source it is procured it always has exactly the same properties, the same colour, the same boiling point, the same freezing point. If different samples of natural water seem to differ one from another, as for instance, rain water from river or sea water, this is merely due to the presence of other substances dissolved in it, and which by well-known methods can be separated from it, leaving the water unchanged.

Another important law in chemistry is the *Law of Multiple* proportions discovered by John Dalton, and illustrated and explained by his famous *Atomic Theory.*

The question whether matter is capable of division and subdivision *ad infinitum*, or whether there is a limit beyond which the particles are so hard that no power in nature is capable of breaking them into smaller pieces, is one which has been debated from the earliest times.

The vague atomic hypothesis of Democritus was the subject of fruitless debate in the Middle Ages. Newton in expounding his gravitation theory, which is applicable to the smallest particles of matter as well as to suns and planets, gave expression to the view that in the beginning matter was formed of "solid, massy, hard, impenetrable, movable particles," and that "those primitive particles being solids are incomparably harder than any porous bodies compounded of them; even so very hard as never to wear or break in pieces." But it was reserved for Dalton to supply that basis of fact without which every hypothesis is useless. How the theory was established is explained in the best text-books. It will be sufficient here to give in Dalton's own words an enunciation of the modern doctrine.[1] "Chemical analysis and synthesis go no farther than to the separation of particles one from another and to their reunion. No new creation or destruction of matter is within the reach of chemical agency. . . . All the changes we can produce consist in separating particles that are in a state of cohesion or combination, and joining those that were previously at a distance." If, therefore, there are two substances which can combine, say nitrogen and oxygen, their union can only occur between whole numbers of atoms, such as one atom of nitrogen to one atom of oxygen, one atom of nitrogen to two atoms of oxygen, or two atoms of nitrogen to one atom of oxygen, etc.

[1] *Dalton's Chemical Philosophy*, 1808, Vol. I, p. 112.

The known compounds of nitrogen with oxygen are represented by the following formulæ in which each of the capital letters represents one atom of the element :—

Nitrous oxide	N_2O
Nitric oxide	NO
Nitrogen trioxide	N_2O_3
Nitrogen tetroxide	N_2O_4
Nitric peroxide	NO_2
Nitrogen pentoxide	N_2O_5

Obviously if the weight of matter represented by each symbol is known, the relative weights in which the two elements combine to form these compounds is also known. The weights attributed to the symbols are called the atomic weights, and what has been learnt about them since Dalton's time will be discussed at length in later pages.

No sooner is the conception of the atom as the ultimate particle of an element firmly established than it becomes obviously desirable to use some other word to designate the pile of atoms, which, according to the theory, is formed when a chemical compound is produced. Such a word is *molecule* (dim. of Latin moles, a heap) which, though introduced into science a century ago, has only become during the last fifty years both familiar and endowed with a precise signification. Dalton himself did not scruple to write of an atom of water, and made no distinction between an atom of an element and an atom of a compound. And in one sense this is justifiable, for what is now called a molecule of water is also an atom (i.e. something indivisible) inasmuch as if further divided it ceases to be water, and becomes an equal mass of mixed oxygen and hydrogen.

The word molecule acquired serious importance when it appeared in the title of a paper published in 1811, which, though it attracted comparatively little notice at the time, was at last recognised by the chemical world so long afterwards as 1860. This was the paper by the Italian physicist, Avogadro, in which is enunciated the hypothesis which bears his name, and which is expressed as follows: "Equal volumes of gases, simple or compound, contain under the same conditions of temperature and pressure the same number of molecules." Hence the weights of gaseous molecules are directly proportional to the specific gravities of the gases. And from the known densities of the elementary gases it follows that many of them consist of molecules containing more than one atom, e.g. hydrogen, oxygen, nitrogen, chlorine contain two atoms each.

The determination of the density of the vapours of a large number of substances which, though not gaseous at common temperatures, are convertible into vapours by heat, provides, therefore, a method very generally applicable to the determination of molecular weights.

A molecule is now always understood to mean the smallest mass of any substance, elementary or compound, which is capable of existing by itself.

At the time of the publication of Avogadro's hypothesis, or law as it is often called, Michael Faraday was a youth just twenty years of age, and as yet following his occupation of bookbinder. It was only little more than twenty years later that, pursuing the study of electro-chemical decomposition or electrolysis, he gave the world the two great quantitative laws which are generally known as Faraday's Laws of Electrolysis. They may be stated as follows in his own words, which will be found in his *Experimental Researches in Electricity* (vol. I, p. 241) :—

I. "The chemical power of a current of electricity is in direct proportion to the absolute quantity of electricity which passes."

II. "Compound bodies may be separated into two great classes, namely, those which are decomposable by the electric current, and those which are not. . . . I propose to call bodies of the decomposable class *electrolytes*. Then again the substances into which these divide under the influence of the electric current form an exceedingly important general class. They are combining bodies, are directly associated with the fundamental parts of the doctrine of chemical affinity, and have each a definite proportion in which they are always evolved during electrolytic action. I have proposed to call these bodies generally *ions*, or particularly *anions* and *cations*, according as they appear at the *anode* or *cathode*, and the numbers representing the proportions in which they are evolved *electro-chemical equivalents*. Thus oxygen, chlorine, iodine, hydrogen, lead, tin are *ions* ; the three former are *anions*, hydrogen and the two metals are *cations*, and 8, 36, 125, 1, 104, 58 are their electro-chemical equivalents nearly."

The theory by which the process of electrolysis is now explained has been very considerably modified within recent times, but Faraday's two quantitative laws remain unaltered, and constitute a firm basis on which researches continued since Faraday's time securely rest.

In the meantime almost before Dalton's Atomic Theory had become familiar to the majority of chemists, and long before Faraday's discoveries in electricity were made known, another discovery was made, the importance of which has in later times been fully recognised. In 1819 the two French physicists, Dulong and Petit, discovered the relation between specific heat and atomic weight, which is expressed by their law, as follows: the specific heat of an element in the solid state is inversely proportional to its atomic weight.

Consequently the number expressing the specific heat, multiplied by the atomic weight, gives a constant which is approximately 6·4 for all temperatures between the freezing and boiling points of water. Four exceptions among the elements, namely, carbon, boron, silicon, and the metal glucinum are known, but are accounted for, and the principle is universally accepted and acted upon for the purpose of regulating the value to be assigned to those atomic weights which cannot be fixed by appeal to other rules, such as that of Avogadro.

Dulong and Petit expressed their law in the following words: "Les atomes de tous les corps simples ont exactement la même capacité pour la chaleur." That is the atoms of all the elements have exactly the same capacity for heat.

An equally remarkable fact was observed some years later by Neumann, who found that there was a similar relation between the specific heats of chemically similar compounds, and the sum of the atomic weights of the elements composing them. Since that day these results have been corrected and extended, so that it may now be said that the specific heat of the molecule of a compound is the sum of the specific heats of the atoms composing it, very approximately, on condition that elements are excluded such as oxygen, hydrogen, nitrogen, of which the specific heat in the solid state cannot be found by experiment. In other words the capacity of the elementary atoms for heat is the same whether they are in the elemental, uncombined state or form part of a chemical compound. The independence of the atom in any condition is the interesting point.

Down to quite the end of the eighteenth century it was supposed that heat was a kind of substance which existed in all sorts of matter and was squeezed out of it when subjected to pressure or friction. The material of heat was called *caloric*, and when associated with certain kinds of matter in sufficient quantity the liquid or gaseous state was produced. Lavoisier, for example, spoke of oxygen gas as made up of the basis of oxygen combined with caloric. But very soon after this time Rumford showed that a given mass of any metal, such as brass, can give out heat in indefinitely large quantity when subjected to friction in the process of boring. Sir Humphry Davy a few years later demonstrated that ice can be melted by merely rubbing it, though kept all the time in an atmosphere below the freezing point of water. Finally, in 1843, Joule of Manchester showed that there is quantitative relation between the work done by a body falling under the influence of gravity and the heat which is produced in the process. Joule's experiments led to the result that a mass of 772 lbs. falling through 1 foot or of 1 lb. falling through 772 feet may produce, by friction or otherwise, heat enough to raise the temperature of 1 lb. of water 1° Fah. This expresses in ordinary English weights and measures the *mechanical equivalent of heat.* It may be expressed in grams, metres, and degrees centigrade, or otherwise, and the work done by the falling mass, attracted by the earth, may be utilised either to produce heat by friction directly, or it may first produce an electric current which may then be converted into heat, but the same quantitative relation is maintained. Hence we have the theory that heat is a "mode of motion" and that this motion is convertible into heat, light, electricity or magnetism, production of steam, and so mechanical force, or finally into chemical action. In any one of these cases the body concerned is said to possess *energy*, and the work it can do while changing its state is a measure of the amount of energy available. The energy of a body in motion is what is called *kinetic energy* (κινέω, to move), while that which it owes to its position or chemical state is called its *potential energy.* The one being convertible into the other, the sum of these two quantities is constant. These facts have been studied from many sides by a large number of physicists and engineers, beside those whose names have already been mentioned. Among those of the past are the names of Carnot and Meyer, while those of Helmholtz, Kelvin,

Maxwell, Clausius, and Willard Gibbs belong to more recent times. And the sum of their work is the principle of the *Conservation of Energy* which, with the *Conservation of Matter*, lies at the foundation of all physical science, including chemistry. These two fundamental propositions are now regarded as practically axiomatic, and are not likely to undergo serious modification or correction as the result of further research. On the other hand, conceptions relating to the Atomic Theory, the process of electrolysis, and the nature of chemical action have undergone very important developments within the present generation, and the general nature of present views will be explained in the following chapters.

CHAPTER V

ELECTRIC DISCHARGE IN GASES—ELECTRONS

THE smallest part of any substance capable of independent existence is called a molecule, and according to Avogadro's law already quoted the number of molecules in a given volume of any gas, under like conditions, is the same, and is independent of the composition of the gas. It requires a little thought to realise how tiny are these particles and how many there are crowded together in any small portion of a gas. In gases also they are much further apart than in a liquid or solid. As to their size no better idea can be conveyed than in the words of Lord Kelvin in a lecture at the Royal Institution in 1883. He says: "To form some conception of the degree of coarse-grainedness indicated imagine a globe of water or glass, as large as a football (or say a globe 16 centimetres diameter) to be magnified up to the size of the earth, each constituent molecule being magnified in the same proportion. The magnified structure would be more coarse-grained than a heap of small shot, but probably less coarse-grained than a heap of footballs."

To express their number in any visible portion of matter is even more difficult. But as the result of the application of a considerable number of different methods it may be stated that it has been estimated that 1 cubic centimetre of air under standard conditions, that is at 0° C., and under a pressure equal to 760 millimetres of mercury, contains $2{\cdot}7 \times 10^{19}$ molecules. The highest attainable vacuum still contains many millions per cubic centimetre.

Fig. 46.
1 Square Centimetre
Contains 10 millimetres on each side, and therefore 100 square millimetres in area. 1 cubic centimetre contains 1000 cubic millimetres.

But the kinetic theory of gases teaches us that in a gas all the molecules are constantly in motion, moving in straight lines, and frequently striking one against another and against the walls of the containing vessel, and so altering their direction. In liquids there is reason to believe that the molecules move but less actively, and clusters of them move in company, while in the solid state the molecules, though not stationary, vibrate more or less rapidly about a mean position, the vibration corresponding to what is called their temperature.

It would be easy to confound the reader with large figures if it were attempted to express the velocity at which molecules travel in a gas, or the mass of an atom of hydrogen or oxygen and so forth, but little would be gained. It is only necessary to remember that gas molecules are far too small to be visible with the aid of any known instrument, that they move with great speed and they collide very frequently. The space between one collision and the next is called the free path of a molecule, and though this varies according to circumstances the *mean* free path can be calculated. In its original form the kinetic theory was not concerned with the form or nature of the molecule itself or of the atoms of which it is composed. But within the last twenty years the experimental researches, especially of Sir Joseph J. Thomson, on the effects of the electric discharge through attenuated gases have supplied information of the most unexpected and startling kind, which may be regarded as giving to physicists and chemists alike an entirely new point of view as to the ultimate constitution of matter.

Soon after the improvement of the induction coil by Ruhmkorff some seventy years ago, experiments on the production of sparks in air and other gases led to the discovery of the beautiful luminous effects which are produced when the discharge passes through gases in an attenuated state. It was discovered, among other things, that the colour and appearance of the light depend not on the substance of the electrodes, but on the nature of the enclosed gas. It was also found by Plücker that the luminous discharge was capable of deflecting a suspended magnetised needle, and is itself acted upon by a magnet.

It will be worth while to begin by a brief account of the principal facts about the electric discharge. If the two terminals of any source of high potential electricity are separated by a gas such as air at common atmospheric pressure, and the voltage is gradually increased, at a certain difference of electric pressure the air is ultimately unable to bear the strain and a current passes momentarily producing a spark. If now the gas contained in the experimental tube is expanded by the use of an air pump, the difference of potential in the two terminals required to cause a discharge is less, and as the pressure on the gas is diminished the character and appearance of the discharge changes. Straight, well-defined sparks are no longer produced,

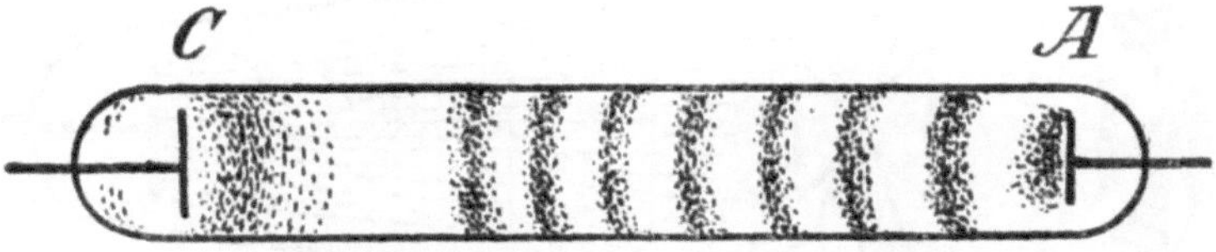

FIG. 47. ELECTRIC DISCHARGE UNDER REDUCED PRESSURE.
C = Cathode. A = Anode.

but a line of light, extending the whole length of the tube, is gradually developed, while the negative pole becomes covered with a violet-coloured glow. If the pressure of the gas is reduced to about half a millimetre of mercury or less, the discharge changes again in appearance and stratification appears, the glow separating into distinct portions with dark spaces between.

Next the cathode or negative electrode there is a non-luminous space, especially noticeable, which is commonly referred to as Crookes' space, as these phenomena have been studied by him for many years. In order to explain some of the phenomena observed in highly exhausted vessels, Crookes attributed them to new properties developed in the gas in consequence of the reduction in the number of molecules present. "The modern idea of the gaseous state is based on the supposition that a given space contains millions of millions of molecules in rapid movement in all directions, each having millions of encounters in a second. In such a case the length of the mean free path of the molecules is exceedingly small as compared with the dimensions of the vessel, and the properties which constitute the ordinary gaseous state of matter, which depend upon constant collisions, are observed. But by great rarefaction the free path is made so long that the hits in a given time may be disregarded in comparison to the misses, in which case the average molecule is allowed to obey its own motions or laws without interference; and if the mean free path is comparable to the dimensions of the vessel, the properties which constitute gaseity are reduced to a minimum, and the matter becomes exalted to an ultra-gaseous state, in which the very decided but hitherto masked properties now under investigation come into play." Matter then, according to Crookes, exists under the circumstances of a very high vacuum in what he regards as a fourth state, which is neither

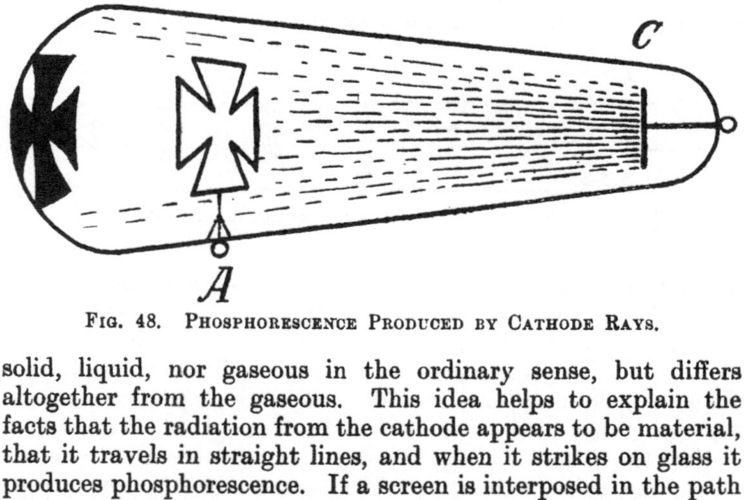

FIG. 48. PHOSPHORESCENCE PRODUCED BY CATHODE RAYS.

solid, liquid, nor gaseous in the ordinary sense, but differs altogether from the gaseous. This idea helps to explain the facts that the radiation from the cathode appears to be material, that it travels in straight lines, and when it strikes on glass it produces phosphorescence. If a screen is interposed in the path of the rays, no phosphorescence is produced within the area of its shadow. If the cathode, instead of being flat, is made concave the rays thrown off from it may be brought to a focus, and in this focus any solid object is heated intensely, and, if fusible, may be melted.

If the degree of exhaustion in the tube is increased beyond a certain point the discharge ultimately refuses to pass. Several other facts also require to be noted. For example, it was found by Lenard twenty years ago that the emanation from the cathode when directed on to a very thin aluminium plate is capable of passing through it, at any rate it appeared to do so, as phosphorescence was excited on a glass surface a short distance from the opposite side. Such a fact was, however, difficult to reconcile with Crookes' hypothesis of electrified particles of the same dimensions as the molecules of ordinary matter. In 1895 the X-rays were discovered by Röntgen, and these X-rays, which are not material, are produced by the cathode rays whenever they strike matter of any kind. The characteristic property of the X-rays is their power to penetrate more or less easily a great variety of substances opaque to ordinary light. It was observed by Röntgen almost immediately that when the hand is held in the path of these rays cast on a fluorescent screen, the shadow of the bones is darker than that of the flesh. Similarly these rays will pass through paper, wood, and metals of small atomic weight, such as aluminium, with little diminution, though they are completely stopped by a comparatively thin layer of a metal of high atomic weight, such as platinum, gold, or lead.

As the particles shot off from the cathode are material and carry an electric charge, the path they follow may be regarded as the path of an electric current, and accordingly they are deflected by the action of a magnetic or electric field. Measurements and observations of the effects produced led Sir J. J. Thomson to conclusions of the greatest importance as to the velocity and mass of these particles. The apparatus used is shown diagrammatically in Fig. 49. Here C is the cathode, A is the anode, and B B′ are two metal plates with a hole through the centre by which a pencil of cathode rays is conducted down the axis of the tube between the two parallel plates of aluminium, D D′, which can be charged as required. A and B B′ are connected to earth. When the discharge is passing the pencil of cathode rays falls on a zinc sulphide screen producing a spot of bright green light at F. If a difference of potential is established between the two plates D and D′ the rays are bent down to some point F′, and the radius of the circle into which they have been deflected can be calculated. The ratio of the mass of the cathode particle to the charge it carries can be calculated from the data thus obtained.

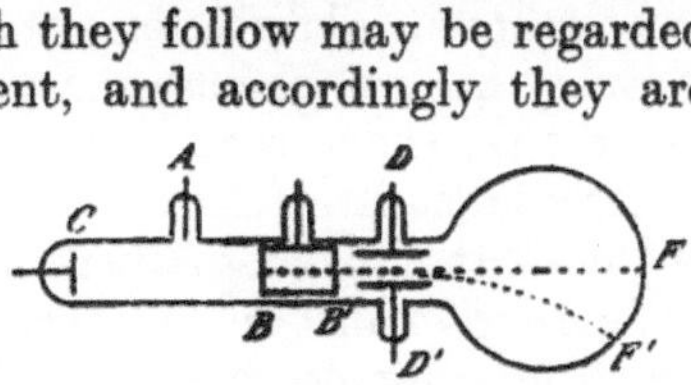

FIG. 49. J. J. THOMSON'S APPARATUS.

The cathode stream was found to consist not of atoms or molecules, but something much smaller which were at first spoken of merely as corpuscles, but the convenient term *electron*, introduced by Johnstone Stoney, was soon adopted, and has been generally used ever since. The word electron was originally employed to designate the atom of electricity or electrolytic unit which is carried by an atom of hydrogen in electrolysis (q.v.).

The mass of each of the cathode particles is about $\frac{1}{1800}$ of the mass of an atom of hydrogen, the smallest of known atoms. But the properties of the electron seem to have no connection with the nature of the gas through which the discharge takes place, nor with the material of the cathode itself. It is now known that similar electrons are emitted not only from ordinary atmospheric gases, but from the inert elements helium, argon, and the rest. Solids of many kinds, red-hot metals or metal surfaces illuminated by ultraviolet rays, as well as such oxides as lime and baryta, when heated also yield electrons. Radium and other radio-active substances emit them, and when derived from this source they are spoken of as β-rays. In fact electrons are to be found in all directions, and probably play an important part in many natural phenomena previously obscure. Thus it is surmised that the aurora seen in the northern sky is produced by electrons discharged from the sun and moving under the influence of the earth's magnetic lines of force.

There remains one important property of the electron to which no reference has so far been made, and that is its power of ionising gases. Under ordinary conditions air and other dry gas is an almost perfect non-conductor of electricity. But the residual gas in a vacuum tube through which a discharge is passing acquires conductivity, and at the same time becomes more or less luminous. Air in the presence of radium, as will be explained later, also acquires the power of conducting electricity. The ions thus produced are some of them positive and some negative; they are thought to be formed by the removal of a negative electron from a molecule of the gas which thus becomes positive, while the negative electron may attach itself to another molecule of the gas forming the negative ion. It seems also possible that they may be formed of fragments of molecules which have been broken up by collisions with the swiftly moving electrons. The latter gradually lose their energy in the process and ultimately lose their cathodic character, possibly entering

J. J. Thomson.

To face page 118.

into combination with some positive ion. The conductivity of the gas soon disappears if the supply of the ionising agent is not kept up, a consequence to be expected from the presence of both positive and negative atomic groups which naturally tend to recombine when they meet.

Thus far the origin and properties of the rays from the cathode have alone been described. But they are accompanied by rays of positive particles which are sometimes called "canal rays" from the manner in which they were first observed. If the cathode is perforated these particles pass through it, and move in straight lines in the opposite direction. The composition of these positive rays is much more complex than that of the cathode rays, for whereas the particles in the cathode rays are all of the same kind, there are in the positive rays many different kinds of particles. They are deflected by strong magnetic and electric fields, but less easily than the cathode rays, from which it is inferred that their mass is greater than that of an electron. They appear to consist of positive ions, derived either from the gas or the electrodes, but they seem to start chiefly from the boundary of the Crookes' dark space. These rays have within the last few years been investigated by Sir J. J. Thomson,[1] who concludes that they have a mass never less than that of a hydrogen atom, anything corresponding to an electron with a positive charge being so far unknown.

The following is a brief description of the apparatus used by Thomson :—

A is a large bulb in which the anode is inserted on one side, the cathode C occupying the opposite neck of the bulb. In the diagram, for the sake of clearness, the tubes are not shown by which the gas is admitted into the bulb, nor the communication with the pump by which the contained gas is kept at the proper low pressure. The cathode has in front an aluminium cap

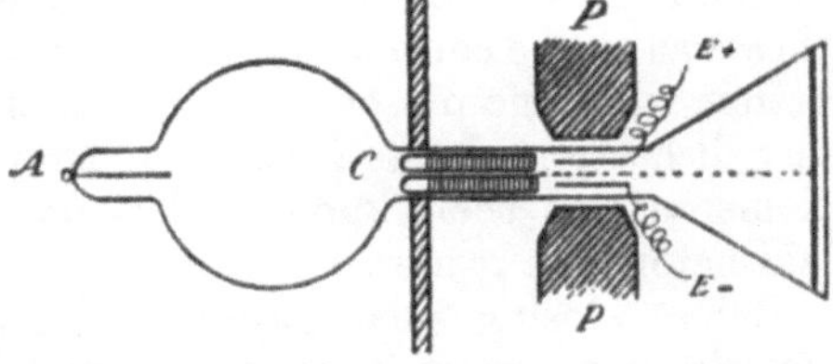

FIG. 50. J. J. THOMSON'S APPARATUS FOR STUDYING POSITIVE RAYS.

[1] See especially "The Bakerian Lecture for 1913" (*Proceedings Royal Society*, Vol. LXXXIX A, pp. 1–20).

which fits on to a cylinder of soft iron with a hole bored along the axis. A very narrow copper tube, only about ·1 mm. to ·5 mm. in diameter, passes through this tube. The particles entering are therefore protected from magnetic forces till emerging from the copper tube, they pass between the poles of the electro magnet PP, or the plates EE, by which an electric field can also be provided. The positive particles then strike on a photographic plate and there record a series of parabolas which depend on the gas or gases present in the tube.

It appears that each of the curves corresponds to a different atomic weight and that these ions are capable of carrying one, two, three or more unit charges. The method therefore can be used as the basis of a new method of analysis, and some very remarkable results have been obtained. Thus among the gases evolved from platinum by bombardment with cathode rays are found not only atoms of hydrogen with one charge and molecules of hydrogen with two charges, but particles of a gas which appears to consist of an element with atomic weight 3, which is either a previously unknown element or consists of hydrogen associated into a molecule H_3.

Another strange result of this investigation is that it appears possible to recognise temporary associations of atoms which are so unstable that they are unknown to the chemist. Thus when marsh gas CH_4 is used in the experimental tube fragments of the molecule CH_3, CH_2, CH and C appear to be able to record their presence on the photographic plate, although the life of each cannot be longer than a very small fraction of a second.

The discoveries which have thus resulted from a close study of the effects of the electric discharge naturally suggest hypotheses as to the constitution of atoms and the nature of chemical action. For the present it will be more convenient to postpone any discussion of this fascinating subject till the history of the elements and their known relations to one another has been laid before the reader.

Before closing this chapter which relates to the action of electricity on gases in general it will be appropriate to give a brief account of some phenomena connected with this subject and brought about by the same agency in connection with the element nitrogen.

The investigation has been carried out by the Hon. R. J. Strutt, professor of physics in the Imperial College of Science at

South Kensington, and has extended over several years. The most important facts were set forth in the Bakerian Lecture given to the Royal Society in 1911, of which a condensed account is given in the following paragraphs.

ACTIVE NITROGEN

It has been known for a long time that vacuum tubes frequently show a luminosity of the contained gas after discharge of electricity through the tube has taken place and has ceased. In the case of air Strutt found that this effect is due to a phosphorescent combustion occurring between nitric oxide and ozone, both formed during the discharge. He also found that other phosphorescent combustions are observed in ozone, notably of sulphur, sulphuretted hydrogen, acetylene, and iodine. With a moderate discharge of electricity it was at first supposed that pure nitrogen gave no afterglow, but when a jar discharge was used with spark gap the glow was readily obtained.

In order to examine the properties of the gas while showing this phenomenon, the vacuum tube through which the discharge passed was connected with an observing vessel, and a current of gas was drawn into the latter by the operation of a powerful air pump. One very remarkable effect on the appearance of the glow is produced by change of temperature. If a long tube, through which a stream of glowing nitrogen passes, is moderately heated the glow is locally extinguished, but the luminosity is recovered as the gas passes on into the cooler part of the tube. If, on the other hand, the gas is led through a tube immersed in liquid air it glows with increased brilliancy where it approaches the liquid air, though the luminosity is finally extinguished when it reaches the coldest part of the tube. It appears then that the change, whatever it is which gives rise to the luminosity, is promoted by cooling and retarded by heating.

The glowing nitrogen has remarkable chemical properties. It combines with common phosphorus at the same time producing much red phosphorus. In this behaviour it resembles the halogens, chlorine, bromine, and iodine. It also combines with sodium, with mercury, and some other metals, in each case developing the line spectrum of the metal concerned.

It attacks nitric oxide, with formation, strangely enough, of nitrogen peroxide, a more oxidised substance. This, however, can be explained by supposing that the active nitrogen prefers to combine with nitrogen, and so withdraws a portion of it from the nitric oxide, leaving more oxygen for the remainder.

It also attacks acetylene and other gases containing carbon, and the result is the production of cyanogen compounds, the presence of which can be demonstrated by shaking up the gas with caustic potash and adding an iron salt in the usual way when Prussian blue is produced.

Glowing nitrogen also exhibits remarkable phenomena when in the presence of iodine. Its normal yellow glow is replaced by a magnificent light blue flame at the place where it mingles with the iodine vapour. At this point a slight rise of temperature is observed. Active nitrogen also attacks mercury, forming with it a compound which explodes when moderately heated.

It appears to be certain that the phosphorescent nitrogen does not owe its activity to a state of condensation corresponding with that of ozone, the molecule of which consists of three atoms O_3, the instability of the molecule being due chiefly to the tendency to the production of the more stable ordinary molecule which contains only two atoms, O_2. Active nitrogen appears rather to consist of separate atoms of the element produced by the dissociation of the ordinary molecules composed of two atoms, N_2. The glowing gas does not appear to owe its peculiarities to the presence of electrified particles, as it is unaffected by passing through an electric field.

CHAPTER VI

THE ELEMENTS OF THE CHEMIST

It is one of the characteristic features of modern physical science, which is not, like the ancient, content with observation of natural phenomena, but depends for progress on the results of experiment, to be perpetually in a state of flux. Its advance is analogous to the ascent of a mountain; the higher the traveller rises the broader is the prospect which becomes visible. He may now and then reach a plateau which tempts him to rest and look backward content for the time with the view, at the same time he knows full well that this resting-place is not the summit and that what he now sees will appear insignificant when a higher altitude is reached. Something of the same kind happens in the evolution of physical science. A theory is formed which for a time seems to provide a satisfactory explanation of all the facts under consideration. But before long some more accurate measurement or the observation of some neglected and apparently trivial circumstance requires a revision of the accepted doctrine or its displacement by a new one.

These remarks apply specially to the case of the chemical elements. The word element has received many applications, and even at the present day in ordinary speech it is used sometimes in a poetical sense, with general allusion to air or water, or it simply means a constituent or ingredient in a mixture of things. Passing over any further reference to popular or ancient usage the word element received for the first time a definition with a scientific character from Robert Boyle in the seventeenth century. And his definition has been current among chemists since his day. An element, according to Boyle, is a substance which resists analysis. It consists of one kind of matter, and by no known process is it possible to extract from it more than one kind of stuff. It is only in the most recent times, namely, since the discovery of radium in 1902 and the related substances, that this definition, accepted as it has been for upwards of two hundred years, can no longer be applied without qualification to many substances which previously would have been included without hesitation under it. On the other hand, we have long since learned that several substances which formerly answered to the definition, having resisted the then known methods of analysis, such as lime, baryta, and the alkalis, are really compound bodies consisting of oxides of metals. But no real advance beyond a position of mere speculation could be accomplished until the phenomena of chemical combination were studied quantitatively and the laws of chemical combination were established. The law of definite proportions, the law of multiple proportions, and the law of reciprocal proportions were enunciated more or less clearly more than a hundred years ago, and all subsequent experiment has only served to establish them the more firmly. Then came in 1808 the Atomic Theory of John Dalton, which at once supplied an explanation of the observed facts. This theory assumes that each element consists of minute separate particles, all alike in size, weight, and chemical properties, and that when chemical combination takes place between any two or more elements to form a compound a definite and limited number of the particles of one kind are intimately associated with a definite and limited number of another kind. Later investigations showed that in every case the associated atoms occupy in space relative positions toward one another which are all definite, and that the properties of the body are connected with and dependent on the configuration of the resulting mass, which is called a molecule. The study of these relationships constitute the department of science known as *stereo-chemistry*, or chemistry in space. It has led to many discoveries in later times, and is still, at the present day, a very important field of investigation.

Dalton himself began attempts to determine the relative masses of the atoms conceived by his hypothesis, but the experimental methods available in his time did not admit of the attainment of accuracy. The subject, however, was pursued with improved methods and greater skill by a number of the most able chemists in the former half of the nineteenth century. The names of Berzelius, Dumas, Gay-Lussac, and Stas are prominent among the workers in that field.

The result of all their labours and that of a host of others was the establishment of a list of substances recognised as elements, in the sense already defined, together with the numbers which represent the relative masses of their atoms or what are called their atomic weights. The revision and criticism of these numbers has for many years past been undertaken by an International Committee of chemists, and a list is issued annually, under the authority of this committee, which represents the latest and best estimates of these important values. Till a few years ago hydrogen, as the lightest body known, was used as the standard for comparison, but after much discussion in the chemical world, which it is not necessary to follow in this place, it appeared more convenient to assume the atomic weight of oxygen to be represented exactly by the whole number 16, and to calculate all the rest accordingly. Hence the atomic weight of hydrogen is not now represented by the number 1·0, which would require the atomic weight of oxgyen to be 15·88, but by the figure 1·008 as given in the following table.

It ought to be understood that the atomic weights given in the list are not all equally trustworthy. Some, such as chlorine, potassium, silver, barium, represent a very high degree of accuracy, while others, for various reasons, such as possible presence of impurities or difficulties in manipulation of the compounds analysed, will probably suffer some slight revision in future years. Lithium, for example, would be a much more difficult case than potassium, and it is still uncertain whether the whole range of rare earth metals, lanthanum, cerium, and the rest, have yet been obtained in the state of purity desirable.

INTERNATIONAL ATOMIC WEIGHTS FOR 1916

Arranged in order of numerical value.

Hydrogen	1·008	Selenion	79·2
Helium	4·0	Bromine	79·92
Lithium	6·94	Krypton	82·92
Glucinum	9·1	Rubidium	85·45
Boron	11·0	Strontium	87·63
Carbon	12·005	Yttrium	88·7
Nitrogen	14·01	Zirconium	90·6
Oxygen	16·0	Columbium	93·5
Fluorine	19·0	Molybdenum	96·0
Neon	20·2	Ruthenium	101·7
Sodium	23·0	Rhodium	102·9
Magnesium	24·32	Palladium	106·7
Aluminium	27·1	Silver	107·88
Silicon	28·3	Cadmium	112·40
Phosphorus	31·04	Indium	114·80
Sulphur	32·06	Tin	118·7
Chlorine	35·46	Antimony	120·2
Argon[1]	39·88	Tellurium[1]	127·5
Potassium	39·10	Iodine	126·92
Calcium	40·07	Xenon	130·2
Scandium	44·1	Cæsium	132·81
Titanium	48·1	Barium	137·37
Vanadium	51·0	Lanthanum	139·0
Chromium	52·0	Cerium	140·25
Manganese	54·93	Praseodymium	140·9
Iron	55·84	Neodymium	144·3
Nickel	58·68	Samarium	150·4
Cobalt	58·97	Europium	152·0
Copper	63·57	Gadolinium	157·3
Zinc	65·37	Terbium	159·2
Gallium	69·9	Dysprosium	162·5
Germanium	72·5	Holmium	163·5
Arsenic	74·96	Erbium	167·7

[1] Placed out of order for reasons which will appear later.

Thulium	168·5	Mercury	200·6
Ytterbium (Neoytterbium)	173·5	Thallium	204·0
Lutecium	175·0	Lead	207·2
Tantalum	181·5	Bismuth	208·0
Tungsten	184·0	Niton (Ra Emanation)	222·4
Osmium	190·9	Radium	226·0
Iridium	193·1	Thorium	232·4
Platinum	195·2	Uranium	238·2
Gold	197·2	Total	83

The numbers which have been adopted in this table have all been selected so as to comply with certain rules long established and fully explained in all the best textbooks of chemistry of the present day. The first of these is known as the law of Avogadro, who proposed the hypothesis on which it is based in 1811. It was not, however, generally recognised till more than fifty years later.[1] The other principle made use of depends on the relation between the specific heat of a solid element and its atomic weight, discovered by the French physicists Dulong and Petit in 1819. In those cases in which both these rules can be applied the result is the same.

As soon as a table such as the one just given could be drawn up a very important discovery was made. In 1863 it was observed by Mr. J. A. R. Newlands that in such a table, imperfect as it was at that time, the properties of the elements are related to their position in the series. Every eighth element in the list, starting from any point, exhibits a revival of the chief properties of the element from which counting is begun. Take, for example, the metal lithium as starting-point, the eighth element following it is sodium, and the eighth following is potassium, and these three elements form a natural family, the members of which are very like to one another in chemical properties, and show a gradation in physical properties. This discovery led to further investigation, and in the end the so-called periodic law of the elements was announced by the late Professor Mendeléeff in 1869.[2] This principle has

[1] The reader who is interested in such matters as the history of Avogadro's doctrine should read the "Memorial Lecture on Cannizzaro," by Sir Wm. Tilden, in the *Transactions of the Chemical Society for 1912*, p. 1677.

[2] For a full account of the origin and history of the conception see the "Memorial Lecture on Mendeléeff," in the *Transactions of the Chemical Society for 1909*.

Series.	Zero group.	Group I.	Group II.	Group III.	Group IV.	Group V.	Group VI.	Group VII.	Group VIII.		
0	*x*	—	—	—	—	—	—	—	—	—	—
1	*y*	Hydrogen, H=1·008	—	—	—	—	—	—	—	—	—
2	Helium, He=4·0	Lithium, Li=7·03	Beryllium, Be=9·1	Boron, B=11·0	Carbon, C=12·0	Nitrogen, N=14·04	Oxygen, O=16·0	Fluorine, F=19·0	—	—	—
3	Neon, Ne=19·9	Sodium, Na=23·05	Magnesium, Mg=24·1	Aluminium, Al=27·0	Silicon, Si=28·4	Phosphorus, P=31·0	Sulphur, S=32·06	Chlorine, Cl=35·45	—	—	—
4	Argon, Ar=38	Potassium, K=39·1	Calcium, Ca=40·1	Scandium, Sc=44·1	Titanium, Ti=48·1	Vanadium, V=51·4	Chromium, Cr=52·1	Manganese, Mn=55·0	Iron, Fe=55·9	Cobalt, Co=59	Nickel, Ni=59 (Cu)
5	—	Copper, Cu=63·6	Zinc, Zn=65·4	Gallium, Ga=70·0	Germanium, Ge=72·3	Arsenic, As=75·0	Selenium, Se=79·0	Bromine, Br=79·95	—	—	—
6	Krypton, Kr=81·1	Rubidium, Rb=85·4	Strontium, Sr=87·6	Yttrium, Y=89·0	Zirconium, Zr=90·6	Niobium, Nb=94·0	Molybdenum, Mo=96·0	—	Ruthenium, Ru=101·7	Rhodium, Rh=103·0	Palladium, Pd=106·5 (Ag)
7	—	Silver, Ag=107·9	Cadmium, Cd=112·4	Indium, In=114·0	Tin, Sn=119·0	Antimony, Sb=120·0	Tellurium, Te=127	Iodine, I=127	—	—	—
8	Xenon, Xe=128	Cæsium, Cs=132·9	Barium, Ba=137·4	Lanthanum, La=139	Cerium, Ce=140	—	—	—	—	—	—
9	—	—	—	—	—	—	—	—	—	—	—
10	—	—	—	Ytterbium, Yb=173	—	Tantalum, Ta=183·0	Tungsten, W=184	—	Osmium, Os=191	Iridium, Ir=193	Platinum, Pt=194·9 (Au)
11	—	Gold, Au=197·2	Mercury, Hg=200·0	Thallium, Tl=204·1	Lead, Pb=206·9	Bismuth. Bi=208	—	—	—	—	—
12	—	—	Radium, Rd=224	—	Thorium, Th=232	—	Uranium, U=239	—	—	—	—

been for the last forty years the most important guide in the prosecution of modern inorganic chemical research.

The last version of his scheme of arrangement of the elements as left in 1904 by Mendeléeff, not long before his death, is shown in the preceding table. Many of the so-called rare earths are omitted owing to the uncertainty which still prevails as to the atomic weights and properties of many members of the group.

A few words of explanation are necessary; y in the table is, according to Mendeléeff, an analogue of helium with a density of about 0·2 and a molecular weight 0·4. He supposed that it might hereafter be identified with *coronium*, a hypothetical element existing in the sun's coronal atmosphere; x is the "ether" of the physicist, for which Mendeléeff, disregarding conventional views, supposed a molecular or atomic structure. It was supposed also to be chemically inert and to have an extremely minute atomic weight.

The spaces left vacant after hydrogen in Series 1 should be occupied, according to Mendeléeff, by elements, at present unknown, having approximately the atomic weights 1·4, 1·8, 2·2, 2·6, 2·8, 3·0, and 3·4. These would be the first members of the Groups II to VIII respectively.

This table is interesting for historical reasons, but the principle of periodicity is more clearly displayed when properties and atomic weights are plotted against each other in a system of rectangular co-ordinates so as to reveal a curve. The first scheme of this kind was published by Professor Lothar Meyer immediately after Mendeléeff's table in 1869. It displays the recurrence of maxima in the values of the atomic volumes when the elements are arranged consecutively in the order of their atomic weights. The physical properties of the elements have been the subject of much study since the time of Mendeléeff and Meyer, and it is now possible to show graphically that not only do atomic volumes wax and wane in following up the series, but other properties such as melting-points and coefficients of expansion follow the same order. The following diagram (Fig. 51) is taken from a paper in the *Journal of the American Chemical Society* (July, 1915) by Professor T. W. Richards, "Concerning the Compressibilities of the Elements and their Relations to other Properties." Here it is interesting to observe how closely the configurations of the several curves agree with one another, the

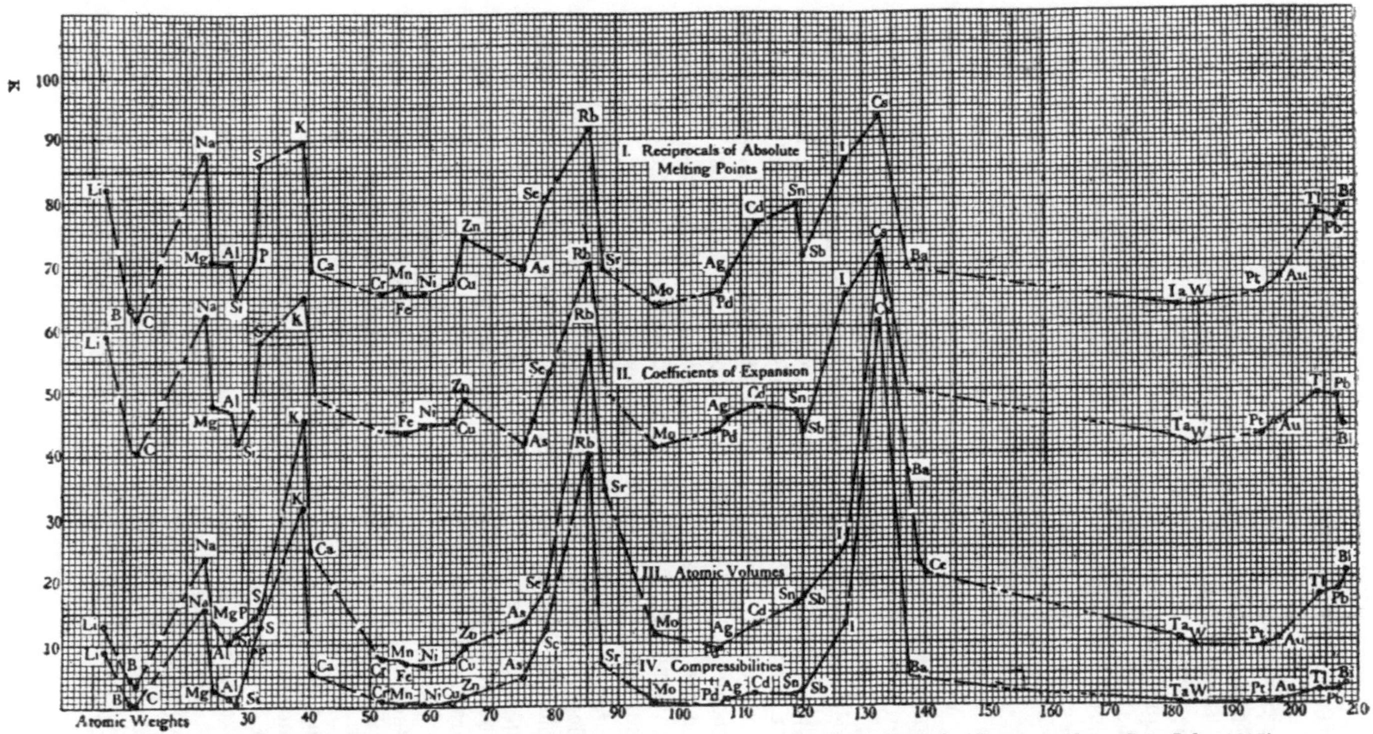

FIG. 51. PROPERTIES OF THIRTY-FIVE SOLID ELEMENTS. After Prof. Richards (*J. Amer. Chem. Soc.*, July, 1915).

maxima and minima in each case corresponding to the same atomic weight.

In the course of experiments on the specific heats of the elements at low temperatures between the boiling-points of liquid nitrogen and hydrogen, Sir James Dewar has observed that the mean atomic heats (at 50° Abs.) of the elements are a periodic function of the atomic weights. The atomic heat is to be understood as the product of multiplying the specific heat by the atomic weight, and at ordinary temperatures from the freezing to the boiling-point of water, the value of this product is almost uniformly 6·2 to 6·4 and periodicity has not been noticed.

At the temperature of only 50° absolute or 223° below centigrade zero the specific heats are very small, that of carbon in the form of diamond being only 0·0028 or $\frac{1}{357}$ of the specific heat of water. When these numbers are multiplied by the atomic weights the product is still of small numerical value. The figures for the atomic heats range from 0·03 for carbon to 6·82 for cæsium. Doubtless the values in some cases will be slightly corrected by future experiment, but in the meantime the recognition of the periodic relation, that is, the rise and fall at regular intervals, is an interesting observation which brings the outstanding property of specific heat into the same category as the rest of the physical properties of the elements.

The diagram published in connection with Dewar's paper in the *Proceedings of the Royal Society*, vol. 89 (1913), p. 169, shows that the rise and fall of the curve follows very nearly the same course as the curve of atomic volumes originally pointed out by Lothar Meyer, and incorporated into the comprehensive diagram already given (page 129).

The importance of the periodic scheme, as arranged by Mendeléeff, deserves a little further notice in view of the influence it has had on the progress of theoretical chemistry. First of all it should be noticed that several of the elements now known, and amounting at the present time to eighty-three, were not known in 1869–70 when the scheme was published. The table of elements first arranged by Mendeléeff in 1869 contained only sixty-three then recognised elements, the atomic weights being in many cases uncertain. But inasmuch as intervals too wide to be accounted for by mere experimental inexactitude occurred in several parts of the table, vacant spaces were left which suggested that substances existed somewhere in nature of

D. Mendeleeff.

To face page 130.

which the chemist had not hitherto taken cognisance. And so it turned out. With most remarkable confidence in the trustworthiness of his scheme as indicating a law of nature, Mendeléeff proceeded to describe in detail the properties, chemical and physical, which would be exhibited by substances corresponding to the vacant places in the table, and in truly prophetic spirit to predict their discovery.

In 1871 the atomic weights represented approximately by the numbers 44, 70, and 72 were not known to belong to any existent element, but to the hypothetical elements expected to fill these places Mendeléeff gave the names ekaboron, ekaluminium, and ekasilicon. In a footnote contained in his famous work on *The Principles of Chemistry*, the English translation of 1891 contains the following words :—

" When in 1871 I wrote a paper on the application of the periodic law to the determination of the properties of yet undiscovered elements, I did not think I should live to see the verification of this consequence of the law, but such was to be the case. Three elements were described—ekaboron, ekaluminium, and ekasilicon[1]—and now, after the lapse of twenty years, I have had the great pleasure of seeing them discovered and named after those three countries where the rare minerals containing them are found, and where they were discovered—Gallia, Scandinavia, and Germany."

Between the elements zinc and arsenic then there were two unoccupied places, and the following are the chief properties which, according to the law, should appertain to them.

The following is the account given by Mendeléeff of the first of these :—

" Zinc, which has an atomic weight 65, should be followed in the III group by an element with an atomic weight about 69. It will be in the same group as aluminium, and should consequently give R_2O_3, RCl_3, $R_2(SO_4)_3$, alums and like compounds analogous to those of aluminium. Its oxide should be more easily reducible to metal than alumina, just as zinc oxide is more easily reducible than magnesia. The oxide R_2O_3 should, like alumina, have feeble but clearly expressed basic properties. The metal reduced from its compounds should have a greater atomic volume than zinc, because in the fifth series, proceeding from zinc to bromine the volume increases. And as the volume

[1] Eka = Sanscrit meaning *one*.

of zinc is 9·2, and of arsenic 18, therefore that of our metal should be near to 12. This is also evident from the fact that the volume of aluminium is 11, and that of indium 14, and our metal is situated in the III group between aluminium and indium. If its volume is 11·5 and its atomic weight be about 69, then its density will be nearly 5·9. The fact of zinc being more volatile than magnesium gives reason for thinking that the metal in question will be more volatile than aluminium, and therefore for expecting its discovery by the aid of the spectroscope, etc."

In 1875 Lecoq de Boisbaudran discovered, by means of the spectroscope, a new metal from the zinc blende in the Pyrenees, which he named gallium. It was found to yield an oxide R_2O_3 and an alum, that is a double sulphate with ammonium and potassium which crystallised in regular octahedrons. Its density was found to be 5·9, and its atomic weight 69·8. The metal was found to possess many of the properties of aluminium, being, however, much more fusible, just as zinc is more fusible than the next metal above it, namely magnesium.

In a similar way the properties of ekasilicon, foreseen by Mendeléeff in 1871, were recognised in the metal *Germanium*, discovered by Clemens Winkler in 1886 in the peculiar silver ore argyrodite. This element stands in Group IV of Mendeléeff's scheme, immediately below titanium, which follows silicon. The missing but expected element was described on the basis of a consideration of the properties of the known elements, silicon, zinc, tin, and arsenic, which in the table are placed at nearly equal distances from the vacant place.

It was expected to have an atomic weight nearly 72 with a higher oxide EsO_2, and a lower oxide EsO, haloid compounds of the type $EsCl_4$ which would boil at about 90°. Its sulphide EsS_2 would resemble tin sulphide SnS_2, and probably dissolve in ammonium sulphide and so forth. Germanium has an atomic weight 72·5, the metal melts at about 900° C. It forms a dioxide GeO_2.

If germanium or its sulphide is heated in a stream of chlorine gas it forms a volatile liquid $GeCl_4$, which boils at 86°, and is decomposed by water after the manner of stannic chloride. In fact the new element was found to possess just the properties to be expected of an element occupying a position intermediate between those of silicon and tin.

A similar correspondence was found to exist between Mendeléeff's ekaboron and the metal scandium when examined by Cleve a few years later.

One other application of the periodic law may be mentioned, and that is the guidance it has afforded in correcting the atomic weights. The element glucinum had formerly the value 13·5 assigned to it. There is, however, no place in the periodic scheme for an element of this atomic weight with properties such as those exhibited by glucinum. A further investigation of its properties and determinations of its specific heat showed that the atomic weight was much lower, and the figure 9·1 entitles it to a place in the table in Group II, next above magnesium, with which it has considerable analogy. Other cases of a similar kind have led to correction which all experience tends to verify. Two elements only present outstanding difficulties. These are the elements argon and tellurium, both of which are placed in the list of elements one step too high in the consecutive order, notwithstanding all the very numerous experimental investigations as to the numerical values of their atomic weights. So strong is the general conviction that their true places in the Mendeléeff scheme are those which have actually been assigned to them, notwithstanding the numerical discrepancy.

Of course it must also be admitted that in Mendeléeff's table satisfactory places have not as yet been found for some of the elements derived from the rare earth minerals, so that the cerium and yttrium and other groups do not quite fall into line.

The element argon has been mentioned, and we must now review as briefly as may be the dramatic story of its discovery.

Previously to 1894 the existence of a group of elements destitute of all power to enter into chemical combination had not been foreseen by Mendeléeff or any of the chemists who had for years made a study of the periodic scheme. But for several years Lord Rayleigh had been engaged in a series of experiments, the object of which was to determine with the utmost possible accuracy, not only the relative densities, but the absolute densities, of the principal gases, that is, to compare their weights with that of an equal bulk of water. The method used was the same as that which had been employed by the French physicist Regnault many years before, and is in principle of the utmost simplicity. It consists in weighing the gas contained in a large glass globe attached to one arm of a balance, using as a counterpoise a similar globe of as nearly as possible the same size, so as to displace, with adjustments, exactly the same volume of air. The weight of the gas contained in the globe could thus be found, and, as the capacity of the globe had been previously determined, the volume of the gas was also known.[1]

Having experimented on hydrogen and oxygen the case of nitrogen came to be considered, and an anomaly was soon discovered.[2] It was found that when the nitrogen had been prepared from atmospheric air by removing the oxygen by any suitable agent, the gas proved to be heavier than the nitrogen made by chemical decomposition of ammonia or one of the oxides of nitrogen. Hence it might be supposed that the atmospheric nitrogen was too heavy on account of imperfect removal of oxygen, or the chemical nitrogen was too light in consequence of its contamination with gases lighter than pure nitrogen. It was proved by direct and laborious experiments that neither of these hypotheses could be adopted.

The following figures represent the weights of nitrogen, made from different materials, with which the experimental globe was filled. (*Proc. Royal Soc.*, vol. 57, p. 267.)

Nitrogen obtained from nitric oxide, NO . .	2·3001	grms.
,, ,, ,, nitrous oxide, N_2O .	2·2990	,,
,, ,, ,, ammonium nitrite, NH_4NO_2	2·2987	,,

Nitrogen was also made from air by first combining it with magnesium to form magnesium nitride, acting on this compound by water so as to produce ammonia and decomposing the ammonia by a hypochlorite. The nitrogen was finally purified by passing it over red-hot copper, and copper oxide.

The weight was then 2·29918 grams, and was therefore practically the same as above. Nitrogen obtained from atmospheric air by removing the oxygen by means of

Red-hot copper . . .	2·3103	grams.
Red-hot iron	2·3100	,,
Ferrous hydrate . . .	2·3102	,,

These figures correspond to the following weights per litre of the gas :—

Chemical nitrogen . . .	1·2505	grams.
Atmospheric nitrogen . .	1·2572	,,

[1] Rayleigh, *Proc. Royal Soc.*, 53 (1893), p. 134.
[2] *Proc. Royal Soc.*, 55 (1894), p. 340.

Yrs very truly
Rayleigh

To face page 134.

Further work led to the conclusion that atmospheric air contains an ingredient hitherto unnoticed by chemists. The announcement made at the Oxford meeting of the British Association in the summer of 1894 was received with a certain amount of incredulity by the chemical world, in view of the immense number of incontestably accurate analyses of air which had been made during the previous half-century. In these, however, the new ingredient, with its characteristic chemical inactivity, had always passed as nitrogen gas.

Lord Rayleigh had by this time secured the co-operation of Professor Ramsay, and the two investigators joined in laying before the Royal Society the results of their work. The interest excited was so great that a special meeting had to be held on January 31, 1895, in the theatre of the University of London with Lord Kelvin, the president, in the chair.

In the paper then published the authors give reasons for suspecting a hitherto undiscovered constituent in air, and, as their statement contains so many interesting features, the following extracts from it will be welcomed by the reader. They say :—

"When the discrepancy of weights was first encountered attempts were naturally made to explain it by contamination with known impurities. Of these the most likely appeared to be hydrogen, present in the lighter gas in spite of the passage over red-hot copper oxide. But inasmuch as the intentional introduction of hydrogen into the heavier gas, afterwards treated in the same way with cupric oxide, had no effect upon its weight, this explanation had to be abandoned, and finally it became clear that the difference could not be accounted for by the presence of any known impurity.

"At this stage it seemed not improbable that the lightness of the gas extracted from chemical compounds was to be explained by partial dissociation of nitrogen molecules, N_2, into detached atoms. In order to test this suggestion both kinds of gas were submitted to the action of the silent electric discharge, with the result that both retained their weights unaltered. This was discouraging, and a further experiment pointed still more markedly in the negative direction. The chemical behaviour of nitrogen is such as to suggest that dissociated atoms would possess a high degree of activity,[1] and that even though they

[1] It is interesting to notice that this hypothesis has been verified fifteen years later by Lord Rayleigh's son, Professor Strutt. See chapter on Active Nitrogen.

might be formed in the first instance their life would probably be short. On standing they might be expected to disappear, in partial analogy with ozone. With this idea in view, a sample of chemically prepared nitrogen was stored for eight months. But at the end of this time the density showed no sign of increase, remaining exactly as at first. . . .

"The simplest explanation in many respects was to admit the existence of a second ingredient in air, from which oxygen, moisture, and carbonic anhydride had already been removed. The proportional amount required was not great. If the density of the supposed gas were double that of nitrogen, $\frac{1}{2}$ per cent only by volume would be needed; or if the density were but half as much again as that of nitrogen, then 1 per cent would still suffice. But in accepting this explanation, even provisionally, we had to face the improbability that a gas surrounding us on all sides, and present in enormous quantities, could have remained so long unsuspected. . . .

"And here the question forced itself upon us as to what really was the evidence in favour of the prevalent doctrine that the inert residue from air, after withdrawal of oxygen, water, and carbonic anhydride, is all of one kind.

"The identification of 'phlogisticated' air[1] with the constituent of nitric acid is due to Cavendish, whose method consisted in operating with electric sparks upon a short column of gas confined with potash over mercury at the upper end of an inverted **U** tube.

"Attempts to repeat Cavendish's experiment in Cavendish's manner have only increased the admiration with which we regard this wonderful investigation. Working on almost microscopical quantities of material, and by operations extending over days and weeks, he thus established one of the most important facts in chemistry. And, what is still more to the purpose, he raises as distinctly as we could do, and to a certain extent resolves, the question above suggested. The passage is so important that it will be desirable to quote it at full length.

"'As far as the experiments hitherto published extend, we scarcely know more of the nature of the phlogisticated part of our atmosphere, than that it is not diminished by lime-water, caustic alkalies, or nitrous air; that it is unfit to support fire,

[1] That is deprived of oxygen. Phlogisticated air is in modern language nitrogen; dephlogisticated air was afterwards named oxygen by Lavoisier.

or maintain life in animals; and that its specific gravity is not much less than that of common air: so that, though the nitrous acid, by being united to phlogiston, is converted into air possessed of these properties, and, consequently, though it was reasonable to suppose that part at least of the phlogisticated air of the atmosphere consists of this acid united to phlogiston, yet it might fairly be doubted whether the whole is of this kind, or whether there are not in reality many different substances confounded together by us under the name of phlogisticated air. I therefore made an experiment to determine whether the whole of a given portion of the phlogisticated air of the atmosphere could be reduced to nitrous acid, or whether there was not a part of a different nature from the rest which would refuse to undergo that change. The foregoing experiments indeed in some measure decided this point, as much the greatest part of the air let up into the tube lost its elasticity; yet, as some remained unabsorbed, it did not appear for certain whether that was of the same nature as the rest or not. For this purpose I diminished a similar mixture of dephlogisticated and common air, in the same manner as before, till it was reduced to a small part of its original bulk. I then, in order to decompound as much as I could of the phlogisticated air which remained in the tube, added some dephlogisticated air to it, and continued the spark till no further diminution took place. Having by these means condensed as much as I could of the phlogisticated air, I let up some solution of liver of sulphur to absorb the dephlogisticated air; after which only a small bubble of air remained unabsorbed, which certainly was not more than $\frac{1}{120}$ of the bulk of the phlogisticated air let up into the tube; so that if there is any part of the phlogisticated air of our atmosphere which differs from the rest, and cannot be reduced to nitrous acid, we may safely conclude that it is not more than $\frac{1}{120}$ part of the whole.'"[1]

The authors repeated this experiment of Cavendish with the advantage of modern apparatus, and they found that on sparking air with added oxygen in the presence of potash the residue which remained unabsorbable was in proportion to the amount of air operated on. An examination of this residue with the spectroscope showed that the gas left was not nitrogen, but had a spectrum of its own.

[1] Cavendish's "Experiments on Air," *Philosophical Transactions of the Royal Soc.*, 1785.

To prepare argon[1] on a large scale air is freed from oxygen by means of red-hot copper. It is then dried by means of soda lime and phosphoric oxide, and the nitrogen is absorbed by passage through a tube packed with magnesium turnings heated to bright redness. The residual gas is then made to circulate through an apparatus containing hot copper, copper oxide, soda lime, and magnesium, by which the last traces of impurities are removed.[2]

Argon is a colourless gas 19·9 times heavier than hydrogen, and therefore nearly 1·4 times heavier than air. It is soluble in water to about the same small extent as oxygen, that is, approximately 4 volumes in 100 of air at common temperatures.

All attempts to induce argon to enter into chemical combination have proved abortive. Most drastic treatment was applied and a great variety of reagents used, but argon remains in all these circumstances unaltered. No substance of this kind having been previously known it will be understood that the ingenuity of the discoverers and the efforts of many other chemists were employed to settle this point conclusively.

In the paper of which an account has just been given the discoverers assume, provisionally, that the gas they had succeeding in isolating was a single substance and not a mixture of gases.

The density of argon being 19·9 the law of Avogadro indicates that its molecular weight is 39·8. The molecule is believed to consist of one atom only, and this is designated by the symbol A.

At the anniversary meeting of the Chemical Society on March 27th, 1895, a fresh surprise awaited the assembled chemists. The discovery of a new element similar in character to argon was announced by Professor Ramsay. In seeking a clue to compounds of argon he was led to repeat experiments of Hillebrand on the rare mineral clèvite, which, when boiled with weak sulphuric acid, gives off a gas hitherto supposed to be nitrogen.

This gas proved to contain very little nitrogen with traces of argon, but examination with the spectroscope showed that the most prominent line was a brilliant yellow one very close to the two lines of sodium, D_1 and D_2, but only known up to this time

[1] The name is derived from the Greek (ἀν privative, ἔργον work), in reference to its chemical inactivity.

[2] Details are described in the original paper entitled "Argon, a New Constituent of the Atmosphere," *Proc. Royal Soc.*, 57 (1895), p. 265.

Yours sincerely,
William Ramsay

To face page 138.

as belonging to a constituent of the sun's chromosphere and designated D_3. This line had been attributed by Professors Janssen and Lockyer, thirty years previously, to a hypothetical solar element which was named by them *helium.*

Clèvite, the mineral originally used, is a variety of uraninite, and contains beside uranium and lead a considerable quantity of the rare earths. The significance of the association of helium with uranium and lead will be referred to on a later page.

Helium has been found to be a constituent of a large number of minerals, but it is especially found in connection with compounds of uranium, thorium, and the rare earths. It is a colourless inert gas like argon, but very much lighter, being, in fact, with the exception of hydrogen, the lightest gas known. Its density is 2·0, and hence its molecular and atomic weight is 4·0. It was found to be very sparingly soluble in water, one volume of the gas being soluble in between 130 and 140 volumes of water, at atmospheric temperatures. This low degree of solubility indicated a probably low boiling-point. The subsequent history of helium confirmed this conjecture, for it was not till 1908 that helium was reduced to the liquid state by Professor Kamerlingh-Onnes of Leiden. The boiling-point was found to be about 4°·5 absolute or 268° to 269° C. below 0° C. This result, by which the last of the known gases was made to change its state, was only accomplished by pressure with the aid of liquid hydrogen as a cooling agent. By causing the liquid helium to boil under reduced pressure the lowest temperature ever attained was reached. This was estimated to be less than 2·5° on the absolute scale.

These two remarkable elements helium and argon presented a difficult problem as to their position in relation to the rest of the chemically active elements.

On reviewing Mendeléeff's periodic scheme of the elements it is noticeable that, if we regard for the moment only the elements known prior to the discovery of helium and argon, the valency or combining capacity of the atoms in Group I is represented by unity. These elements are univalent. The same is the characteristic valency of the halogens, fluorine, chlorine, bromine, and iodine, which stand, along with manganese, somewhat doubtfully in the Group VII. If then, following the order from left to right, it is found that the valency steadily increases as shown below, it appears that if a column is provided to the left of

Group I any elements finding a position therein would have a valency one unit less than 1 or 0.

Valency	0	1	2	3	4	5 or 3	6 or 2	7 or 1
	He 4	Li 6·94	Gl 9·1	B 11·	C 12·0	N 14·01	O 16·0	F 19·0
		Na 23·0	Mg 24·3	Al 27·1	Si 28·3	P 31·04	S 32·06	Cl 35·46
	Ar 39·88	K 39·1	Ca 40·07	Sc 44·1	Ti 48·1	V 51·0	Cr 52·0	

The differences between the first and second lines are pretty constantly equal to very nearly 16 units, thus 23–6·94 = 16·06, etc. The differences between the second and third lines are at first about 16, but increase in passing from left to right. If helium and argon are introduced into such a table it is at once observed that there is an interval between them which would apparently require an element having an atomic weight 16 units greater than that of helium, or approximately 20. The question then arises does such an element exist ?

This question Ramsay set to work to investigate. Many fruitless experiments were undertaken on the gases obtained from minerals and in attempts to separate helium and argon into two gases by process of diffusion. The gases found in many mineral waters, such as the hot springs at Bath, and the waters of Cauterets in the Pyrenees in which argon and helium had been found. The production of liquid air on a large scale provided the material from which the first success was obtained ; and, after allowing about a litre of it to evaporate away, the last portions were found to contain a gas having about twice the density of argon to which the name *Krypton* (Gr. hidden) was given. Very shortly afterwards, by liquefying a large quantity of "argon," obtained from atmospheric air, and in a similar manner allowing the liquid to evaporate, the successive fractions yielded two other gases. The one lighter than argon fitted the place already prepared for it in the periodic table, the other had a density of 64, and was found in small quantity in the least volatile portions of the liquid, and was named Xenon (the stranger). The complete story of these wonderful researches by Lord Rayleigh and Professor Ramsay deserves to be read in the original papers by every serious student of chemistry. The details of the countless experiments, the ingenuity in devising apparatus, the skill involved in its manipulation, and the knowledge which could turn to account so many physical facts and methods have never been surpassed and perhaps not equalled in the history of physical science.

There is one other point worthy of notice. The periodic scheme is justified completely, notwithstanding necessary qualifications, by the results of this work, for by its use discoveries have been made which, without such a guide, would probably have remained unsuspected for ever.

The new elements are as follows, and they will be found in the table (p. 127) in their several positions.

Name.	Symbol.	Atomic or molecular weight.	
Helium	He	4·0	Most recent estimates.
Neon	Ne	20·2	
Argon	A	39·88	
Krypton	Kr	82·92	
Xenon	Xe	130·2	

CHAPTER VII

DISCOVERY AND PROPERTIES OF RADIUM

WHILE all this work was going on in England researches were proceeding in France which were destined to lead to other surprising discoveries. The X-rays had been discovered by Professor Röntgen of Würzburg in the autumn of 1895, and early in 1896 Henri Becquerel, Professor of Physics in the Museum of Natural History in Paris, set to work to examine the radiations emitted by phosphorescent bodies of all kinds, in the expectation that the luminous rays might be accompanied by invisible but penetrating radiation, identical with or similar to the Röntgen rays. Observations on the radiating properties of the salts of uranium led him to the discovery that these substances possess the property of affecting a photographic plate. In the first experiments the uranium compound was exposed to the rays of the sun while the sensitised plate was protected by black paper and a sheet of metallic aluminium.

A few days later he discovered, almost by accident, that exposure of the whole to light was unnecessary, and that if the crystal was kept attached to the plate long enough a very strong impression was obtained in the dark. It thus appeared that the phenomenon could not be attributed to luminous radiations emitted by reason of phosphorescence, since at the end of one hundredth of a second phosphorescence becomes so feeble as to be imperceptible. It appeared then that the photographic cation produced by an uranium compound was a phenomenon of a completely new order. In order to ascertain whether it was due to energy stored up in the crystal and that the effect would disappear or diminish with time, some years would have been necessary. Fortunately Becquerel discovered almost immediately after the photographic experiments just mentioned, that the new radiation had the property of rendering the surrounding medium an electrical conductor, and consequently of discharging electrified bodies when brought near them. He used a gold leaf electroscope, of which a common form is shown in the figure. A couple of strips of gold leaf are attached to the end of a brass rod having a brass disc at the top. The leaves must be protected from draughts of air, etc., and they are therefore suspended inside a glass case, the rod being electrically insulated by means of ebonite or some good non-conductor. When the plate is electrified the gold leaves receive part of the charge, and both being alike either positive or negative are repelled by each other and so diverge. Under ordinary circumstances in air the charge slowly leaks away and the leaves collapse again. In the presence of a radio-active substance the leaves of the electroscope collapse more rapidly, at a rate which can be measured by having a scale on the glass case.

FIG. 52. GOLD LEAF ELECTROSCOPE.

Like the X-rays those discovered by Becquerel travel in straight lines and are capable of traversing wood, paper, and the less dense metals such as aluminium. The radio-activity of a uranium compound is, however, tested more conveniently by the use of the electroscope, and it was by the systematic use of this instrument that Madame Curie, assisted by her husband, the late Professor Curie, was led to the discovery of radium.

By the examination of a large number of uranium and thorium minerals it was found that the electrical conductivity of the air induced by the rays from an uranium compound varies directly with the amount of this element present in the mineral. All uranium compounds are active, and the metal itself more so than any of its salts, except pitchblende or uraninite, U_3O_8, and native chalcolite (copper uranyl-phosphate). The latter substance, when prepared artificially, was found to be less active than the metal. Hence it appeared that the natural minerals contained a substance far more active than uranium. Thorium compounds were found to be active, the action of some of them being actually more pronounced than that of uranium.

A specimen of pitchblende possessing 2½ times the activity of uranium was examined chemically with the object of isolating the radio-active substance. The mineral dissolved in acids was brought into contact with sulphuretted hydrogen, and it was found that the uranium and thorium remaining in solution, the active substance was precipitated with the sulphides insoluble in ammonium sulphide. After separating these in the usual manner it was found that the active substance remained with the bismuth. An extremely active substance was also obtained from pitchblende by sublimation, and when the sulphides were heated in a vacuum at 700° a sublimate was obtained possessing an activity 400 times that of uranium. By further treatment a very active substance was obtained to which the name *polonium* was given in honour of Madame Curie's native country.

The chemical examination of the uraniferous minerals studied by the Curies proceeded nearly on the lines of the ordinary method of qualitative analysis. In the sulphides obtained polonium had been recognised, but another very active substance was found to be associated with the barium also present. Barium is easily separated from solution by sulphuric acid, which causes it to be thrown down as a white insoluble precipitate consisting of the sulphate. This sulphate was accompanied by the active substance. The sulphate was converted into the chloride, and it was then found that a partial separation of the inactive barium from the attendant active substance could be effected by taking advantage of the greater degree of solubility of the barium chloride in water, alcohol, or hydrochloric acid. Ultimately it was found that separation was more easily effected by fractional crystallisation of the bromides. The new substance was the bromide of the previously unknown element about to become so famous under the name of *radium*.

According to Professor Rutherford the amount of radium present in a mineral is uniformly about 3·4 parts for 10,000,000 parts of the uranium present. Consequently in a mineral containing 3 kilograms of uranium there is present about 1 milligram of radium.

The process of extracting radium on a large scale from pitchblende residues which contain barium and radium together is extremely expensive and laborious. A complete account of the operations required as well as of the properties of radium examined and recorded up to that date is provided in the Thesis presented by Madame Curie to the Faculty of Sciences of the University of Paris in 1903. This is printed *in extenso* in the *Annales de Chimie et de Physique*.

The nature of the process has already been sufficiently stated. The first supplies of material were given by the Austrian Government from the residues left after the extraction of uranium from the pitchblende in the State mine at Joachimsthal in Bohemia. But a company has been formed to work the pitchblende found in certain Cornish mines. The most productive sources of radiferous ores are at present found in the United States, but as the search for uranium is now proceeding in many countries other minerals will probably be found at least as good as those already known, and perhaps more abundant. The case is parallel to that of the rare earths which fifty years ago were known almost exclusively in connection with Swedish minerals but are now obtained from copious deposits on the other side of the Atlantic. One of the most promising of uranium minerals appears to be the substance called Carnotite, which is a complex vanadate of uranium.

Radium bromide is still very costly, the price being about £15 per milligram at the present time. The enquirer must therefore not expect to see anything more than what appears as a contemptible little grain of salt at the bottom of a small glass tube perhaps an inch long. There is, however, a reason other than the cost, which would preclude the exhibition of any large quantity such as a quarter of a pound if at any time so much should become available. Its physiological effects are so powerful that any large quantity is dangerous to handle.

A small tube containing only a few milligrams was long ago

M. Curie

To face page 144.

found to produce a sore on the body if carried in the pocket. It is this caustic effect of the radiations from radium which is being used experimentally for medical purposes on cancers and other malignant growths in the human tissues.

A specimen of radium then looks like a few grains of common salt, which, however, is slightly luminous, and therefore visible in the dark. But one of its most striking properties is the power it possesses of exciting phosphorescence in other substances brought near it. Thus all diamonds gives out light of various tints and intensity in the presence of radium, and a certain variety of blende (native zinc sulphide) lights up brilliantly. On examining the light given forth by the zinc sulphide by means of a magnifying glass it was observed to be due to brilliant separate flashes which are more numerous as the radium is nearer to the screen and so less numerous as it is further away, that the sparks, which appear like stars on a black sky, may be counted.

This effect was discovered by Sir William Crookes, who has arranged a simple apparatus called the spinthariscope (Gr. *σπινθαρίς*, a spark) for observing it. This consists of a tube about two inches long, having a zinc sulphide screen at one end with a small surface coated with a radium salt near it. At the other end is a low-power lens through which the sparks can be seen.

Contact with a radium salt is followed in some cases with remarkable changes of colour. Sir William Crookes possesses a diamond which, having been embedded in radium bromide for some months, has assumed an olive-green colour though unchanged in other respects. This colour is persistent, and cannot be removed by boiling the stone in acids or other chemical agents. The glass tubes in which radium salts have been kept always become discoloured, generally assuming pretty rapidly a purplish tint. Sir William Crookes has also shown recently that the diamond may acquire and retain indefinitely the property of radio-activity. In the Philosophical Transactions of the Royal Society for 1914 he thus describes a case: "A large brilliant-cut diamond of pure water assumed a fine green colour after having been kept for sixteen months (from May, 1904, to September, 1905) in a bottle and covered with powdered radium bromide. At the end of that time it was highly radioactive. This diamond has been carried about in my pocket, off and on, since 1905, and has been tested on a sensitive photographic film at intervals of a year or more. No appreciable difference in its radio-activity can be detected from that which it possessed when first removed from the radium bromide in September, 1905. Examined at the present time, nine years after its removal from the bottle of radium bromide, it is luminous in the dark, it rapidly discharges a sensitive electroscope when held near it, and produces scintillations on a zinc sulphide screen as if it were a radium compound."

Another very remarkable fact about radium (discovered by P. Curie and Laborde in 1903) is that a mass of the salt is always at a temperature several degrees above that of the surrounding atmosphere. Obviously the exact difference will depend upon circumstances, but this spontaneous liberation of energy was made the subject of many later experiments, and among other facts it was found that the rate of emission depended on the age of the specimen. The quantitative estimation of the heating effect by Rutherford and Barnes in 1904 led to the result that one gram of radium bromide gives out 110 gram calories per hour.

The element radium was obtained in the metallic state in 1910 by Madame Curie and M. Debierne. It is a white metal which melts at about 700° C., and which dissolves in water, decomposing it and forming the alkaline hydroxide, hydrogen gas being given off. The salts of radium are very similar to those of barium, and it agrees in general properties, valency, etc., with the metals of the alkaline earths, and is consequently placed in the periodic table below barium. But while the other members of the same series are destitute of radio-active properties the activity of radium as measured by the electroscope is about 2,000,000 times that of uranium.

The atomic weight of radium has been the subject of much careful experiment. The conditions necessary to ensure the utmost possible accuracy in such determinations, and the chief considerations involved have been so admirably exposed in the paper by Sir William Ramsay and Dr. R. Whytlaw Gray on "The Atomic Weight of Radium," in the Proceedings of the Royal Society for 1912, p. 270, that we cannot do better than quote the greater part of the introductory portions of this paper. It not only gives the history of the important question as to the atomic weight of radium, but it affords very instructive information as to the procedure in work of this kind in general.

"The essentials in determining the correct equivalent of an element are :—

"(1) A pure compound of the element and sufficient evidence of purity.

"(2) An advantageous transformation in which the weight of the element or elements combined with the one of which the equivalent is to be determined is as large as possible.

"(3) If possible no transference, and no operation which necessitates the use of reagents which can convey into the solution matter which may be absorbed.

"(4) A quantity sufficient in amount to make it possible with the balance at disposal to determine its weight to, at least, 1 part in 20,000.

"(5) Resistant vessels which will not themselves give up any material to the substance and so make its purification difficult.

"Determinations of the equivalent of radium have been made by Madame Curie, by Sir Edward Thorpe, and by O. Hönigschmid. Madame Curie's first determination, made in 1902, may be taken as avowedly only a rough approximation. Using 90 mgrm. of chloride she found the atomic weight to be 225, assuming, no doubt with justice, that radium is a diad. Her second determination employed the same method, viz., precipitation and weighing of silver chloride from a known weight of anhydrous radium chloride. Madame Curie, in her earlier work, proceeded to the ultimate atomic weight progressively, raising the number from 140 to 146, then 174, then greater than 220, and in 1902 to 223·3 ; finally, with 90 mgrm., she obtained the figures 225·5, 226·0, and 224·2 ; mean 225·2, which she regarded as accurate within a unit.

"The method of crystallisation described in her later paper is merely indicated. . . . The samples were tested spectroscopically for barium. . . . The amount taken was about 0·40 grm. After deducting the weight of the filter ash the figures 226·62, 226·31, and 226·42 were obtained, the values for Ag= 107·93 and Cl=35·45 having been taken. Substituting 107·88 and 35·46 the figures are less by 0·09 or 226·53, 226·22, and 226·33 ; the mean of these is 226·36. . . .

"In his Bakerian Lecture for 1907, Sir Edward Thorpe described experiments on the equivalent of radium. His raw material, placed by the courtesy of the Austrian Government at the disposal of the Royal Society, was 'about 500 kgrm.,' or, say, half a ton of pitchblende residues from Joachimsthal. These residues were delivered by the Austrian Government to M. Armet de Lisle in Paris for preliminary extraction. . . .

"Thorpe received from Paris 413 grams of mixed chlorides of barium and radium, the radio-activity of which was 560 times that of uranium. . . .

"The method of separation of radium and barium followed by Thorpe was substantially the same as that adopted by Madame Curie; 9400 recrystallisations of the chlorides were carried out, towards the end in silica vessels. . . .

"The analytical process was also identical with that employed by Madame Curie, namely, precipitation of silver chloride from the dissolved radium chloride acidified with nitric acid, subsidence, washing with distilled water six times, drying at 160°, and weighing on an assay balance sensitive to 0·1 mgrm. . . .

"The results of Thorpe's determinations are:—

I. 226·7 II. 225·6 III. 227·6

"A new set of determinations has been made by Hönigschmid (1911) in which quantities somewhat exceeding 1 gram were used. The method of purification was again that employed by Madame Curie and Thorpe, viz., repeated crystallisation of the chlorides from hydrochloric acid and precipitation of the aqueous solution of the salt with alcohol. The equivalent was not altered after 50 such crystallisations and 13 precipitations with alcohol, and that material was regarded as pure, and employed in the final determinations. The method, too, was the same as that described, but in two cases the chloride of silver was reduced and the weight of the silver ascertained. The mean result, taking Cl=35·46 and Ag=107·88, was 225·95. The extremes in seven determinations were 225·92 and 225·97.

"Whilst these researches show the approximate atomic weight of radium it cannot be said that the results must be accepted as final, for they are lacking in several of the conditions stated at the beginning of this paper. There is the possibility of contamination of the solutions with the reagents used; transference was necessary in all the experiments; both Madame Curie and Sir Edward Thorpe were troubled with insoluble deposits; and the accuracy of weighing was in the former case only 1 in 8000, and in the latter only 1 in 700. These disadvantages were absent from the method which we employed, viz. the conversion of radium chloride into radium bromide by heating it in a current of hydrogen bromide and *vice versâ*; there was no transference and only gaseous reagents were used."

The material used by Ramsay and Gray was obtained from Cornish pitchblende through the Radium Corporation, and consisted of 330 mgrms. of anhydrous radium barium bromide, containing about 70 per cent of radium bromide. These bromides were purified by the authors by treatment first with sulphuretted hydrogen whereby a small black precipitate, probably lead sulphide, was formed and removed. The solution was then acidified with sulphuric acid so as to precipitate the sulphates, which were dried and heated to redness in a mixture of carbon tetrachloride vapour and gaseous hydrogen chloride, whereby they were converted into chlorides. The chlorides were next converted into bromides by heating to redness for some hours in a current of hydrogen bromide.

The bromides thus obtained were submitted to fractional crystallisation whereby the barium was removed and the purified radium bromide was divided into a number of separate portions which were used for determining the equivalent. This was done as already indicated by determining the loss[1] of weight which ensued on converting a weighed quantity of bromide into chloride, and the gain[1] on converting the chloride into bromide. As the result of all this experiment, with the calculations following, the final result for the atomic weight of radium was 226·36, which is identical with the number found by Madame Curie.

These then are some of the most striking facts which have become known to us about radium, and during the first year or two after the isolation of this curious substance no explanation was forthcoming which seemed to satisfy both chemists and physicists, for the simple reason that nothing of the kind had previously been dreamt of in their philosophy.

The results of Madame Curie's work showed that the phenomena exhibited by radium salts were due not to the molecule as a whole, but to the atom of the new element independently of the other elements with which it was associated. Thus equivalent quantities of chloride, bromide, and sulphate of radium show equal activity electrically.

It was immediately found that the radiations of radium are

[1] The quantities to be weighed being very small, usually 2 to 3 mgrm., a special balance was used, one form of which was described, p. 78, Part I.

very complex, and unlike the cathode rays, some are not affected by a magnetic or electric field and they differ in the extent to which they may be stopped by metallic or other screens interposed in their path.

According to Professor Rutherford[1] radio-active substances afford three types of radiation which he distinguishes by the Greek letters, alpha α, beta β, and gamma γ.

The α-rays are readily stopped by tinfoil or a sheet of writing-paper, and travel only a short distance even through air, but are little influenced in direction by a magnet.

The β-rays are similar in character to the cathode rays produced in a vacuum tube (see p. 116).

The γ-rays resemble Röntgen or X-rays.

At this point it will be well to explain briefly the principles of the methods which are employed in studying these radio-active substances. Three general methods have been used, and reference to them has already been made in the preceding account of the discovery of radium. The first depends on the action of the radiation on a photographic plate. The second on the luminosity produced when the rays strike the surface of certain substances, such as zinc sulphide (blende) or certain platinocyanides. The third process is the most important, as it lends itself to the purposes of exact measurement more readily than the other two. This is electrical and rests on the property possessed by the radiations of ionizing the gas or gases through which they pass. Ionization means the production of positively and negatively electrified particles, which act as carriers of electricity and remove the charge on gold leaves or other charged surfaces exposed to contact with them.

An electroscope which has been much used by Professor Rutherford is shown in the accompanying figure. Within a brass case, provided with a window W, is suspended a brass plate B connected by a rod D with the gold leaf C. It is supported by the plug of sulphur S, which is a very perfect insulator. One side of the lower box opens on a hinge, and access is thus gained to the lower plate A, which is connected through the case to earth. The gold leaf is charged to a suitable potential to give it a deflection of about 40°, and the cap E is then placed over

[1] For the information contained in this much condensed account of the radiations from radio-active bodies the author is indebted chiefly to Rutherford's "Radio-active Substances and their Radiation" (*Cambridge Univ. Press*, 1913), to which work the reader is referred for further detail.

the end of the rod. The active matter to be tested is placed on the plate A, and the rate at which the gold leaf falls is observed by means of a telescope having a scale of divisions in the eye-piece. The time taken for the leaf to pass between two points on the scale is noted by a stop-watch, and the average rate of movement per minute can be determined. The average rate of movement per minute is directly proportional to the ionization current between the two plates A and B, that is to the intensity of the radiation emitted by the active substance on the lower plate. There is a small natural leak due to atmospheric ions, for which allowance is always made.

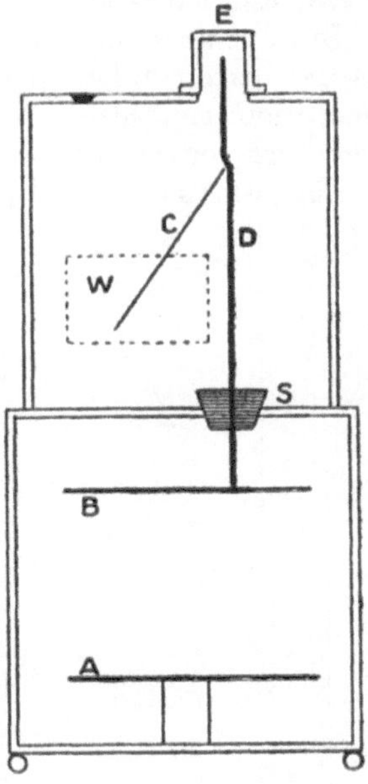

Fig. 53.
RUTHERFORD'S ELECTROSCOPE.

As to the α-rays some doubt was for a long time experienced as to their nature. But physicists were agreed that they consisted of particles of matter charged with electricity and projected with great velocity. The discovery of helium in association with the minerals showing radio-activity and from which radium was extracted, led Professors Rutherford and Soddy to suggest in 1902 that helium might be a product of the disintegration of the radio-elements. Soon after this Sir William Ramsay and Professor Soddy discovered that helium is contained in the gases, oxygen and hydrogen, which are set free on dissolving radium bromide in water. And later it was found that radium bromide heated in a vacuous tube gives off helium.

The α-rays, therefore, are believed to consist of atoms of helium positively charged and ejected from the atom of radium with a velocity about $\frac{1}{15}$ the velocity of light.

The β-rays were discovered in consequence of the fact that they are drawn aside by a magnetic field. If an active preparation is placed in a short narrow lead tube it will cause a fluorescent patch to become visible on a screen coated with a platino-cyanide held above it. If the poles of an electro-magnet placed on either side are excited the phosphorescent patch is broadened out on one side showing that the rays causing the fluorescence have been deflected.

By reversing the magnetic field the broadening of the fluorescent patch takes place in the opposite direction. The deflection can also be shown by taking advantage of the photographic properties of these rays.

The β-rays consist of particles (electrons, see p. 118) carrying a negative charge. This can be observed by means of an ingenious device arranged by Strutt and called the "radium clock." It consists of a glass tube evacuated as completely as possible by means of a mercury pump, and partially lined with tinfoil connected with earth. In the vertical axis of this vessel is suspended, by a quartz rod, a small tube containing a radium salt in metallic connection with a pair of gold leaves attached to the lower end by means of a brass cap. The lower part of the tube containing the radium is smeared with phosphoric acid to render it conducting. While the negative β-rays are discharged into the glass of the tube, the gold leaves gradually acquire a positive charge, which they retain, if the vacuum is good, till they diverge sufficiently to touch the tinfoil lining of the bulb, when they instantly collapse. They then gradually get recharged, and the operation is repeated at intervals, the frequency of which depends on the amount and activity of the substance connected with the gold leaves. (*Philosophical Magazine*, 1903, p. 588.)

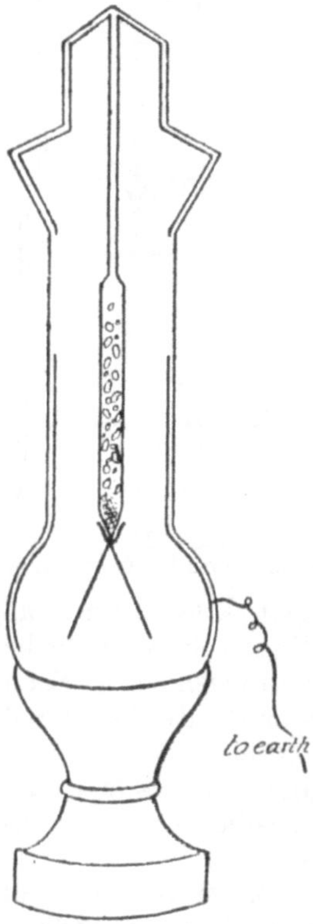

FIG. 54.
STRUTT'S RADIUM CLOCK.

But the radiations of thorium and radium designated α, β, and γ, are accompanied by an active substance which was found to be carried off in a current of air, and which, though it would pass slowly through paper, could be prevented from escaping by a thin sheet of mica. This seemed to agree with the properties of a gas, and it was proved by special experiments that the observed activity was not due to particles of dust. Further investigation showed that the "emanation," as it is called, possesses the properties of a chemically inert gas. This substance has since been named "niton" by Sir William Ramsay, who, notwithstanding its instability, places it among the argon group of elements in the periodic scheme.

According to Rutherford's disintegration theory of radio-active change a definite number of atoms of radium break up per second, each atom evolving an α particle which ultimately becomes a helium atom, leaving behind the residue of the atom which forms the gas known as the "emanation" or niton. It seems to be agreed that the radium atom on disintegration to niton splits up into two parts only, one of which is the α particle. The atomic weight of the resulting niton must be therefore the atomic weight of radium *minus* the atomic weight of helium or $226{\cdot}4 - 4 = 222{\cdot}4$.

This question, however, as Sir William Ramsay remarks, can only be settled by appeal to experiment, and in 1910, in association with Dr. Whytlaw Gray, and with the aid of a balance constructed on the same principle as the balance used for the estimation of the atomic weight of radium he proceeded to determine the density of niton.[1] To appreciate the extreme delicacy and difficulty of the operations involved it is necessary to read the original memoir in its entirety. It will be sufficient to state in this place a few facts connected with the enquiry.

To determine the density of a gas, four separate measurements are essential,—the volume, the temperature, the pressure, and the weight of the gas. In the present case the volume of niton which accumulates in a given time from a known weight of radium is a constant quantity and has been repeatedly measured. In the present case the total volume of niton obtainable for weighing scarcely exceeded 0·1 cubic mm. The weight of this volume on the assumption that the atomic weight is 222 is less than $\frac{1}{1400}$ mgrm. It is therefore evident that in order to weigh this minute quantity of gas with sufficient exactness a balance turning with a load not greater than $\frac{1}{100,000}$ mgrm. is a necessity. This seems an almost inconceivably small weight to

[1] "The Density of Niton (Radium Emanation) and the Disintegration Theory," *Proc. Royal Soc.*, vol. 84 (1911), p. 536.

attempt to measure when one considers that the limit of sensibility of a delicate assay balance is about $\frac{1}{200}$ mgrm.

The atomic weight of niton deduced from five series of experiments was as follows :—

I.	II.	III.	IV.	V.	MEAN.
227	226	225	220	218	223

The determinations of the atomic weight of radium show that it is almost exactly 226·4, and the loss of one helium atom leads to the value 222·4 for the atomic weight of niton, a value which is established by these experiments.

The same authors have shown that niton is liquefiable by cold, and its boiling-point under atmospheric pressure is $-62°$ or 211° absolute. The critical temperature, that is the temperature at which the gas cannot be liquefied by pressure, is 104·5° or 377·5° absolute.

The liquid emanation is colourless, it causes the glass of the containing tube to phosphoresce brightly. On further cooling it sets into a solid which melts at $-71°$.

The emanation or niton, as it may now be called, though so definite a substance is even less permanent than radium itself, for while the life of radium extends to thousands of years, the period of half-decay being calculated as 2000 years, the half-period of transformation of niton is only 3·85 days. Its immediate products of disintegration are α-rays which escape and a solid active deposit which is still more evanescent, having a half-transformation period of only 3 minutes. This substance, which has been labelled RaA, is transformed again in six stages till a product is obtained which has been identified with the *polonium* separated from uranium residues by Madame Curie in her original investigation. It is believed that the final product of the disintegration of polonium is lead.

As already mentioned the proportion of radium found in a mineral bears a constant ratio to the amount of uranium present, and this seems to suggest that radium is formed by the disintegration of uranium. If this is the case it is obvious that if a specimen of a uranium compound were prepared free from radium it should be possible in course of time to recognise the production of radium in such a material. Experiments of this kind have been made, but the results showed that the change was too slow to admit the supposition of an immediate connection between radium and uranium. An intermediate substance has been detected, and it has been called *ionium*, but it has, according to Professor Soddy, an average life of about 30,000 years. It is not possible as yet to establish a direct relation between radium and pure uranium, but it appears that ionium is the parent of radium. While therefore much has been accomplished the history of radium is by no means yet complete. The story is further complicated by the fact that radium is not the only radio-active substance obtainable from uranium minerals. A substance called *actinium* was early recognised among these products. Further it was discovered by Sir William Crookes that by adding excess of ammonium carbonate to the solution of an active uranium salt, a precipitate is formed which redissolves in excess of the reagent, leaving a small insoluble residue in which all the activity is concentrated. This contains a substance to which the symbol UrX has been assigned. The inactive part of the uranium regains its activity by keeping.

The following table provides a summary of the changes which have been traced in the uranium-radium series of disintegration products. It should be added by way of explanation that there is reason to believe that at some stages in the process the disintegration of the active substance may take place in two ways so that a second series of products may arise.

Name.	Atomic weight.	Time of ½ transformation.	Rays emitted.	Chemical character.
Uranium 1	238·5	5×10^9 yrs.	α	Non-separable
Uranium 2	234·5	10^6 yrs. ?	α	
Uranium Y	230·5	1·5 days	β	Separated from U with ferric hydroxide
Uranium X	230·5	24·6 days	$\beta+\gamma$	Do., chemical properties allied to Th, both derived from U 2
Ionium .	230·5	2×10^5 yrs.	α	Non-separable from thorium
Radium .	226	2000 yrs.	$\alpha+\beta$	Resembles barium
Niton .	222	3·85 days	α	Inert gas, density 111
Radium A	218	3 mins.	α	Solid, with positive charge; volatilisable
Radium B	214	26·8 mins.	$\beta+\gamma$	Volatilisable; precip. on zinc
Radium C	214	19·5 mins.	$\alpha+\beta+\gamma$	Volatilisable
Radium C_2	210 ?	1·4 mins.	β	Probably derived from C
Radium D	210	16·5 yrs.	soft β	Non-separable from lead
Radium E	210	5·0 days	$\beta+\gamma$	From soln. of Ra D on nickel and by electrolysis
Radium F (Polonium)	210	136 days	α	Resembles bismuth Probably converted into lead

Another point to remember is that the α particles, when they give up the two unit positive charges carried by each leave an atom of helium of which the atomic weight is 4. Hence the loss of an atom of helium corresponds to a diminution in the atomic weight of the residue to the extent of 4 units. Thus the change from Ra=226 (approx.) to one α particle and one atom of niton leaves the atomic weight of the latter 222 (approx.).

The number of radio-active substances is not limited to the uranium-radium series. The minerals containing thorium were early found to exhibit properties similar to those of uranium.

An examination of Ceylon thorianite, by Dr. O. Hahn in Sir William Ramsay's laboratory in 1905, led to the discovery of a substance which evolves the same emanation as thorium, and a year or two later two products, *meso-thorium* and *radio-thorium*, were separated.

Meso-thorium has properties similar to those of radium, and is prepared commercially in Germany as a substitute for that substance.

It will be sufficient here to show in the following tables the successive products which result from the disintegration of thorium and actinium, from which it will be seen that the phenomena are similar to those observed in the case of radium, the various stages representing very different degrees of stability.

Element.	Radiation.	Half-value period.
Thorium	α	3×10^{10} years
Meso-thorium 1	no rays	5·5 years
Meso-thorium 2	$\beta+\gamma$	6·2 hours
Radio-thorium	α	2 years
Thorium X	$\alpha+\beta$	3·64 days
Emanation	α	54 seconds
Thorium A	α	0·14 second
Thorium B	slow β	10·6 hours
Thorium C_1	α	60 minutes
Thorium C_2	α	very rapid ?
Thorium D	$\beta+\gamma$	3·1 minutes

Element.	Radiation.	Half-value period.
Actinium	no rays	?
Radio-actinium	$\alpha+\beta$	19·5 days
Actinium X	α	10·5 days
Emanation	α	3·9 seconds
Actinium A	α	·002 second
Actinium B	slow β	36 minutes
Actinium C	α	2·1 minutes
Actinium D	$\beta+\gamma$	3·47 minutes

From Rutherford's " Radio-Active Substances, etc.," 1913.

CHAPTER VIII

GENESIS AND TRANSMUTATIONS OF THE ELEMENTS

It has already been stated that the α particle is an atom of helium, the atomic weight of which is 4, and that it carries 2-unit charges of positive electricity. When an element loses an alpha particle by radio-active change its atomic weight, therefore, is reduced by 4 units as shown in many cases in the preceding tables. It also changes in valency to the extent of 2 units; as in the case of radium which, like barium, is bivalent, and forms a dichloride or dibromide, but when it changes into niton the valency disappears and niton is found among the non-valent or chemically inactive gases.

On the other hand the separation of a β particle from a radio-active element corresponds to a loss of 1 unit of negative electricity, corresponding to 1 unit of chemical valency without appreciable loss of atomic weight; the β particles being regarded as electrons with an atom about $\frac{1}{1800}$ of the mass of a hydrogen atom.

If now the radio elements and their products of disintegration are traced through the periodic table it appears that three changes, occurring in any order, of which one is attended by expulsion of one alpha particle, and two by the expulsion each of a β particle, may result in the formation of a product which remains in the same group as the parent though not in the same series. The consequence of such changes is that the radio-active elements and their ultimately inactive products are found in one or other of the last twelve places of the periodic table from thallium to uranium, and though not identical in atomic weight with the long known elements are indistinguishable from them. Thus it is believed that the final product of the decay of thorium and the final product of radium is lead, but the lead derived from thorium has a calculated atomic weight 208·4, while the lead from radium has an atomic weight 206. These figures are derived from the atomic weights 232·4 for thorium and 226 from radium, in consequence of the loss of six and five atoms of helium respectively.

Estimations of the atomic weight of lead from uraniferous and thoriferous minerals lend some support to this view. Thus the following values for the atomic weight of lead, derived from the radio-active minerals named in the list, were published in 1914 by Professor Richards and Mr. Lembert, both experienced in the work of atomic weight determinations.

Lead from N. Carolina uraninite . .	206·40
,, ,, Joachimsthal pitchblende .	206·57
,, ,, Colorado carnotite . .	206·59
,, ,, Ceylon thorianite . . .	206·82
,, ,, Cornish pitchblende . .	206·86
Common Lead	207·15

It deserves to be mentioned that earlier experiments made at Harvard with the object of testing this question, namely, the possible variation of atomic weights in the cases of copper, calcium, sodium, and iron, all gave negative results.

The evidence so far, therefore, tends to sustain the hypothesis, but the conclusion cannot yet be considered satisfactorily established in view of the fundamental importance of the issues involved. For it is obvious that these conclusions, if accepted, involve an entirely new view of the nature of the elements and the constitution of matter. If as is suggested there may be two or more different atoms with atomic weights divergent from each other to the extent of several units, but which are indistinguishable from each other by chemical or physical properties, and which exhibit, it is said, the same spectrum, the properties of the elements are not wholly dependent on their atomic weights as the periodic law prescribes. It may be that hereafter the

T. W. Richards

To face page 158.

methods of analytical chemistry will be so far improved as to enable the chemist to discriminate between the several assumed varieties of such an element as lead.

Whether that will come about or not it is clear that the supposed "isotopes,"[1] as they have been called by Professor Soddy, are much more like each other than any others of the elements standing equally close together in respect to atomic weight. Take cobalt and nickel, for example, with atomic weights differing from each other by little more than a quarter of a unit; Co = 58·97; Ni = 58·68; difference = 0·29. These two metals resemble each other very closely, and if it had not been for the fact that the salts of cobalt are generally red, while those of nickel are green, it is quite possible that if they had both given compounds of the same colour they might have been for a long time confused together. They are not distinguished by any great differences in properties which do not involve colour or degree of solubility, etc., till they are put through some of the less common transformations, e.g. the production by cobalt of a complicated series of ammoniacal compounds for which there is no analogy among the compounds of nickel. If the final products of radio-active disintegration are ever obtained in appreciable quantities, it may turn out that they do present differences of character though not recognisable by the instruments or agencies at present at our disposal.

It will have been noticed by the reader that the radio-active elements thus far referred to are found among the last few members of the series in the periodic table possessing the highest atomic weights known.[2] A very minute activity is exhibited by various ordinary materials, but this appears to be attributable to the wide distribution of such substances as radium and thorium in extremely attenuated quantities. The only elements of relatively small atomic weight which have been found to exhibit an appreciable activity, though weak, are potassium and rubidium. According to Professor Rutherford no α rays are emitted but only β rays, and "the activity due to a potassium salt is not more than $\frac{1}{1000}$ of the activity of the β rays due to a equal weight of uranium."

[1] From Greek ἶσος equal, τοπος a place.

[2] On the other hand, it may be remarked that the commonest materials of the earth's crust, the ocean, and the atmosphere are formed of elements of relatively small atomic weight.

For the present it is impossible to do more than mention these facts, which may turn out on further investigation to have a special significance.

But the elements highest in the scale are all undergoing transformations, attended by radio-active phenomena, into products which are all of smaller atomic weight. The question arises whether uranium with its atomic weight 238·2, is really the limit, or whether there may not be elsewhere in the cosmos elementary substances more complicated in structure and having a greater atomic mass. So far as at present known the evidence is against such a supposition assuming the present conditions prevailing in our solar system. No new elements have been discovered in the meteorites which have fallen on the earth's surface, and the spectroscope directed to the sun, stars, and nebulæ, though indicating the presence of elements unknown on the earth, also indicates that they are probably of lower atomic weight than any terrestrial constituent. Nevertheless the possibility of substances of higher atomic weight than uranium being formed, and existing temporarily under other cosmical conditions is not excluded. Helium is known to be expelled by radium and thorium, and helium is the first term of the series of inactive gases, helium, neon, argon, krypton, xenon, with atomic weights gradually increasing up to 130. If helium is the product of the disintegration of uranium it is at least conceivable that the other gases of like character may have resulted from the breaking up of elements of higher atomic weight, which, though temporarily existent, became extinct before the present order of things settled down to a comparatively stable condition. Uranium, on this view, represents the limit of mass for an atom under present circumstances.

Such considerations as these lead us to the contemplation of the questions which have so long had a supreme fascination for chemists: How did the present elements come into existence? Is there a common material out of which all have been evolved —an "urstoff" or "protyl"—and if so what is its relation to electricity? And what is electricity, is it a state of matter or matter itself?

To such questions there have been many attempts to provide at least partial answers.

The essential unity of all matter is no modern idea, but the discoveries of the last few years have supplied material which has stimulated speculative discussion to a degree previously unknown.

The ancient Greek philosophers might conceive systems of the universe, based on ideas of motion or of matter, but as their actual knowledge of nature was both limited to mere observation and the results of observation often mistaken and always very imperfect, their theories had no secure foundation. And even down to the comparatively modern times of the eighteenth century philosophical writers, Kant and Leibnitz and Laplace, there was little to consolidate or support conclusions as to the origin of matter and the cosmos. But through all the preceding centuries it is remarkable how frequently the idea of an original primal stuff, called by Aristotle ὕλη, appears in the works of philosophical writers. When, however, the definite discoveries in chemistry made in the latter half of the eighteenth century, and when the properties of hydrogen, especially its lightness, became familiar it appeared to many chemists of that day as though some of the speculations of the ancients were likely to be fulfilled.

The first, the crudest, and yet one of the most famous of the modern forms of this idea is what appears in chemical literature as "Prout's Hypothesis." William Prout was a physician and chemist who lived from 1785 to 1850. He held the view that the atomic weights of all the elements are integral multiples of the atomic weight of hydrogen. But in the first place the atomic weights in his day were but very inaccurately determined, and the atomic weight of chlorine has always been found to approximate to 35½ when hydrogen is 1. He says in one of his essays: "we may almost consider the πρώτη ὕλη of the ancients to be realised in hydrogen, an opinion, by the way, not altogether new."

Some form of this hypothesis has doubtless hovered in the minds of many chemists since that time, but in its original form it has no basis in fact.

The positive knowledge now derived from the discoveries of J. J. Thomson as to the electric discharge in gases, and the work of Rutherford, Soddy, Ramsay, and others on radio-active substances seem to afford a justification for a reconsideration of the subject. The world has also been influenced in many directions by the use of ideas of evolution borrowed, in the first instance, from the biological sciences though substantially modified. So that in our own times speculation as to the origin of the elements and their relations to one another has grown bolder. The discovery of the periodic relations of the elements by Newlands, Mendeléeff, and Meyer may be regarded as the starting-point of much of this modern hypothesis, though, strangely enough, not shared by Mendeléeff, who devoted the greater part of his life to the subject.

The earliest serious treatment of the periodic relations of the elements as the basis of ideas as to their origin we owe to Sir William Crookes. His first exposition of his views was given in 1886 at the Birmingham meeting of the British Association, but he has on several occasions revived the hypothesis then brought forward, with modifications suggested by successive discoveries already described. The last formal utterance of his views was addressed to the International Chemical Congress at the meeting in Berlin in 1903. Briefly his ideas on "The Genesis of the Elements" are based on the assumption of a "protyle," as he calls it, in the form of a mist of minute particles, which gradually accreting into larger and larger clusters gave rise to elemental atoms, which became assorted by a selective process dependent on the tendency of particles with the same kind and rate of motion to separate from a crowd and keep together. In order to explain the production of the chemical elements, as known, it must be supposed that in this process of accretion some clusters of particles are more stable than others, and therefore survive the changes of conditions (temperature, pressure, etc. ?) to which they are exposed throughout the ages which have elapsed since they were formed. To do justice to his views it is necessary to quote Crookes' own words, taken from the Proceedings of the Royal Society (vol. 63, p. 409. 1898) in which he describes a model designed to represent diagrammatically the evolution of the elements.

"I take any arbitrary and convenient figure of eight without reference to its exact nature; I divide each of the loops into eight equal parts, and then drop from these points ordinates corresponding to the atomic weights of the first cycle of elements. In the model the elements are supposed to follow one another at equal distances along the figure of eight spiral, a gap of one division being left at the point of crossing. The vertical height is divided into 240 equal parts on which the atomic weights are plotted from H=1 to Ur=239·59.[1] Each black disc represents

[1] Since corrected to 238·2.

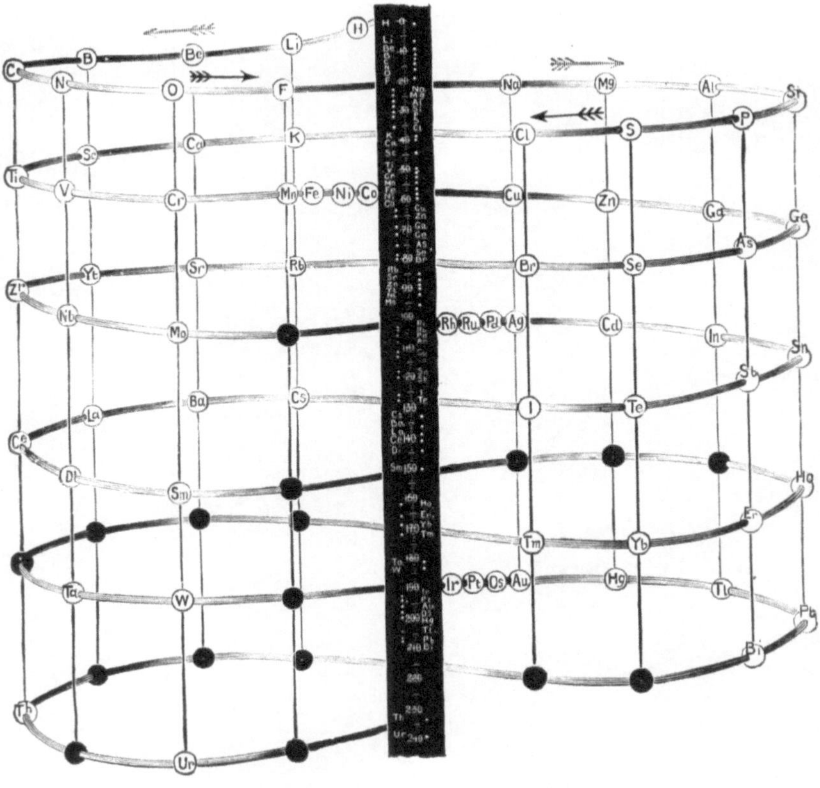

FIG. 55.—EVOLUTION OF THE ELEMENTS (CROOKES).

To face page 162.

an element, and is accurately on a level with its atomic weight on the vertical scale. . . . Let me suppose that at the birth of the elements, as we now know them, the action of the *vis generatrix* might be diagrammatically represented by a journey to and fro in cycles along a figure of eight path, while simultaneously time is flowing on, and some circumstance by which the element-forming cause is conditioned (e.g. temperature) is declining (variations which I have endeavoured to represent by the downward slope). The result of the first cycle may be represented in the diagram (Fig. 55) by supposing that the unknown formative cause has scattered along its journey the groupings now called hydrogen, lithium, glucinum, boron, carbon, nitrogen, oxygen, fluorine, sodium, magnesium, aluminium, silicon, phosphorus, sulphur, and chlorine. But the swing of the pendulum is not arrested at the end of the first round. It still proceeds on its journey, and had the conditions remained constant, the next elementary grouping generated would again be lithium, and the original cycle would eternally reappear producing again and again the same fourteen elements. But the conditions are not quite the same. Those represented by the two mutually rectangular horizontal components of the motion (say chemical and electrical energy) are not materially modified ; that to which the vertical component corresponds has lessened, and so, instead of lithium being repeated by lithium the groupings which form the commencement of the second cycle are not lithium but its lineal descendant potassium."[1]

In this model a glance will satisfy the chemical reader that the majority of the elements fall into natural order in the series which fall vertically under one another. We find, for example, the helium-argon-krypton group in order, with neon close by, the place for xenon, 130, being vacant at the time the model was designed. The triplets :—

P	S	Cl	K	Ca
As	Se	Br	Rb	Sr
Sb	Te	I	Cs	Ba

[1] Sir William Crookes writes to the author January 29, 1916, as follows: "Since giving my account of it (the model) to the Chemical Society I have not materially modified my views on the subject. But had I to make the model again I should turn it upside down, and put H at the bottom, one millim. above the level of the board on which the spiral stands. Then the position vertically of each element would be its atomic weight measured from the base-board in millims. Uranium would then stand at the top 238·5 millims. above the base."

and some others also fall together. But objection might very well be raised to the position in which hydrogen is placed above the halogens, also to the position of fluorine above the iron group, the separation of sodium from lithium, and the close association of the latter with potassium, beside other anomalies. The model, however, shows very clearly the fundamental idea in the hypothesis as to the origin and growth of the elements. They are supposed to be formed successively by the condensation of the primal matter or protyle, but it is not supposed that they were generated one from the other, helium from preformed hydrogen, lithium from preformed helium and so on.

The hypothesis makes use of a sort of uniformitarian principle in the assumption that the whole of the elements have been brought into being by the operation of the same physical conditions acting uniformly from first to last. We now know from the products of radio-active change that some elements must have been formed not by a process of condensation of protyle or anything else, but by the reverse operation, namely, the disintegration of more complex masses. It must also be borne in mind as one of the results of modern work that the elemental atoms appear to contain in every case positive and negative constituents which are held together by their mutual attraction. It does not appear therefore that the simple condensation or polymerisation of a mist-like material is a sufficient account of what must have been a more complex process.

Of the known elements the metals have more characters in common than any of the others ; they all appear at the cathode when they are deposited from their compounds in the process of electrolysis, they are good conductors of heat and electricity, and are generally malleable and ductile. They vary in density from lithium, which is the lightest solid known, to gold, platinum, iridium, and osmium, which are at the other end of the scale. This is nearly in the order of their atomic weights. Moreover in many families the metallic positive chemical character develops and becomes more evident in passing from the members of lowest atomic weight to those of higher atomic weight. This is illustrated in such cases as the alkali metals potassium, rubidium, cæsium, and the metals of the alkaline earths, calcium, strontium, barium. The family likeness is so strong as to be suggestive of the existence in the metals of some common constituent or a similar arrangement of the particles or corpuscles of which we are led to believe their atoms consist. The non-metallic substances placed in Groups V, VI, and VII of the periodic scheme, on the other hand, are very dissimilar from one another in obvious properties; fluorine and chlorine, nitrogen and oxygen are gases at common temperatures, while bromine is a liquid, and phosphorus, sulphur, and iodine are solids. They form well-defined groups, but instead of their negative activity increasing in proportion as the atomic weight increases, they lose much of this character, and gradually assume more or less completely the appearance and properties of metals. Moreover the number of non-metals is relatively small when compared with the elements which exhibit more or less definitely the characteristics of the metals. Leaving aside the five indifferent gases of the argon family and hydrogen, we may count thirteen non-metals in the list of elements, while all the rest, numbering upwards of sixty, are metals.

The discovery of helium as a product of the disintegration of the radio-active elements is suggestive when coupled with the fact that in passing from series to series in the periodic table the difference in the atomic weights of the common elements is approximately a multiple of 4. Take the first three rows of elements after hydrogen; they stand as follows:—

He = 4	Li = 6·9	Be = 9·1	Bo = 11	C = 12	N = 14	O = 16	F = 19
Diff. 16·2	16·1	15·2	16·1	16·3	17	16	16·4
Ne = 20·2	Na = 23·	Mg = 24·3	Al = 27·1	Si = 28·3	P = 31	S = 32	Cl = 35·4
Diff. 19·7	16·1	15·7	17	19·8	20	20	18·5
A = 39·9	K = 39·1	Ca = 40·	Sc = 44·1	Ti = 48·1	V = 51	Cr = 52	Mn = 54·9

Reviewing these figures it will be seen at once that the difference between the first and second row of elements is uniformly very close to 16. The differences between the second and third rows vary somewhat, inasmuch as three are very near to 16, while in four cases they are very near to 20. Both these numbers are multiples of 4, which is the atomic weight of helium.

The view which appears to be held, by those who have devoted themselves to the study of radio-active phenomena, concerning the constitution of the atom is somewhat as follows: The mass of the atom is associated with positive electricity which is neutralised by electrons, each carrying unit charge of negative electricity. There is a difference of opinion as to whether the positive mass is a kind of shell or skeleton in which the electrons are imbedded, or whether the positive charge is concentrated at the centre while the electrons are distributed round it. In either case the number of electrons is not indefinitely large, as it has been shown by Sir J. J. Thomson that only a minute fraction of the mass of an atom consists of electrons, which in *number* are about three times the atomic weight of the element. The positive mass is supposed to be, at least in part, made up of helium atoms, associated with hydrogen, especially among the elements of high atomic weight.

Since the electrons are all negative they must repel one another, and the helium and hydrogen particles being all positive also repel one another. The electrons must therefore be distributed in such a manner throughout the structure that electrical neutrality is preserved. In any of the more complicated cases the structure must be unstable, and under the action of forces the nature of which is unknown disruption occurs, a helium atom escapes and with it two electrons, giving rise to the phenomena of radio-activity. On the other hand, the stability and chemical indifference of helium and its allies are probably due to simplicity of structure and the presence in the atom of the least possible number of electrons.

Crookes' speculations concerning the genesis of the elements have already been recounted and some criticisms formulated. It seems to the writer that for many reasons the assumption that all the elements from first to last were evolved from one protyl by a process essentially the same throughout is untenable. In every atom it seems to have been established that there are two principles associated respectively with positive and negative electricity. Though the properties of the elements are in some way connected with the masses of the atoms of which they consist it seems improbable that atomic mass alone determines properties. There are too many anomalies in the periodic scheme itself. There can be little doubt that their properties are dependent in no small degree on the arrangement of the constituent particles, positive and negative, within the atom. Like properties indicate a like constitution, and the conclusion seems irresistible that in certain cases the analogy between a family of elements and a homologous series of carbon compounds expresses the physical fact.

Take the alkali metals for example, the formula a+nd, in

which a is the atomic weight of the first member and d a common difference, seems to represent their true relations to one another, thus :—

$$\begin{aligned} a &= 7 &&= \mathrm{Li} \\ a+d &= 7+16=23 &&= \mathrm{Na} \\ a+2d &= 23+16=39 &&= \mathrm{K}, \text{ etc.} \end{aligned}$$

This was the idea of Dumas three-quarters of a century ago.

In any discussion as to the mode in which the elements may have been evolved this principle must be borne in mind. It leads to the conclusion that in certain families of the elements the successive terms have been produced by a process which, after the formation of the first term, is continued by successive condensations of similar matter on the first as a nucleus.

In other cases it may be supposed that the process was disturbed in some way before completion, so that the result may be carried along two or more independent lines. The case of the metals allied to iron may be taken as an example. They are probably all derived from aluminium as a basis somewhat as follows :—

Al (27)

Sc (44)	Mn (55)	Fe (56)	Ga (70)
Y (89)	Co (59)	Ni (58·7)	In (115)
La (139)	?	Cu (63)	Tl (204)

As to the radio-active elements some have been formed by disintegration of others of greater atomic weight, while some may have been produced by condensation of a less deliberate character than usual, so that an unstable internal arrangement of electrons resulted with production of an endothermic atom.

Helium is known to be derived from radium. Hence it seems reasonable to suppose that argon and the rest which are found only in very minute quantity in the atmosphere, and not in minerals, may have originated in a corresponding disruption of more complex atoms of elements which had only a temporary existence, and have long disappeared from among terrestrial materials.

It is obvious, however, that here we reach almost the limit of legitimate speculation. There can be no doubt that much further light may be expected as the result of the experimental researches which are going on, and to allow the imagination to run too far in advance of positive knowledge is not good physical philosophy. The difficulties of investigation in this field, however, are likely not to diminish but rather to multiply. This is partly due to the excessively minute quantities of matter which have in some cases to be dealt with. An illustration of this occurs in connection with the chemical action of the rays from radium. Thus water is decomposed with production of oxygen, hydrogen peroxide, and hydrogen gas, and it has long been known that the glass of tubes containing the salts of radio-active elements becomes discoloured, assuming a purplish tint after a time, and this appears to be due to the liberation of sodium or potassium or possibly some other element contained in the ingredients of the glass.

These observations led not unnaturally to the idea that under the influence of electrons from radio-active matter some of the common elements might be broken up into fragments so that, for example, copper might be degraded into sodium and lithium. Experiments announcing this change were published, but other chemists could not get the same result, and at the present time the general opinion would not be in favour of such a conclusion.

Somewhat similarly it has been asserted by Professor Collie and Mr. Patterson that the cathode rays are capable of producing helium and neon out of gases and other matters in which they are believed not to exist as such, that is to say, by disintegration of one kind of atom other kinds are produced. But the same effects are not always producible even by the original observers, and by others are altogether denied. In such cases there is no alternative but to " wait and see."

THE AGE OF THE EARTH

A very interesting question has been raised which, though not definitely chemical in its character, may be mentioned here in consequence of its connection with the study of the properties and distribution of radium and other radio-active substances. It relates to estimates concerning the probable age of the earth.

The earth was once a hot fluid mass which is supposed to have resulted with the sun and planets from the condensation of nebulous matter. The question arises how long is it since the globe of the earth assumed approximately the condition in which it now exists, since the solidification of the crust and the separation of the waters from the dry land, and in fact the establishment of conditions which would fit it to be the abode of life ?

Attempts to answer this question have assumed a variety of forms, and very diverse estimates have been arrived at.

From inferences as to the rate of cooling of the earth deduced from observations of the temperatures at different depths in the crust Lord Kelvin estimated that not more than one hundred million years were required for the earth to cool down from the temperature when in the molten state to its present state.

Many geologists and physicists have studied the question and have arrived at different conclusions. The *History of the Geological Society of London*, published in 1907, on the occasion of the Centenary of the Society, contains the following passage (p. 227): "At present it may be said that an estimate of one hundred million years, for the period during which life has existed on the earth, is regarded as fairly approximate." Such is the view of the geologists, but as all such estimates are based on very insecure foundations they can be regarded only as mere speculations which serve to show how incomplete is our knowledge of the world in which we live.

The properties of radium have provided a new starting-point which, however, cannot be regarded as more secure than the rest. Radio-active matter is found everywhere in the rocks of the earth's crust as well as in the ocean and the gases of the atmosphere. The estimation of the amount of radium in the rocks is beset with difficulties, and it is certain that the amount varies enormously from rock to rock, being probably about fifteen times as much in some of the old granites as in basalts and some other rocks which are presumably less ancient.

Radium in its process of spontaneous disintegration gives out heat continually, and this heat developed in the substance of the earth might counteract the effect of radiation by which the earth is continually parting with heat and tending to cool down.

Now it has been estimated by Professor Rutherford that if one gram of radium emits 100 calories per hour "the presence of $4{\cdot}6 \times 10^{-14}$ gram of radium per gram uniformly distributed throughout the volume of the earth would produce as much heat as that lost from the earth by conduction to the surface. In other words, with such a distribution of radium, the temperature gradient of the earth would remain constant."

Now the average amount of radium found in rocks is about

twenty times the amount required by this hypothesis ; therefore after all the temperature of the earth may be actually rising and not falling, and the consequence is that evidence as to the age of the earth is still very nebulous.

CHAPTER IX

SOLUTIONS

PROBABLY few persons as they drop the sugar into a cup of tea concern themselves with the problem which is presented by the disappearance of the lump and the fact that in a few minutes, even without stirring, the taste of the sugar can be recognised in every part of the liquid. Suppose the sugar to be immersed in water the change can be watched more readily, and it is at once seen that the crystalline mass falls quickly asunder, while a dense syrupy liquid streams away from it. After a time, if sugar is added in successive portions to the same quantity of water, the process of dissolution slackens, the lumps crumble away less rapidly, and ultimately they undergo no change, the liquid being then, to use the common expression, saturated.

But every cook knows that if heat is applied and the temperature of the liquid raised more sugar will dissolve until another point of saturation is reached as the liquid boils. If such a liquid is then allowed to cool to the temperature of the air a portion of the sugar soon begins to separate from the liquid in the form of crystals the size of which depends on the volume of liquid and the condition whether it is stirred about or left at rest.

A lump of white marble looks to the unaided eye so much like loaf sugar that it might easily be mistaken for that substance, but if a lump of white marble is placed in hot tea or cold water no change would be observed ; it would not dissolve. But now suppose that the lump of white marble is immersed in water to which some nitric or hydrochloric acid has been added, there will be a great effervescence, bubbles of gas (carbon dioxide) escape, and the marble rapidly disappears, forming a clear colourless solution. This solution, however, contains something different from marble, and if the liquid is duly concentrated it will yield crystals, quite unlike marble, which consist of the nitrate or chloride of calcium, containing also a certain proportion of water. The case of sugar in water differs from that of marble in acid, therefore, in the fundamental fact that the sugar can be recovered unchanged in properties, while the marble cannot be recovered from the liquid because a chemical change has taken place, and part of its components has been lost.

Why does the sugar dissolve in water while the marble does not dissolve except on condition of undergoing chemical change? These are questions to which the physicist and chemist can give as yet only partial and imperfect answers. We may try to follow in imagination the change in the sugar. First we must recall the fact that the molecules of gases are free from each other and every one, according to the kinetic theory, moves about rapidly and independently of the rest, only knocking up against them and continually altering the direction of its course. In liquids we must believe, from the phenomena of diffusion, that something of the same kind is continually going on with this important difference, that there are relatively few separate and independent single molecules. A large proportion of the molecules move together in clusters or companies, which are larger at low temperatures and smaller if the temperature is raised. Thus the molecule of water in the state of gas, that is superheated steam, consists of two atoms of hydrogen and one atom of oxygen, or expressed in symbols H_2O. In the liquid state at the common temperature of the air these join together or at any rate move together in parties of two or more molecules, such as $(H_2O)_2$, $(H_2O)_3$, etc.

A minute quantity of the compound is also probably in a state of dissociation, being resolved into ions H and HO. This will be explained later and does not concern for the moment the consideration of the question relating to sugar. We must suppose then that a crystal of sugar immersed in water is exposed to a shower of blows from the moving molecules of the water which are sufficiently strong to detach separate molecules of the sugar from the surface and cause them to move about in the liquid in the same manner as molecules of the solvent itself. They are thus made to behave as they would do if converted into gas by the application of heat. Sugar cannot be gasified in this way because its atoms separate from each other and form new combinations, that is chemical decomposition takes place, before the necessary temperature is reached.

Why the molecules of calcium carbonate are not separable from one another in a similar way by the action of water is usually explained by saying that the cohesion between them is greater than the cohesion between molecules of sugar. That, however, is merely a word and not an explanation, and serves as an admission of ignorance as to the nature of the difference between the two substances.

Sugar dissolved in water is then assumed to be in a condition comparable with that of a gas, only the spaces between the molecules are occupied by moving molecules of another kind. The molecules of gases exert pressure which is regulated by temperature, and the number of molecules in a given volume conforms to the law of Avogadro. In the solution of sugar an analogous condition prevails. The boundary of a mass of liquid is determined by what is called "surface tension," which acts in such a way that the liquid, at any temperature much below its boiling-point, behaves as though it were confined within an elastic skin which always tends to squeeze it into the smallest possible space. This is shown by the spheroidal form of detached drops. Within this bounding surface the sugar and the water exert a pressure which is called the "osmotic pressure." The earliest observations of this pressure and measurements of it were made by Pfeffer, professor of botany at Bâle some thirty years ago. But the interpretation of his results led the Dutch professor Van 't Hoff, in 1887, to formulate the theory of solution which has just been briefly explained. Van 't Hoff found from Pfeffer's measurements the existence of a complete parallelism between the osmotic pressure of a dissolved substance and the laws which govern gas pressures.

In the first place the osmotic pressure, at any rate in dilute solutions, is in direct proportion to the strength of the solution, that is, to the amount of dissolved substance in unit volume of the liquid, and this is equivalent to saying that the osmotic pressure is inversely proportional to the volume, which is one form of Boyle's Law.

Pfeffer also found that osmotic pressure increases with temperature and that the increase is in harmony with Gay Lussac's law for gases, which states that the volume of a gas is directly proportional to the absolute temperature.

Further, when solutions containing different substances of the same chemical character are compared together and the quantities of the different substances are in the proportion of their *molecular weights* dissolved in equal volumes of the same solvent, they exert the same osmotic pressure. This is parallel with the law of Avogadro. The qualification here mentioned must, however, be observed, that is, sugars, alcohols, and neutral substances generally which are not electrolytes may be compared together, while acids, bases, salts may be compared together, but for reasons to be explained presently, the osmotic pressures given by such substances, which conduct an electric current, and are decomposed by it, are not comparable with those of substances like sugar. Their osmotic pressures are in general much greater.

Pfeffer's method of experiment was based on the employment of a membrane which allows water to pass through, but which does not allow the dissolved substance, such as sugar, to pass. If now we imagine a small vessel composed of this *semipermeable* material, filled completely with a solution of sugar and connected with a manometer, any change in volume of the liquid will be indicated by the manometer. If the semipermeable vessel is then immersed in pure water an increase of pressure soon begins to be manifest by the movement of the mercury in the manometer, and this is due to the passage of water from without inwards through the membrane. The maximum pressure indicated by the manometer is the osmotic pressure. Pfeffer's methods have been improved upon in more recent years, and measurements have been made by Morse and Frazer in America up to 20 atmospheres and more, also by means of a special apparatus devised by Lord Berkeley and Mr. Hartley, pressures have been recorded for strong sugar solutions up to more than 100 atmospheres.

The nature of the membranes which have been used is a matter of interest on account of physiological considerations. The material most commonly used in the earlier experiments was a film of copper ferrocyanide, which is formed when a drop of solution of copper sulphate is brought into contact with a solution of potassium ferrocyanide, but measurement of pressures could only be made possible when this material is deposited in the pores of unglazed china-ware, of which the experimental vessel to hold the solution is made. Through this material water passes freely, but neither copper sulphate nor potassium ferrocyanide, common sugar nor dextrose. Other artificial membranes have been prepared for use in the experimental study of the phenomena, but there are many interesting cases of natural membranes, of which one has been carefully studied by Professor Adrian Brown within recent years. He finds that the barley grain is covered with a membrane of this kind, which, so long as the seed is uninjured, allows water from any solution in which it is immersed to pass into the interior of the grain, but it completely excludes sulphuric acid, common salt, and many other substances. And even grains which have been boiled in water so as to destroy their vitality retain this selective power, and thus show that this power is due to a physical property of the membrane and is not a physiological effect of living matter. The importance of discoveries of this kind is obvious in connection with the changes which go on in both vegetable and animal tissues. There must be many different semipermeable membranes existent in such tissues to account for the remarkable and rapid exchanges between fluids contained in adjacent cells. To cite one instance the ascent of the sap from the root to the stem and often distant branches of a tree must be dependent on action of this kind in which water passes freely, while the soluble contents of the cells forming the wood and leaf are not suffered to be lost or washed away by rain.

If the principle is accepted that different substances taken in the proportions of their molecular weights in equal volumes of the solution have the same osmotic pressure, it is obvious that by the determination of the osmotic pressure a method is provided for the determination of molecular weight. But the experimental determination of osmotic pressure is not an easy matter, and it is therefore more practicable to use for comparison a solution of known osmotic pressure. Liquids which have the same osmotic pressure are usually described as *isotonic*, and the comparison may be made by observing whether or not water passes from the solution to be tested into the standard solution contained in a natural or artificial cell membrane.

It is unnecessary in this place to pursue the subject further, especially as it is treated very fully in all the best books on physical chemistry. Enough, however, has been said to show that the physics and chemistry of a cup of tea are more complicated than is perhaps commonly supposed.

The subject, however, is by no means exhausted. There is reason to believe that in the act of dissolution in a liquid many substances unite chemically with a portion of the solvent. Evidence of this is derived from the fact that heat is evolved when many substances are placed in contact with a liquid capable of dissolving them, the evolution of heat being a common sign of chemical union.

In many cases also changes of colour are observed, and when, from the resulting solution, crystals are deposited, these crystals contain definite molecular proportion of water or alcohol or other liquid, in the midst of which they have been formed. An example may be found in the crystals of common washing soda, which is represented by the formula $Na_2CO_3.10H_2O$. If some of these crystals placed in a saucer are gently heated they melt easily and soon give off steam. After a time the liquid dries up to a white powder which consists of the soda without the water. If the dry powder is now allowed to become quite cold and an equal weight of cold water is poured on it the mass becomes hot, and no solution is produced, for the water unites with the salt to reproduce the original substance. If more water is now added the salt passes again into solution, and after some time, if too much water has not been added, crystals appear having the same composition and properties as the original washing soda. In most cases the solubility of a salt in water is, like that of sugar, continuously greater with rise of temperature up to the boiling-point of the solution or even much beyond that point. But this is not always the case. Glauber's salt (sodium sulphate) is an excellent example. If some of this salt, which is sold in small crystals for medicinal use, is mixed with less than its own weight of water and very gently warmed it dissolves freely, forming a clear solution. If the latter is then heated to the boiling-point a shower of small crystals will be observed falling within the solution, and these crystals consist of sodium sulphate, Na_2SO_4, in the anhydrous state, that is, without combined water. The crystals of Glauber's salt contain ten molecules of water, $Na_2SO_4.10H_2O$, and the solubility of these increases with rise of temperature up to 32°·5 C., when a portion of this hydrate splits up into water and the anhydrous salt which is less soluble, and consequently a portion of it is deposited from the liquid. Such a separation of two components of a compound is called "dissociation," and in such a liquid as the solution referred to the process is doubtless continuous with rise of temperature. It certainly cannot be inferred that below the critical temperature

32°, the salt is wholly in the hydrated state, and that above that temperature it is wholly anhydrous. It is pretty certain that part of the salt is in both conditions throughout the whole range of temperature. Both the two compounds just referred to are salts, and as already mentioned such substances differ from sugar in the fact that when an electric current is passed through them in a state of solution, or when melted they undergo a peculiar decomposition called "electrolysis."

This subject is so important and has of late years occupied so much of the attention of chemists that it deserves a somewhat close consideration.

Electricity has long been known to be capable of passing through matter in two ways. Metals, and some other bodies in a less degree, allow a current to pass through them with comparatively little loss. They are called "conductors," and silver and copper are among the best conductors, while platinum, tin, and lead are less good. When a metal like platinum or a non-metal such as carbon, as in an Edi-swan incandescent lamp, is used to carry a current of electricity a part of the energy disappears as electricity, and appears as heat, but no chemical change is produced. On the other hand, if a current is passed through a solution of an acid, a base, or a salt the compound is resolved into two parts which are liberated at the opposite poles or electrodes. Such substances are called "electrolytes." Sugar, alcohol, and neutral substances generally, such as chloroform, ether, etc., are not electrolytes, and offer great resistance to the current. The older theory of electrolysis assumed that the two components of the electrolyte were torn from each other by the force of the current. According to the modern doctrine, on the other hand, the assumption is that the separation of the ions begins, and in all cases of good electrolytes is carried to a considerable extent *when the electrolyte is dissolved* in water or other appropriate liquid. Take the case of common salt for example; when dissolved in water this theory requires us to believe that a considerable proportion of the salt is no longer sodium chloride, the molecules of which are each constituted of an atom of sodium combined with an atom of chlorine as in the original solid salt. It is believed on the contrary that most of the molecules are broken up into ions of positive sodium, represented by the symbol $\overset{+}{\mathrm{Na}}$, and ions of negative chlorine, $\overset{-}{\mathrm{Cl}}$. These ions move about in the liquid as independent particles, their freedom of movement being complete in every sense but one. Any attempt to remove one kind of ion is fruitless, for if the liquid is evaporated and crystallisation of solid salt ensues every positive ion unites with a negative ion to reproduce molecules of the ordinary neutral kind. And when a current is passed through the liquid for every positive ion removed at one electrode a negative is withdrawn at the other.

Neither is it possible by the process of liquid diffusion through a membrane or a porous partition to separate sodium ions from chlorine ions even temporarily.

If a solution of common salt is placed in a porous pot or bag of parchment paper which is immersed in pure water, a portion of the salt will diffuse into the water. It is an eminently good diffuser, as compared with sugar and especially with gum or albumen, all of which move in a similar way, but far more slowly. But the salt which passes through to the water as the result of this spontaneous movement consists of sodium ions and chlorine ions in exactly the same proportions as in the salt which remains behind, so far as chemical analysis can determine. No doubt, as will be mentioned presently, all sorts of ions do not move about at the same rate, some being very much faster than others, but there are electrostatic forces at work between the positive and negative particles which prevent them from wandering beyond the range of each other's attraction, and thus becoming separable to anything more than an almost infinitesimal extent far beyond recognition by chemical analysis. There is in fact no reason to suppose that it will ever be possible to separate a neutral salt solution into a positive portion and a negative portion. There may be membranes in animal and vegetable tissues which possess to a small extent this kind of semipermeable property, but very little is definitely known in this direction. At the same time it is possible that some of the electrical effects observed in leaves and other living parts may be traced to this cause. Evidence of the existence of free ions is obtained from other considerations.

If this hypothesis is adopted it will be apparent why salts in general exert an osmotic pressure so much greater than that which is observed in the case of sugar. The following figures show some of the results of Pfeffer's work with different solutions :—

SUGAR IN WATER

Percentage of sugar, C.		Osmotic pressure P in mm. of mercury.		$\frac{P}{C}$
1	..	535	..	535
2	..	1016	..	508
2·74	..	1513	..	554
4	..	2082	..	521
6	..	3075	..	513

It is evident in this case that, allowing for reasonable experimental errors, the osmotic pressure is in the case of sugar directly proportional to the amount of dissolved substance per unit volume.

POTASSIUM NITRATE IN WATER

Percentage of nitrate, C.		Osmotic pressure, P.		$\frac{P}{C}$
0·8	..	1304	..	1630
1·43	..	2185	..	1530
3·33	..	4368	..	1330

In this case the ratio $\frac{P}{C}$ declines somewhat as the concentration of the liquid is increased, but this is evidently due to the passage of a small quantity of the salt through the membrane.

When a substance of any kind is dissolved in, say, water the properties of the liquid are modified. Thus it is a matter of common knowledge that sea water does not freeze so easily as fresh water, and the practice is familiar of strewing salt on a frozen surface to induce thaw, that is to form a liquid which remains liquid, while water at the same temperature is ice. It may not be so commonly known that the boiling-point of water is raised many degrees by the addition of common salt, but this is familiar to the practical chemist, who makes use of the fact when he requires a bath for experimental purposes somewhat hotter than boiling water.

About 1883 the first tolerably accurate estimations of the effect of dissolved salts in lowering the freezing-point of water were published by the late Professor Raoult of Grenoble. It was previously known that a determinate quantity of the same substance dissolved in the same quantity of water always reduced the freezing-point by the same number of degrees, and that when equal quantities of different substances were dissolved there was a simple relation of some kind among their molecular weights. But it was only after a long series of experiments that Raoult succeeded in establishing his important generalisations.

When known quantities of the same substance are successively dissolved in the same portion of a solvent on which it has no chemical action, there is a progressive lowering of the freezing-point of the solution, which is proportional to the weight of substance dissolved in a constant weight of the solvent. But it is not possible to carry the process very far, as deviations from this rule occur when strong solutions are used. But the relation of the depression produced by a small quantity of the substance to its molecular weight enabled Raoult to arrive at an important conclusion.

The following are some of his experimental results :—

Name of dissolved substance.	Molecular weight.	Depression of freezing-point by 1 gram of substance in 100 gr. water.	Product of depression and molecular weight.
Methyl alcohol	32	–0·541	17·3
Ethyl alcohol	46	0·376	17·3
Butyl alcohol	74	0·232	17·2
Glycerine	92	0·186	17·1
Mannite	182	0·099	18·0
Invert Sugar	180	0·107	19·3
Milk Sugar	360	0·050	18·1
Cane Sugar	342	0·054	18·5
Salicine	286	0·060	17·2
Phenol	94	0·165	15·5
Pyrogallol	126	0·129	16·3
Acetone	58	0·294	17·1
Ether	74	0·224	16·6
Ethyl acetate	88	0·202	17·8
Acetamide	59	0·301	17·8
Urea	60	0·286	17·2
Ammonia	17	1·117	19·9
Ethylamine	45	0·411	18·5

A glance at these figures is sufficient to show that the depression produced by 1 gram of the substance is inversely as the molecular weight, and hence that the product of the two is a constant. Of course the product varies a little owing to experimental difficulties, but the rule has been since well established by the results of hundreds of experiments made by other chemists in all the laboratories in the world. And in fact determinations of molecular weights by observation of the freezing-point of solutions is now one of the commonest and most useful operations in the course of research into the composition and character of new compounds of all kinds.

Suppose the weight P in grams of such a substance as sugar is dissolved in 100 grams of water and the number of degrees below 0° at which the solution begins to freeze be expressed by C. Then $\frac{C}{P}$ will represent the depression which would be produced by 1 gram of substance dissolved in 100 grams of water. If this expression is multiplied by the molecular weight of the substance dissolved, in this case sugar, the product is the depression which would be theoretically produced by the molecular weight M in grams of sugar dissolved in 100 grams of the solvent. This cannot be directly determined as it is necessary to work with rather weak solutions which give a depression of only 1° or 2° C. This molecular depression may be expressed by K, and is equal to $\frac{C}{P} \times M$.

When different substances of the same neutral character as sugar are dissolved in the same solvent the value of K is found to be the same. Hence a method is provided whereby if K is known for the solvent chosen, and for compounds analogous to the one under investigation, the molecular weight of the latter can be determined, or rather, from the possible molecular weights that value is chosen which comes nearest to the value of M in the formula $M = \frac{P \times K}{C}$. The solvents most commonly used are water, acetic acid, and benzene, for which the values of K now adopted are respectively 19, 39, and 49.

The method outlined here is the most accurate and the most commonly used method for determining the molecular weight of soluble substances which are not volatile without decomposition and of which the vapour density cannot therefore be determined.

The apparatus used in such work is described in all the best text-books, where also an account will be found of the method based on the observation of the boiling-points of solutions. The latter method is a little more difficult to carry out, and is not so commonly resorted to as the freezing process. Obviously the boiling-point of a liquid is directly related to its vapour pressure, for a liquid boils when the pressure of the vapour produced just exceeds the pressure of the atmosphere on the surface. Another method is occasionally used which consists in observing the change in the vapour pressure of the solvent at a fixed temperature when a known quantity of a substance is dissolved in it. And by another formula based on Raoult's work the molecular weight of the dissolved substance can be determined. The rules relating to the vapour pressures of dilute solutions are similar to those relating to freezing-points.

An important point in regard to these practical experimental methods is the relation in which they all stand to osmotic pressure, for molecular proportions of different substances when dissolved in the same quantity of the same solvent exert equal osmotic pressures, and raise the boiling-point or lower the freezing-point and the vapour pressure to the same extent. These effects are dependent on the number of particles present without regard to their composition.

Many other problems connected with solutions might be discussed if space permitted. The application of the process of dialysis or selective diffusion of dissolved substances discovered by Graham more than half a century ago has been utilised in the extraction of sugar from the beet. At the present time there is much discussion about the cultivation of the sugar beet in England. In France, Belgium, and Germany the crop and the production of sugar have long been matters of great national importance, while England, relying largely on Colonial sugar, has been obliged also to purchase large quantities of beet sugar from other countries.

Beet root contains 14 to 18 per cent of cane sugar together with a variety of other substances such as vegetable acids, asparagine and albuminous matter. In order to separate the sugar as much as possible from these substances, which if expressed along with it would interfere with its crystallisation when the juice is evaporated, the beet roots after being washed are cut into very thin slices, which are laid in warm water.

During its immersion a process of dialysis goes on, each cell of which the tissue consists acting as a membranous bag through which the sugar and any salts present pass pretty rapidly, while the non-crystalline albuminous and gummy matters remain behind. A series of tanks is employed into which the water is pumped in regular order so that the fresh water is added to the already partly exhausted pulp, while the extract is passed on to the tanks containing fresh beet, and so a fairly concentrated solution of sugar is ultimately obtained. This solution is then treated with lime and afterwards with carbon dioxide or sulphur dioxide gas.

The insoluble precipitate, which contains a considerable amount of organic matter, is then passed through a filter press, and the clear liquid evaporated in vacuum pans till it begins to crystallise. This method of extraction, by taking advantage of the process of diffusion, is indispensable and could not be replaced without great disadvantage by any process of extracting the juice by pressure.

The study of aqueous solutions has also been applied to the elucidation of problems connected with the formation of mineral deposits. The ocean is the recipient of all the very numerous substances washed out of the land by the action of rain, and the consequent delivery of these substances by streams and rivers into the sea. From the sea the water evaporates and passes invisibly into the atmosphere, but there can be no return of the dissolved salts to the land except in the form of spray carried by the wind. In countries like England with an extensive coast the amount of salt thus returned is considerable, as fine spray is carried by strong winds a long distance inland.

Deposits of rock salt have been formed by slow evaporation of the water of such enclosed basins as the Great Salt Lake in Utah, the Dead Sea, etc., into which streams bring soluble matter and from which there is no exit. The consequence of these conditions is that in the course of ages such water becomes highly saline and deposits are formed at the bottom and round the shores of such lakes. Naturally the salts which are least soluble in water are deposited first if they are present in appreciable quantities. Thus a bed of gypsum (hydrated calcium sulphate) is usually present below deposits of rock salt, but as salts of potassium and magnesium, partly in the form of sulphate, are associated with the sodium chloride the products may be very numerous. A very important research was undertaken a few years ago on the formation of oceanic salt deposits by Professor Van 't Hoff, and carried on by him with the aid of his students for many years, till shortly before his death in 1911. This had special reference to the famous salt deposits at Stassfurth in Prussia, from which supplies of potassium and magnesium salts have been distributed in large quantities during many years past, and the lack of which has caused some inconvenience outside Germany during the war. The salts from Stassfurth which have become familiar in commerce are :—

Carnalite . .	$KCl, MgCl_2, 6\,H_2O$,
Kainite . .	$KCl, MgSO_4, 3\,H_2O$,
Kieserite . .	$MgSO_4, H_2O$,
Sylvite . .	KCl.

There are many other double salts in these deposits, but the result of Van 't Hoff's work has been to explain how it came about that these compounds were formed and in what order. Temperature has a good deal to do with it, but pressure is also a condition which may modify the composition of some of the salts in the solid state as they occur in the veins underground.

Another important question which has been studied within recent years is that which relates to the formation of the double carbonate of magnesium and calcium which constitutes the mineral called *dolomite*, which is so abundant as to form whole mountains in some parts of the world, the Eastern Alps for example. The substitution of magnesium carbonate for calcium carbonate in such rocks is the problem which has occupied chemical geologists without the discovery of a definite answer in each case. The examination and analysis of the core obtained from a bore hole drilled into the atoll of Funafuti led Professor Judd in 1904 to the conclusion that the original calcium carbonate, secreted by the corals, has been partly replaced by magnesium carbonate after the death of the organisms. While the proportion of magnesium carbonate near the surface is from 12 to 16 per cent, at a depth of 637 to 1114 feet, it increases and is maintained with some variations at 40 per cent, a proportion which approaches the amount required (45·65) to form dolomite. The true explanation of this change is yet to be sought, but a further attack on the problem, with the aid of the knowledge acquired by a study of Van 't Hoff's researches, is much to be desired, and would probably lead to interesting results.

CHAPTER X

ELECTROLYSIS

It is little more than a hundred years since the decomposition of water by an electric current was first seen by Nicholson and Carlisle. A few years later Davy, making use of the same agency, isolated potassium and sodium, and with the aid of a battery consisting of 2000 plates first showed to an audience at the Royal Institution the arc light between points of charcoal. Faraday succeeded Davy, and within ten years after the death of the latter the quantitative laws, relating to electro-chemical decomposition, were established by him. The electric deposition of metals from solution is the basis of the beautiful art which gives to every household its silver-plated spoons and forks, and supplies the means of copying with the most minute detail any surface which is, or can be rendered, conductive of electricity. Since the days thus briefly referred to the means of generating electric currents have developed chiefly out of the discoveries of Faraday. The application of the current to the production of heat or motion is familiar but cannot be further described at this point. The question which has occupied chemists for a hundred years is what is the nature of the process of electrolysis; why does the electric current so easily decompose a solution of common salt or acidified water, while it has scarcely any effect on a solution of pure sugar or alcohol?

Around questions like these a revived discussion has raged for more than twenty years, and though practical unanimity on the main part of the modern theory has been reached, many accessory questions have yet to be settled and probably will remain unsolved for many years to come.

First of all it will be useful to recall the hypothesis which was accepted for the greater part of the nineteenth century. It will be the easier to perceive the profound nature of the change which has been introduced. Imagine, for example, that a solution of hydrochloric acid is submitted to the action of a current. If the solution is moderately strong hydrogen gas is evolved from one pole, and chlorine gas from the other.[1] These two substances Faraday called the " ions " of hydrochloric acid. According to the older hypothesis introduced by Grotthus in 1805 it was supposed that throughout the liquid, between the poles or electrodes while in action, the molecules were ranged in a series of polar chains, the positive constituent, in this case the hydrogen, facing in one direction toward the negative plate or cathode, while the negative constituent, in this case chlorine, was drawn toward the positive plate or anode. In the process of electrolysis the positive hydrogen atoms moved from molecule to molecule along the chain until at the end, at the surface of the cathode, they were attracted away and then appeared in the free state in the form of hydrogen gas. The negative atoms of chlorine were supposed to move in a similar manner along the chain in the opposite direction. This may be represented in the following diagram :—

Immediately before Decomposition

$$\underbrace{HCl}\quad \underbrace{HCl}\quad \underbrace{HCl}\quad \underbrace{HCl}\quad \underbrace{HCl}\quad \underbrace{HCl}$$

During electrolysis. First phase

$$H\quad \underbrace{ClH}\quad \underbrace{ClH}\quad \underbrace{ClH}\quad \underbrace{ClH}\quad \underbrace{ClH}\quad Cl$$

In order to explain the continuance of the process it is then necessary to assume that the new molecules turn round so that their ions face the electrodes to which they then move.

During electrolysis. Second phase

$$H\quad \underbrace{HCl}\quad \underbrace{HCl}\quad \underbrace{HCl}\quad \underbrace{HCl}\quad \underbrace{HCl}\quad Cl$$

According to this idea the effect of the current is first to cause all the molecules between the electrodes to arrange themselves in lines having the ionic constituents facing in opposite directions. This would appear to involve the setting up of a peculiar structure or at least a tactical arrangement of the particles in the liquid which should have some effect on its optical or other properties. Nothing of the kind can, however, be detected in the space occupied by the liquid between the electrodes. Secondly, the

[1] If weak the liquid gives hydrogen at the cathode and hydrochloric acid with oxygen at the anode.

theory assumes that the molecules of the electrolyte, though moving about in the liquid, are all complete and entire until the moment when the electro-motive force is applied, and that then, and not sooner, they are separated by the opposite electrical attraction of the cathode and anode respectively.

In such operations if the same current passes successively through solutions of different electrolytes the quantities of substance liberated at each pole is in proportion to the chemical equivalents of these substances. That is if 1 part by weight of hydrogen is liberated in the cell charged with hydrochloric acid, 108 parts of silver would be deposited from a solution of a silver salt included in the same circuit. Similarly $32\frac{1}{2}$ parts of zinc, $103\frac{1}{2}$ parts of lead, $31\frac{3}{4}$ parts of copper would be liberated for 1 part of hydrogen. At the anode where the chlorine appears for every $35\frac{1}{2}$ parts of that element set free, 8 parts of oxygen, or 80 parts of bromine, or 49 parts of sulphuric acid would appear at the same surface. This is in accordance with Faraday's law. It appears that all soluble or fusible compounds are decomposed by the same current in chemically equivalent quantities, however different they may be in chemical constitution. It is also a fact that the smallest current sent through an electrolyte produces chemical decomposition in proportion to its strength in accordance with Faraday's first law (p. 110). If molecules, in the fluid state of a substance, are broken up by the current it might be expected that while some would be easily decomposed, others would be broken up with greater difficulty, and one or other of Faraday's laws would not be valid.

In one of his lectures on "Theories of Chemistry," given in 1904, Professor Arrhenius refers to the beginning of his own work on the subject in the following words: "In the year 1883 I carried out an investigation on the conductivities of different electrolytes, and was thereby led to the conclusion that all the molecules of an electrolyte do not conduct the electric current. The molecules were therefore divided into two classes, active and inactive. At high dilutions all the molecules were supposed to be transformed into the active state. The number of electrically active molecules in a solution (e.g. of an acid) was measured by its conductivity.

"Now the order of different acids, as regards their power of displacing one another from their salts, was known from thermochemical measurements. This order was exactly the same as

To face page 186.

that of their conductivities in equivalent solutions. This circumstance led me to suppose that chemically active molecules are identical with electrically active ones, and therefore the conductivity of an acid was regarded as a measure of its strength. In consequence it was argued that the velocity of a reaction, which may be brought about by different acids, is proportional to the conductivity of the acid used. . . . Generally speaking, there seems to be a certain parallelism between electrical conductivity and chemical reactivity. Gore found that pure anhydrous hydrochloric acid does not (appreciably) attack oxides and carbonates ; also it is practically a non-conductor of electricity. Similarly, one can understand why concentrated sulphuric acid may be transported in iron vessels, whereas diluted sulphuric acid attacks them very rapidly."

He then goes on to show that electrolytes differ from non-electrolytes in giving, when in solution, greater osmotic pressures, and greater effect on the freezing, boiling-points, and vapour pressures. This has already been sufficiently explained in the preceding chapter on "Solutions."

An idea of great importance was introduced into chemistry by the late Professor Alexander Williamson about 1850, in the endeavour to explain the remarkable process by which alcohol and sulphuric acid react to form ether. He says,[1] "We are forced to admit that in an aggregate of molecules of any compound there is an exchange constantly going on between the elements which are contained in it. For instance, a drop of hydrochloric acid being supposed to be made up of a great number of molecules of the composition ClH, the proposition at which we have just arrived would lead us to believe that each atom of hydrogen does not remain quietly in juxtaposition with the atom of chlorine with which it first united, but on the contrary is constantly changing places with other atoms of hydrogen, or what is the same thing, changing chlorine. Of course this change is not directly sensible to us, because one atom of hydrochloric acid is like another ; but suppose we mix with the hydrochloric acid some sulphate of copper (of which the component atoms are undergoing a similar change of place), the basylous elements, hydrogen and copper, do not limit their change of place to the circle of the atoms with which they were at first combined,—the hydrogen does not merely move from

[1] *Quarterly J. Chem. Soc.*, vol. 4, p. 111.

one atom of chlorine to another, but in its turn also replaces an atom of copper, forming chloride of copper and sulphuric acid," and so forth. It is not necessary to quote further, but this idea, which we owe to Williamson, of atomic motion and exchange of partners in a liquid or gas, is now acknowledged to lie at the root of all modern ideas of chemical action.

The same idea was made use of in 1857 by the German physicist Clausius, in order to explain electrolysis, but he could not answer the question as to the proportion of the dissolved substance which must be supposed to be in the act of exchanging constituents, and therefore dissociated into ions. To Arrhenius we owe the knowledge that from the conductivity and the osmotic pressure of the solution of an electrolyte the relative proportions of the undissociated and dissociated molecules can be calculated. It is well known that water is an extraordinarily bad electrolyte. Experiments originally made by Davy a hundred years ago, and repeated by numerous experimenters down to our own day, have shown that in proportion as the dissolved impurities in ordinary water are removed or excluded the conducting power, small at any time, is steadily reduced, and at one time it was supposed that *pure* water was a non-electrolyte and non-conductor. What is called " conductivity " water, that is water specially distilled under special precautions, does possess a very small conductivity which is attributed to the presence of minute quantities of hydrogen ions and hydroxyl, HO, ions. It will therefore behave as an extremely feeble acid or base.

Hydrogen chloride or hydrochloric acid, in the liquid state but absolutely free from water, is a colourless liquid, but it exhibits none of the characteristic properties of water as a solvent. It is a non-conductor. It has none of the properties of an acid, for it does not alter the colour of litmus, nor does it act on alkaline oxides or carbonates. If, however, this liquid is mixed with water the solution is one of the best electrolytes known, and with the development of conductivity the chemical activity commonly attributed to hydrochloric acid at once becomes manifest. It is obvious, therefore, that some change has taken place in the constitution of both water and hydrogen chloride when in presence of each other.

It appears probable that the ions available in the liquid in a moderately strong solution are chiefly those of hydrochloric acid, inasmuch as the products actually evolved are hydrogen and chlorine. There are, however, also the ions, hydrogen and hydroxyl, resulting from an increased dissociation of water, the latter interacting with chlorine to reproduce hydrogen chloride and oxygen gas. Water seems to have a special power of inducing ionisation, superior to that of other liquids which have been tried as solvents.

All these and many other facts are accounted for by the doctrine of Arrhenius. The current passing through a dissolved or melted electrolyte does not tear the molecules asunder, but simply directs the ions according to the electrical charges with which they are already invested, and causes them to be discharged and so restored to the common molecular neutral state at the surfaces of the electrodes where they appear.

The laws of electrolysis indicate the relation of electrolytic decomposition to the ordinary chemical doctrine of valency. A molecule of hydrogen chloride, HCl, is resolved into 1 part by weight of hydrogen and $35\frac{1}{2}$ parts by weight of chlorine. According to the law of electro-chemical equivalents, if the same current passes through a solution of common salt, NaCl, and then through solutions of calcium chloride, gold chloride, and tin chloride, for every $35\frac{1}{2}$ parts of chlorine set free in the first cell, the same amount of chlorine would be liberated in each of the others. But while $35\frac{1}{2}$ parts of chlorine combine with 23 parts of sodium, which is the atomic weight of that metal, the same quantity combines with 20 parts of calcium, with $65\frac{3}{4}$ parts of gold, and with $29\frac{3}{4}$ parts of tin, and these are respectively $\frac{1}{2}$, $\frac{1}{3}$, and $\frac{1}{4}$ the atomic weights of those metals. But since the atomic weights are known from other considerations we must suppose that an atom of calcium combines with twice as much chlorine as an atom of sodium, an atom of gold with three times as much, and an atom of tin with four times as much. And these metals are respectively said to be univalent, bivalent, trivalent, and quadrivalent. Chemically equivalent atoms carry in electrolysis the same electric charge, and the charge carried by one atom of hydrogen or one atom of chlorine is called the unit charge, and is represented by 96,500 coulombs of electricity for 1 gram of hydrogen.[1] The atom of calcium, therefore, carries two such atomic charges, an atom of gold three, and an atom of tin four atomic charges of electricity.

[1] No explanation can be given here of the electrical units, definitions of which will be found in every text-book of physics.

The name *electron* is now used to signify one atomic charge, and from the account already given of these minute masses it must be inferred that the addition of one electron to an atomic mass converts it into a univalent negative ion. The removal of an electron from an atom converts it into a positive ion. When electrolysis of hydrogen chloride, for example, occurs each atom of chlorine set free receives one electron, while each atom of ionised hydrogen loses an electron, the other elements gaining or losing two, three, or four electrons according to their valency or combining capacity. According to this idea the movable electron by its presence or absence determines the possibility of chemical combination between atoms. It has been considered as an element by Sir William Ramsay, who assigns to it the symbol E, and has discussed the use of such formulæ as NaECl for ionisable compounds.

In the story of the action of the electric discharge on gases the electronic constitution of atoms has been described. If the hypothesis just explained is accepted as to the process of electrolysis, it must be admitted that while the electrons when free are all of the same kind, no difference of mass among them having been observed, their relation to atoms must be of two kinds. It would appear that the majority enter into the permanent constitution of the atom, while others, usually from one to four and never more than eight, are capable of being detached in chemical exchanges. These correspond to the valency of the atom.

In considering such hypotheses we may observe that among the elements known there are at least three kinds of molecules.

(1) There are the elements of the argon group whose molecules consist of one atom only which is electrically neutral and incapable of taking up either a positive or negative charge, and hence are chemically inactive and are unknown in the form of any chemical compound either among themselves or with other elements.

(2) The elements of the zinc, cadmium, mercury group have also monatomic molecules, but they are fairly active as chemical agents. Zinc, for instance, dissolves readily in all acids, and mercury, though not so easily attacked by acids, combines with the halogens and even with many metals. An atom of zinc therefore must be constituted internally quite differently from, say, an atom of argon.

(3) The great majority of the elements have polyatomic mole-

cules, H_2, O_2, S_2, S_4, S_6, P_4, N_2, etc. This appears again to indicate that the atoms of these elements must be capable of presenting opposite charges to each other, in virtue of which they combine together to form molecules, thus $\overset{+}{H}\overset{-}{H}$ or $\overset{+}{Cl}\overset{-}{Cl}$. So that their positive and negative constituents are both capable of changing their attitude according to circumstances. There is, however, no condition known in which hydrogen ions go to the anode or chlorine atoms to the cathode in the process of electrolysis.

Here we are on difficult ground. Chemists are far from general agreement as to the valency of the elements and the cause of the differences which are observed in so many cases between the principal or ordinary valency, which seems to rule the composition of the most stable compounds, and the extra valency which is frequently developed. By the latter they endeavour to account for the composition of what are still frequently called molecular compounds.

The measure of the principal valency is the behaviour of the compound in electrolysis. Thus there is no difference of opinion as to the constitution of common salt, which whether in the fused state or dissolved in water, yields the ions $\overset{+}{Na}$ and $\overset{-}{Cl}$. But when a saturated solution of common salt in water is cooled prismatic crystals are formed which contain $NaCl2H_2O$. An immense number of compounds of a similar character are known containing this " water of crystallisation " ; for example :—

Washing soda . .	$Na_2CO_3.10H_2O$
Glauber's salt . .	$Na_2SO_4.10H_2O$
Epsom salt . .	$MgSO_4.7H_2O$, etc.

What is the nature of the bond which holds together the molecule of the salt and the molecules of the water superadded no theory yet explains in a manner generally accepted by chemists. If we look through the periodic scheme of the elements many of them are found to exhibit two degrees of valency ; thus carbon appears to be bivalent in carbonic oxide, CO, while it is quadrivalent in marsh gas, CH_4, and in nearly all other compounds. Nitrogen appears to be bivalent in nitric oxide, NO, while it is trivalent in ammonia, NH_3, and probably quinquevalent in sal-ammoniac, NH_4Cl, and in nitric acid, $HO.N.O_2$, etc. Oxygen is usually bivalent, but it is now agreed that in certain cases it develops two more units of valency and becomes quadrivalent.

If in common salt the chlorine atom owes its univalency to the presence of one electron, and in water the oxygen owes its bivalency to two electrons, it appears that the combination of the chloride with water must be due to some other agency than the presence or absence of electrons. This question is still the subject of active debate, and restrictions as to the number of valencies which may be assumed have of late been considerably relaxed.

The most interesting and practically useful assumptions are those which within the last few years have been proposed by Professor Alfred Werner of Zürich, as the result of investigations into the composition and properties of the complex compounds formed by the union of the salts of certain metals, such as platinum and especially cobalt, with ammonia. Many of these have long been known, and their constitution has remained a puzzle. Werner has shown that only a portion of the groups of atoms attached to the central metallic atom are capable of forming ions, so becoming chemically reactive. Those which are not ionisable, are connected by a peculiar subsidiary valency and never exceed six. The number of these is known as the *coordination number.*

It is possible that the case of water of crystallisation, already referred to, may be found to come within Werner's hypothesis.

The question arises therefore whether valency, that is, capacity to enter into chemical combination, is in all cases electrical, and due to the intervention of electrons. It is of course well known that some elements usually very readily ionisable show no signs of the same property in certain forms of combination. Thus chlorine when in the form of the chloride of hydrogen or a metal is immediately reactive when brought into contact with other ionised elements or groups of elements. But the compounds,

$$CH_3Cl,\ CH_2Cl_2,\ CHCl_3,\ \text{and}\ CCl_4,$$

and many others in which the halogen is directly attached to carbon, show no signs of response to the common reagents. Thus pure chloroform, $CHCl_3$, does not give, like other chlorides, a white precipitate of silver chloride. It does, however, interact with ammonia, exchanging its three atoms of chlorine for one atom of nitrogen, producing hydrogen chloride and hydrogen cyanide. The electrolytic test cannot be applied to compounds such as those mentioned, and in the case of carbon compounds generally there are difficulties which attach to the view that the constituent atoms are held together by electrons. It is only necessary to inspect the formula of such a compound as one of the sugars.

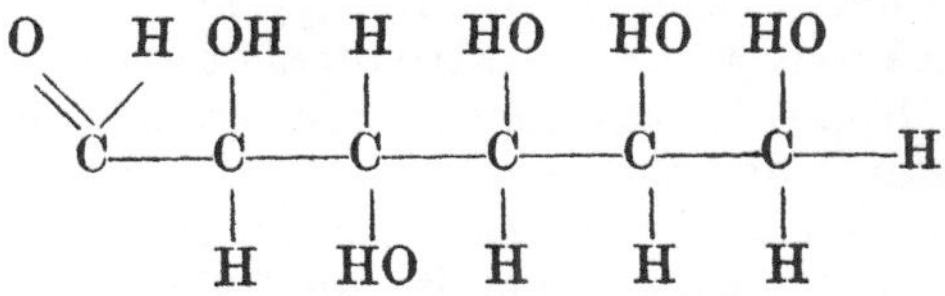

Here are five hydroxyl groups, HO, and more than an equivalent number of hydrogen atoms, but neither hydroxyl nor hydrogen is capable of separating from the carbon so as to form an ion.

Ionic dissociation in a liquid has been sometimes regarded as similar in nature to the dissociation of many volatile compounds when converted into vapour by heat. This, however, is a case of imperfect analogy. When, for example, ordinary ammonium chloride is heated it can easily be shown that the vapour contains both ammonia and hydrogen chloride, for on passing through it a pipe of porous clay, such as a tobacco pipe stem, and blowing air through the pipe, the issuing gas is alkaline, owing to the presence in it of an excess of ammonia. This is explained by the escape through the pores of the clay of both ammonia and hydrogen chloride, but in accordance with the ordinary law of gaseous diffusion, the ammonia being the lighter gas, diffuses most rapidly. The law states that gases diffuse at rates which are inversely proportional to the square roots of their densities. The density of ammonia (compared with hydrogen) is 8·5, while that of hydrogen chloride is 18·25; hence their relative rates of diffusion are $\sqrt{18{\cdot}25}$ and $\sqrt{8{\cdot}5}$, so that ammonia gas passes through a porous partition nearly $1\frac{1}{2}$ times faster than hydrogen chloride.

If an electrolyte, such as sodium chloride, is dissolved in water, and a layer of pure water floated above it, it might be expected that if the ions sodium and chlorine into which much of it is supposed to be resolved are free it should be possible to detect the chlorine in the watery upper layer in excess of the sodium, while in the lower layer there would remain an excess of sodium. But as explained in the previous chapter the ions retain their opposite electric charges, they are prevented from travelling far from each other, and the minute quantity which may be supposed to escape, according to calculations by Professor Nernst, is too small to be recognisable by chemical tests.

The theory of electrolytic dissociation is still beset with a few difficulties which, however, will probably disappear in time, as a certain proportion of them are due to experimental inaccuracies and to complications which are attributable to the formation of compounds, possessing electrolytic properties, between the dissolved substance and the solvent.

Another kind of difficulty is met with in the case of the compounds which are known as "amphoteric" electrolytes. These are substances which are capable of behaving as acids or bases according to the nature of the substance with which they are associated. Amino-acetic acid or glycine, $NH_2.CH_2.COOH$, and amino-succinic or aspartic acid, $COOH.CH(NH_2).CH_2.COOH$, are examples of these compounds. These two compounds contain the group NH_2 in virtue of which they are feeble bases, while they both contain the carboxylic group CO.OH, which confers acid properties. In either case, acting as both acid and base, they must give rise to the hydrion $\overset{+}{H}$ and hydroxidion $\overset{-}{HO}$, the latter resulting from the ionisation of the hydrated form of the substance; amino-acetic acid becoming $OH.NH_3.CH_2.COOH$. From this if the substance acts as base, the ions formed are $NH_3.CH_2.COOH$ and $\overset{-}{OH}$, while if it acts as acid the ions are $OH.NH_3.CH_2.COO$ and H^+, or the corresponding anhydrous ion $NH_2.CH_2.COO^-$. Combination may take place between two molecules of the compound, the one acting as base, the other as acid, or the acid and basic portions of one and the same molecule may act on each other. In any case, according to the theory, all these ions must exist together in an aqueous solution of glycine, and the experimental enquiry into the electrical conductivities of such solutions is therefore difficult, but there is no reason to suppose that they offer any obstacle to the acceptance of the hypothesis.

Industrial processes based on electrolysis are now common. Such are, for example, the modern methods by which such metals as sodium and aluminium and zinc are obtained, copper is refined, and electro-plating with silver and gold (a long-established method) with nickel, zinc, and brass. In addition to the deposition of metals there is also the liberation of hydrogen, now required on a very large scale, by electrolysis of alkaline solutions. The liquid round the anode in any electrolytic arrangement is exposed to oxidising influence, while the cathode provides a means of reduction, and both these effects are now turned to account for manufacturing a considerable number of salts and other compounds formerly procured by purely chemical processes. Falling water is usually the source of the energy which is transmuted through the current into chemical energy and heat in the cells, and at Niagara, for example, there has been a large development of electrical industries on both the Canadian and American sides.

Potassium chlorate is an important compound formerly made by passing chlorine gas into alkaline solutions. It is now made almost exclusively by the electrolysis of potassium chloride solution, keeping the liquid at about 70° C., at which temperature the hypochlorite formed at lower temperature is changed into a mixture of chlorate and chloride. When the electrolytic cell is divided by a diaphragm so that the electrodes are kept separate the electrolysis of sodium chloride may be arranged to yield caustic soda and hydrogen gas at the cathode, with chlorine at the anode. The gases may be led off and utilised in any way desired. The caustic soda is in the solution, and when the decomposition is effected in the Castner cell it is obtained free from common salt.

Another important product is permanganate, which was formerly prepared by fusing together black oxide of manganese and potassium hydroxide or carbonate, whereby a green manganate, K_2MnO_4, is formed. By passing carbon dioxide through the solution one-third of the manganese was precipitated as dioxide MnO_2, while the potassium carbonate and permanganate were left in solution. In order to prevent waste of potash the manganate may be dissolved from the fused mass by water, and submitted to electrolysis with iron electrodes at a temperature of about 60°, when the manganate is oxidised to permanganate. Other salts, such as perchlorates and persulphates, are obtained in a similar manner.

On the other hand, the electrolytic method is applied to reduction, and a good example is afforded by the preparation of hydroxylamine from nitric acid. The electrolyte is sulphuric acid of 40 per cent strength, to which nitric acid is slowly added while the whole is kept cool. The hydroxylamine crystallises from the liquid in the form of sulphate and the yield is almost theoretical.

By the use of a cathode cell and appropriate current density, temperature, and dilution many organic compounds may be similarly produced with results, as to yield, which in many cases are superior to the older methods of reduction by means of sodium, sodium amalgam, or zinc dust and acid.

CHAPTER XI

CATALYSIS AND CATALYSTS

THESE look like very hard words, but as *catalyst* is derived from the Greek, which merely means an agent which *unloosens* or sets free something else, they can be regarded as reasonable in their application, and not, as appears in some other cases, a device for concealing ignorance.

An example will make the matter clear. Oxygen mixed with twice its volume of hydrogen forms a mixture which is well known to explode on approach of a flame or an electric spark, or when heated strongly enough in any other way. This mixture of gases may be kept indefinitely in a closed vessel in the dark or in sunlight without the production of water or any other sign of chemical combination. If, however, a perfectly clean piece of platinum foil is introduced into the mixture combination between the oxygen and hydrogen immediately begins, and often proceeds so rapidly that the metal becomes red hot and ultimately the residual gas explodes. But when the action is all over the platinum betrays no sign of having had anything to do with the matter. It is unaltered in weight and appearance and, if unsoiled by handling or otherwise, it retains the peculiar catalytic property which it has just manifested.

The facts just related were discovered so long ago as 1817 by Sir Humphry Davy, who, with the aid of a spiral of platinum wire suspended in a glass containing a little alcohol or ether, demonstrated the union of the vapours with the oxygen of the air on the surface of the metal, producing what he called his "lamp without flame." Many observations of the same kind made by later experimenters proved that the effect was producible by many substances beside platinum.

Catalysis, then, is a process in which a chemical change, which without assistance proceeds either not at all or very slowly, is greatly accelerated by contact of the materials with a small quantity of some agent, called the catalyst, which remains after the reaction undiminished.

The catalyst in some cases retains in its appearance some evidence of having suffered change, but in most cases this can be fairly attributed to the action of the heat which is developed during the reaction. It is, however, possible frequently to account for the starting and continuance of the observed chemical action by the hypothesis of the alternate formation and decomposition of a compound of the catalyst with some constituent of the materials engaged in the change. Thus in the process formerly employed for procuring oxygen by heating potassium chlorate with a relatively small quantity of manganic oxide, there can be little doubt that the action consisted essentially in the formation of an oxide of manganese containing a larger proportion of oxygen than the dioxide employed, and its subsequent decomposition so that the dioxide remained at the end.

In this case evidence is derived from the altered appearance of the oxide left behind, and also from the two facts that the oxygen thus obtained always contains a trace of chlorine, while the residual potassium chloride is alkaline from the formation of a trace of potassium oxide.

An example of this kind seems to differ from that of the first mentioned, namely, the action of platinum on oxyhydrogen gas. And on enquiry among the numerous and various instances of catalytic action, among carbon compounds especially, it appears that no explanation can be found which is applicable generally. All that can be said is that the action is in the majority of cases chemical and that it involves the alternate formation and destruction of an unstable compound.

As an example of the effect of a very small quantity of acid may be cited the action of almost any acid on common sugar, whereby it is converted into "invert" sugar, a mixture of equal quantities of dextro- and lævo-glucose, or the action of a few drops of sulphuric acid on aldehyde whereby in a few minutes it is converted, with evolution of heat, into paraldehyde, a compound having the same composition as aldehyde but three times the molecular weight. The preservative effect may also be cited of a minute quantity of sulphuric or phosphoric acid on hydrocyanic (prussic) acid, which in its absence passes quickly into a mixture of ammonium formate and a brown substance.

One very interesting case in which a minute quantity of a third substance affects the mutual behaviour of two others is provided in those numerous instances in which the presence of a minute quantity of water seems to be essential to interaction. It certainly appears to be so when chlorine is brought into contact with metals, for even metallic sodium may be preserved for years in contact with chlorine gas at common temperatures if the latter is perfectly dry, and the indifference of combustible substances such as carbon monoxide gas, charcoal, and phosphorus to oxygen gas when all are free from moisture has been the subject of much experimental enquiry.

So striking are these phenomena that it has even been supposed that chemical action cannot take place except in the presence of a small quantity of some electrolyte. This generalisation is, however, for the present too wide and many facts are known which seem to oppose it.

Catalytic agents are employed in many industrial operations, and since the publication of the researches of Professor Sabatier of Toulouse, less than twenty years ago, a stimulus has been applied to the utilisation of catalytic change for manufacturing purposes.

An example of catalytic effect is to be found in the long-established lead-chamber process for making sulphuric acid, in which sulphur dioxide, water, and atmospheric oxygen are enabled to interact rapidly in the presence of a relatively small proportion of nitric peroxide NO_2. Here several intermediate nitrogenous compounds are undoubtedly formed, but whether they are essential stages in the process by which sulphuric acid is ultimately produced from its dissociated constituents is a question which cannot be regarded as even yet finally settled. The lead chamber survives, but of late years has found a serious rival in the "Contact" process, which is merely the outcome of a long known catalytic operation based on the use of finely divided platinum. Theoretically sulphur dioxide requires one atom of oxygen to convert it into sulphur trioxide, $SO_2+O=SO_3$. This combination is attended by the evolution of a considerable amount of heat, which, if allowed to accumulate, so as to raise too high the temperature of the tubes containing the contact substance, will partly undo the result of the combination, and the sulphur trioxide is destroyed. Another point to attend to is the necessity for providing a considerable excess of oxygen in the form of atmospheric air, the nitrogen of which takes no part in the change. The platinum is used in the form of a deposit of fine particles on asbestos fibre, which is easily produced by soaking the asbestos in a solution of platinic chloride, drying it and then exposing to a low red heat by which the chloride is completely decomposed. Since platinum has become so costly as it now is, many attempts have been made to replace it by other substances, and many patents have been taken out. It appears, so far as can be ascertained, that ferric oxide (red oxide of iron) has met with some success, but has not served to replace platinum.

In connection with the use of platinum a discovery was made in the earliest days of this process which for a time checked its development and even threatened failure. The fact came out that minute quantities of certain substances have the property of "poisoning" the catalyst, so that its activity pretty rapidly declines, and it becomes "dead." Fortunately this was traced to the impurities which accompany the sulphur dioxide produced by roasting iron pyrites. Of these the most important is arsenic. By cooling and spraying with water the gases brought from the pyrites ovens these impurities can be removed and the gas, cleared of mist, can be safely delivered into the series of pipes charged with the platinised asbestos. Many other practical points require attention to secure success in the operation. Thus it is found that at a temperature of 400° to 430°, 98 to 99 per cent of the sulphur dioxide is converted into the trioxide, while a further rise of temperature reduces the yield. It is therefore necessary to provide the means of cooling by a draught of air when the temperature tends to rise too high.

The Claus kiln affords another example of a catalytic process which has been turned to account for industrial purposes. In the production of soda, that is sodium carbonate, from common salt by the Leblanc process which has been in operation for more than a century, the accumulation of impure calcium sulphide in the form of alkali-maker's waste was for generations a source of loss to the manufacturer, and of annoyance to the district.

Small mountains of this material are to be seen in the "Black Country" of Staffordshire and in parts of Lancashire. Attempts to deal with it proved unsuccessful till about the year 1887 when Messrs. Chance, alkali makers, of Oldbury, succeeded in overcoming the difficulties which had previously stood in the way of success. Their process consists essentially in decomposing the wet tank waste or impure calcium sulphide with carbon dioxide, obtained from limekiln gases, in such a way as to obtain a gas very rich in sulphuretted hydrogen, the residue being almost inodorous and harmless. The problem then is to get the sulphur out of this gas in a convenient form, so as at once to get rid of the offensive smell and obtain a marketable product. This is done by the use of the catalytic action of ferric oxide, which in the presence of a mixture of sulphuretted hydrogen and a limited quantity of air, slightly warmed, causes the hydrogen to unite with the oxygen while sulphur is set free. The Claus kiln is a cylindrical brick chamber having a perforated bottom on which is laid first a quantity of broken fire brick, upon which a layer of peroxide of iron in the form of some suitably porous ore is laid. The mixture of about 4 volumes of air with 5 volumes of sulphuretted gas (containing 38 per cent of H_2S, the rest being chiefly nitrogen) is passed through this bed of oxide, and the water vapour and sulphur vapour pass into adjoining chambers, where some of the sulphur is collected in the fluid state and some in the form of crystalline powder or flowers of sulphur.

The action of the oxide of iron seems to consist in local reduction to a lower oxide and reoxidation by the passing air. A somewhat similar action occurs in the oxide of iron purifier employed in the gas works for the removal of the last portions of sulphuretted hydrogen from coal-gas.

The combination of nitrogen with hydrogen so as to produce ammonia has long been a desideratum. The passage of electric sparks through such a mixture gives rise to the formation of a minute amount of the compound, but inasmuch as ammonia is decomposed by heat the process soon reaches a stage at which equilibrium is established, the ammonia being destroyed as fast as it is formed. Pressure has been found to promote the combination of the two elementary gases in accordance with the rule that pressure in a mixture of gases tends to facilitate the formation of that substance which occupies the smallest volume, in this case the ammonia, as shown by the equation,

$$N_2+3H_2=2NH_3.$$

The introduction of a catalyst in the form of an active metal greatly increases the yield of ammonia, and when the ammonia formed is withdrawn from the mixture as quickly as possible, by absorption by an acid or otherwise, the action is still further promoted.

It is only recently that proposals to use the nitrogen of the air for the manufacture of ammonia have taken a practical shape.

Experiments carried on during the last seven years, under the direction of Professor Haber of Karlsruhe, with the support of the Badische Colour Company at Ludwigshafen, have resulted in the production of ammonia on an industrial scale.

The arrangements are understood to be somewhat as follows: A mixture of nitrogen with three times its volume of hydrogen gas is passed under pressure of about 150 atmospheres into a tube which contains the catalysing substance, and which is maintained by means of an electric coil at a temperature between 500° and 700° C. To collect the ammonia thus produced the mixed gases return through a coil surrounded by liquid air where the ammonia condenses.[1] Now as the nitrogen and hydrogen which are returned along with a fresh supply of the mixed gases to the contact chamber are at a low temperature they have to be warmed by passing through a heat exchanging coil. Here they receive heat from the mixture of gases as they leave the tube containing the catalyst.

The most active catalyst appears to be metallic osmium, but in view of its costliness and the very limited supply of this rare metal, many other materials have been tried, and if patents are to be regarded as any indication of what is going on pure iron appears to be the favourite.

The nitrogen required in the process is obtained by fractional evaporation of liquid air to be explained in a later chapter, and thus the process is made continuous. Whether Haber's process is destined to take a permanent place among chemical industries

[1] The temperature must not be much below −75° or the ammonia may freeze and stop up the pipes.

is as yet uncertain, but the continual increase in the demand for ammonia, not only for agricultural but for manufacturing purposes, coupled with the idea of utilising this compound as a step toward the production of nitric acid and nitrates, shows that this possibility is not to be lightly ignored.

A very pretty and interesting experiment sometimes exhibited in the lecture room consists in bubbling oxygen through a little moderately strong solution of ammonia contained in a flask, in which is suspended a coil of clean platinum wire. White fumes of ammonium nitrate and nitrite appear in the neighbourhood of the coil, and if the supply of oxygen is rapid the flask will become filled with orange coloured peroxide of nitrogen, and the bubbles of gas, containing as they do an explosive mixture of oxygen and ammonia, often burn as they escape from the surface of the liquid and are ignited by the now red-hot metal. In this experiment air may be substituted for oxygen with similar though more moderate effects.

It is perhaps remarkable that these phenomena should have remained unnoticed by the industrial chemist till quite recent times. Early in the present century, however, the conditions of the reaction between ammonia and air in the presence of platinum were investigated by Professor W. Ostwald of Leipzig, and it was found that the yield of nitric acid amounted to something like 85 per cent of the theoretical. The ammonia mixed with a relatively large volume of air is passed through a layer of platinum coated with the spongy metal, or other less expensive catalysts such as one of several metallic oxides, and maintained at a temperature of 300° C. or somewhat higher.

It appears that in this process the purity of the ammonia is not a matter of much importance, and even the gas from crude gas liquor may be used. Of course ammonia from any source may be employed, and this method of producing nitric acid has latterly been associated with the process of Frank and Caro in which synthetical calcium cyanamide is decomposed by steam :

$$CaCN_2 + 3H_2O = CaCO_3 + 2NH_3.$$

The catalytic processes which have become most familiar are those in which the addition of oxygen is the object in view, and the idea of adding hydrogen to nitrogen as a practical means of obtaining ammonia is still in the early stages of development.

But the researches of Professor Sabatier of Toulouse on combinations of hydrogen effected by the agency of a catalyst date from the extreme end of the last century, and have excited not only great interest among those occupied in scientific chemistry, but have led to unexpected applications to industrial purposes which already have assumed a position of great practical importance.

These researches seem to have originated in attempts, known to have been made by Moissan, to contrive the direct union of acetylene with certain metals such as copper, nickel, and iron. The expected fixation did not take place, but Sabatier found that ethylene, as well as acetylene, when directed on finely divided metals at a temperature of only 300° C. produces incandescence with a deposit of carbon, the escaping gas consisting of hydrogen mixed with ethane. This seemed to indicate that hydrogen had been added to the elements of the ethylene, and by further experiments it was found that a mixture of ethylene and hydrogen passing through a column of reduced nickel is changed into ethane by combination of the two gases : $C_2H_4+H_2=C_2H_6$.

In association with M. Senderens further research enabled the enquirers to generalise this result. Even at common temperatures acetylene mixed with excess of hydrogen, in contact with the metal, is completely converted into ethane, without destruction of any portion of the hydrocarbon and without formation of secondary products.

These experiments carried out in 1899 showed that nickel freshly reduced from its oxide possesses this catalytic power in relation to hydrogen in a peculiar degree. Reduced cobalt, iron, and copper, as well as spongy platinum partake of this property more or less.

Various modifications of the process have since been devised, especially with the object of operating on materials in the state of liquid, without resorting to the process of converting into vapour or gas. At the temperatures necessary for this purpose many carbon compounds are destroyed or seriously altered in composition. These modifications are in many cases successful, but the interaction takes place much more slowly.

As to the chemical changes which are brought about by the process of hydrogenation they may be ranged into several classes. When oxygen is present water is formed in some cases, while in others hydrogen is simply added on. In other cases the molecule is broken up as when benzene, C_6H_6, is converted by addition of hydrogen into methane or marsh-gas, CH_4. Aniline is also converted by similar action into hydrocarbons and ammonia which is, of course, produced from the nitrogen detached.

Professor Sabatier explains the action of these metals by the hypothesis of the formation of an unstable temporary hydride of the metal formed by combination of hydrogen with the superficial layers. Such hydride would be easily dissociable, and the hydrogen, therefore, is easily removed by contact with unsaturated compounds.

This hypothesis is in harmony with the views which seem to prevail about catalytic processes in general. In all cases which have been sufficiently investigated there appears to be formed a small, often minute quantity, of an unstable and often merely temporary compound which seems to carry its effect from molecule to molecule throughout the mass, sometimes remaining recognisable at the end or merely reverting to the condition of the original catalytic agent introduced. It is, however, often difficult to determine in special cases whether the action is attributable to the formation of a definite though unstable chemical compound, or whether it is to be included among those still obscure cases of physical condensation which come under the modern designation "adsorption."

It has long been known that metallic palladium can "occlude," to use the common expression, several hundred times its volume of hydrogen gas. The metal retains its ordinary appearance after being charged either by acting as the cathode in an electrolytic cell decomposing acidified water, or by heating the metal in hydrogen gas. This palladium-hydrogen immersed in a solution of ferric chloride reduces it to the state of ferrous chloride, again without visible change in the palladium; is this to be regarded as a compound of palladium and hydrogen in which the hydrogen can become active in consequence of being ionisable?

Hydrogen gas, under ordinary circumstances, is without action on a ferric salt, but under considerable pressure it produces reducing effects. If materials capable of generating hydrogen, such as zinc and dilute sulphuric acid, are brought into contact with such a compound as ferric chloride, reduction occurs, and the result is commonly spoken of as an effect of "nascent" hydrogen, which is then supposed to be in the state of free atoms electrically charged. There is apparently a little conflict of evidence here, and time only will show what further light can be obtained in this direction by research.

We may now turn to the consideration of the important practical applications which have arisen out of these apparently recondite investigations. In the first place it is perhaps not inappropriate to observe that they supply a satisfactory answer to those persons who are often disposed to enquire as to the utility of this or that piece of pure scientific work which seeks to extend knowledge without reference to the further use of it. Scientific literature abounds with examples, but this deserves remark, because it is so recent as to be still in process of development. But there is another reason for noticing the present case attentively, and that is that it serves as an example of the common failure of the discoverer to participate in the commercial profit which is made of his discovery. Professor Sabatier has shown the chemist and manufacturer how the element hydrogen may be made to unite with a great diversity of substances, by a process which is easily carried out and which involves the use of no costly materials.

A large proportion of vegetable oils and some animal fats are liquid at the common temperature of the air. They are, therefore, of smaller value for many purposes than the fats which are solid under the same conditions. Consequently many attempts have already been made to act on the oils in such a way as to convert them more or less completely into solid substances. It should be explained that most of the oils consist essentially of a compound called *olein*, which is the ester or compound ether of glycerine with oleic acid. The solid fats are similar compounds derived from stearic or palmitic acid. Stearic acid is so named from the Greek word *στέαρ*, tallow. Palmitic acid occurs abundantly in palm oil. Now these latter acids are what the chemist calls saturated compounds, that is they contain the carbon, hydrogen, and oxygen, of which they are composed, in such a condition that they fully satisfy their mutual attractions and cannot enter directly into any further chemical union. But olein is unsaturated and can unite with two atoms of hydrogen, forming stearic acid.

The most practical of the older attempts to produce a solid fat from oil was based on the fact, discovered long ago, that in contact with nitrous acid olein and oleic acid are converted into solid compounds, called respectively elaïdin and elaïdic acid.

Another method depends on the action of strong sulphuric acid on oleic acid, whereby it is converted into a mixture of solid compounds which require subsequent distillation under reduced pressure.

No sooner had Sabatier made known the nature of his method than numerous patents were taken out by other people with the object of applying his principle to the hydrogenation of unsaturated fats.

The practical feature of these patents consists in the fact that it is only necessary to add the catalyst, usually porous metallic nickel, to the oil, to heat it to a moderate temperature, namely from 200° to 250° C., and to inject hydrogen gas into the mixture. The result is that the unsaturated oils present combine more or less completely with hydrogen to form the corresponding saturated fat, and thus at the end of the process a product is obtained which has a melting-point considerably above the melting-point of the material operated on. In fact an oily substance is thus hardened into a fat which is solid at common temperatures. The resulting hardened fats are of great commercial importance, being largely employed especially in soap and candle making. The following extract from a recent trade report for the year 1913[1] indicates the manufacturer's view of the position :—

"In the year 1913 the imports of the hitherto customary raw materials for soap-making, such as tallow, palm oil, and cocoanut oil, showed a drop of more than 6000 tons. But it must not be concluded from this that the production of British soaps decreased in that year, because the new hardening process has also given to certain other fatty substances, such as whale oil and linseed, which formerly were scarcely of any account, a great importance for the soap industry. This process is all the more important for the British soap industry because through the establishment of numerous soap factories in countries which formerly supplied basic materials (South Africa, Australia, Argentina, Japan, etc.), these materials are now employed locally instead of being sent to Great Britain. Thus the hardening process has prevented an enhancement of the prices of the basic materials, and of the soap itself, which would have considerably restricted both the consumption and the production."

The production of the enormous quantities of hydrogen

[1] Messrs. Bigland Sons and Jeffreys, of Liverpool.

required in these operations is a question of great practical importance which will be dealt with in a later chapter. It is only necessary to say here that the hydrogen employed must be approximately pure, as the presence of small quantities of sulphur or arsenic compounds serves to diminish the activity of the catalyst, and ultimately to destroy it, as in other cases already referred to. Though the weight of hydrogen actually absorbed by the oil is relatively small, the volume of gas, by reason of its lightness, assumes enormous proportions. One ton of oleic acid requires roughly 79,000 litres or 2800 cubic feet, and one ton of ordinary olein requires 75,900 litres or 2680 cubic feet of hydrogen gas.

It is also of the utmost importance to prepare the catalyst, usually reduced nickel, in such a way as to avoid the introduction of impurities. Thus it is found that when the hydroxide of nickel from which the metal is to be reduced has been made from the sulphate, it is impossible to avoid the presence of minute quantities of basic sulphate, and when this is heated in hydrogen it is reduced to sulphide and the resulting metal is ruined so far as its catalytic power is concerned.

An interesting method of introducing a very active form of catalyst is based on the remarkable power possessed by nickel of uniting with carbonic oxide to form a volatile compound. According to this process water gas, a mixture of hydrogen and carbonic oxide, is passed over nickel ore at about 100° C. The carbonic oxide unites with the nickel, leaving other metals behind, and nickel tetracarbonyl, $Ni(CO)_4$, is produced in any required amount. This compound in vapour together with the excess of hydrogen is then passed into the liquid to be hydrogenated at a temperature between 200° and 240° C., when the nickel carbonyl is decomposed, depositing the metal in a finely divided and very active state. It is probably at the moment when it is liberated from the compound that the nickel causes the union of the hydrogen with the fat, as it is stated that if the carbonyl compound is first passed in so that the nickel is liberated and the hydrogen is then supplied no practical result is obtained.

The application of the process is not confined to soap and candle making. Already various inferior oils are being converted into solid fats which, after due purification, are transformed into edible products, and in all probability considerable additions to the supply of margarine from the chemical factory may be expected. The only point which seems to require further investigation is the effect which the minute quantities of nickel present in these foods may have on the human consumer. There is at present no indication of serious consequences.

In this great development the discoverer of the principle and its application has, it is understood, no share. It is gratifying, therefore, to record the recent award of a Nobel Prize for chemistry to Professor Sabatier.

The process of hydrogenation has been supplemented in an interesting way by the discovery that the same catalytic agents are capable in certain cases of inverting the process, and so causing a disruption of the substance into hydrogen and a residual compound. The alcohols, for example, may be resolved in the presence of copper into hydrogen and aldehyde. The process may be turned to industrial use for the production of formaldehyde from methyl alcohol.

Other catalytic agents have been found among the metallic oxides by the Russian chemists Gregorieff and Ipatieff. Alumina, for instance, at a temperature near 300°, causes generally the dissociation of the primary alcohols into water and the corresponding ethylenic hydrocarbon.

But the process of dehydration seems to be always accompanied by dehydrogenation with the production of an aldehyde and hydrogen gas. But whether the one or the other of these changes predominates depends on the nature of the oxide. Thus thoria, alumina, and tungstic pentoxide are very active in decomposing the vapour of ethyl alcohol at 340° to 350° C., and the gas evolved consists almost entirely of ethylene. Oxides of zinc, manganese, and vanadium, on the other hand, are much less active, and the gas produced is chiefly hydrogen. Alumina at the lower temperature of 240° to 260° splits ethyl alcohol almost entirely into ether and water.

The action of various surfaces in promoting chemical action, especially combustion, is illustrated by the important work accomplished during recent years by Professor W. A. Bone of the Imperial College of Science and Technology at South Kensington. A complete account up to that date was the subject of a lecture given in November, 1912, to the German Chemical Society in Berlin. A further summary was given at the Royal Institution in London on 27th February, 1914, and in the Howard Lectures to the Royal Society of Arts in March, 1914.

To these publications readers interested in the matter must be referred for details.

The following abbreviated version contains the most prominent and important facts :—

Mr. Thomas Fletcher in 1887 showed that when a mixture of ignited gas and air is directed on to a large ball of iron wire so as to heat it to the necessary temperature, and the current of gas is then momentarily interrupted, the ball will continue to glow with great increase of temperature, but without any sign of flame.

Bone began investigations in 1902 as to the influence of various hot surfaces on the combustion of hydrogen and carbon monoxide, and has arrived at the following general conclusions. (1) The power of accelerating gaseous combustion is possessed by *all* surfaces at temperatures below the igniting point in varying degrees, depending on their chemical characters and physical texture. (2) Such an acceleration of surface combustion is dependent on an absorption of the combustible gas and probably also of the oxygen by the surface, whereby it becomes activated (probably ionised) by association with the surface; and (3) the surface itself becomes electrically charged during the process.

It also appears that while hot surfaces possess the power of accelerating gaseous combustion at temperatures below or near to the igniting point, the same power is manifested in an increasing degree as the temperature rises. And there is experimental evidence that the differences manifested by different surfaces at low temperatures practically disappear when the temperature of the surface reaches bright incandescence.

Incandescent surface combustion has been applied to a number of practical purposes, such as heating rooms and providing a hot surface suitable for many cooking operations such as grilling or roasting. It has also been applied on a large scale to raising steam, melting metals and alloys, and other practical purposes.

In the former case a diaphragm is prepared of granulated fire brick or other material bound together into a block and fitted into a suitable frame, which provides a space at the back into which the gas and air mixture can be fed (Fig. 56, p. 210). The gas being first turned on and lighted, air is then gradually added till a fully aerated mixture is obtained.

The flame soon becomes non-luminous and diminishes in size;

a moment later it retreats on to the surface of the diaphragm, which at once assumes a bluish appearance. Finally, all signs of flame disappear and there remains an intensely glowing surface. The temperature thus attained is high enough even to melt platinum. Consequently in applying the principle to the construction of a furnace for fusion or other purposes in which a high temperature is required the choice of the contact material is necessarily rather limited. Practically all solids except calcined magnesia and carborundum are excluded.

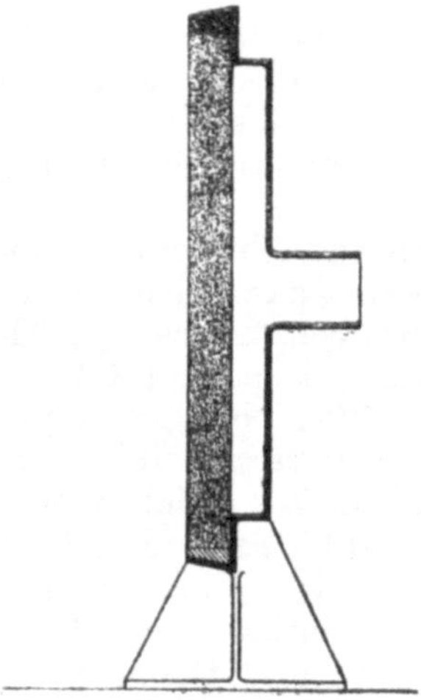

FIG. 56.
SECTION OF DIAPHRAGM FOR SURFACE COMBUSTION (BONE).

In raising steam three forms of apparatus may be adopted. In the case of a multitubular boiler in which the heating tubes pass through the body of the boiler containing the water, the tubes are packed with the refractory contact material in a granular state. The combustible mixture of air and gas passes through these tubes, and the control of the heat communicated to the boiler and the amount of steam raised, is effected either by adjustment of the amount of gas admitted or by working the tubes in groups, so that any number can be brought into action as required.

The diagrammatic section of an experimental boiler on this construction with ten tubes is shown in the figure. The tubes are 3 feet long and 3 inches in diameter, fixed in a cylindrical steel shell capable of withstanding a pressure of over 200 lbs. per square inch. Three only of the tubes are shown here. The gaseous mixture was forced through the tubes from a special feeding chamber attached to the front plate of the boiler; the products of combustion after leaving the boiler passed through a small heater containing nine tubes by which the water before entering the boiler could be warmed. The feed water on entering was at the temperature 5°·5 C., or nearly 42° F., on passing to the boiler the temperature was 58° C., or about 136° F. The successful results obtained with this apparatus led to the erection

Fig. 57.—BOILER WITH 110 TUBES AT SKINNINGROVE IRON WORKS. (BONECOURT SYSTEM.)

To face page 210.

of a boiler of about ten times its capacity by the Skinningrove Iron Company, Ltd., to be fired by the waste gas from an Otto by-product coking plant. The boiler is shown in the illustration. It consists of a cylindrical drum 10 feet in diameter and 4 feet from back to front, traversed by 110 steel tubes 3 inches in diameter packed with granular refractory material. In front of the boiler is a specially designed feeding chamber which delivers washed coke-oven gas under a pressure of 1 to 2 inches of water.

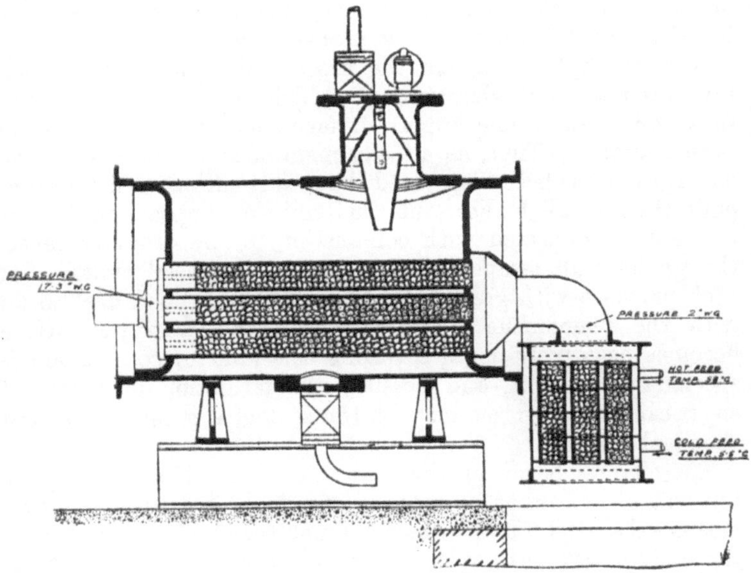

FIG. 58. EXPERIMENTAL TUBULAR BOILER FOR SURFACE COMBUSTION (BONE).

This gas with a regulated supply of air is drawn, by suction from a fan, through a short mixing tube, into each of the combustion tubes, where it burns without flame. The products of combustion pass outwards into a semicircular chamber at the back of the boiler, and thence to the tubular feed-water heater. The fan, which is attached just beyond this heater, is driven by an electric motor, sucks out the cooled products at a temperature of 100° C. (212° F.) or under and discharges them into the atmosphere.

A second kind of arrangement provides for the use of liquid

fuel in which the liquid is first burnt in a separate space under the boiler, and the imperfectly burnt products are carried with the requisite proportion of air through the tubes containing the granular contact substance.

In the third arrangement the granular material is placed in trays beneath the boiler.

The new method admits of the employment of almost any form of combustible gas, such as waste gases from the blast furnace or coke-oven, producer-gas of any kind, as well as water-gas and coal-gas. A high efficiency has been obtained up to 92–94 per cent in the most favourable cases.

Any attempt to explain the process of surface combustion involves many considerations for which at present we are not fully prepared. Some important facts, however, have already been revealed. Thus, as already mentioned in the chapter on electrons, it has been discovered that many surfaces when heated emit these small bodies, and this emission appears to be not necessarily connected with combustion in the ordinary sense, though at high temperatures a greatly increased emission of such particles with great velocity may have a good deal to do with the phenomena. The fact that the catalytic surface becomes negatively charged during this kind of combustion is specially significant, and possibly the formation of layers of electrically charged gas may be the seat of a greatly increased chemical activity.

The chapter began on the subject of catalysis and catalysts, but though it cannot be extended further the subject is by no means exhausted. Everywhere through the literature of chemistry, old or new, examples occur of processes which must be regarded as operating under the influence of these mysterious agencies. Hence to be exhaustive the whole known range of chemical changes would have to be reviewed.

Among processes long recognised as catalytic is the Deacon process for obtaining chlorine. This depends on the interaction of hydrogen chloride gas with the oxygen of air in the presence of cupric chloride. A temperature somewhat over 400° C. is required. The process was introduced more than forty years ago, and will be found described in all the principal manuals of chemistry.

There are many other cases in which a small quantity of a substance is sufficient to initiate or promote a chemical change otherwise not to be accomplished. These are to be found especially in connection with laboratory processes employed in the study of organic compounds. The remarkable characters and properties of enzymes will be described in a later chapter.

CHAPTER XII

ARCHITECTURE OF MOLECULES

STEREO-CHEMISTRY AND THE NATURE OF VALENCY

We may begin by recalling the fact that Dalton and several of the early promoters of his atomic theory were led to consider, though without giving the subject much attention, the question of the arrangement which chemically combined atoms assume in space of three dimensions. If a detached molecule could be seen, what would it look like? There is a great deal of evidence, some of which is indicated in a preceding chapter, that each atom retains its independence, so that there is a certain rough analogy between the bricks in a wall and the atoms in a molecule. Dalton, referring to the diagrammatic representations of atoms in his *Chemical Philosophy*, Part I (1808), says :—

"The combinations consist in the juxtaposition of two or more of these (atoms); when three or more particles of elastic fluids are combined together in one it is to be supposed that the particles of the same kind repel each other and therefore take their stations accordingly."

Dalton also gives diagrams showing the arrangements which he supposed might exist in a number of different compounds, including a substance so complex as alum.

Wollaston about the same time recognised that it would be necessary "to acquire a geometrical conception of their relative arrangements in all the three dimensions of solid extension."

It was nearly fifty years before the germ of the doctrine of valency was recognised by Frankland, but it has been a subject of constant enquiry, experiment, and discussion down to the present day.

In the chapter on Electrolysis a brief indication was given of one, and perhaps the most important method, of measuring this property, though for obvious reasons it cannot be applied in all cases. By valency must be understood the habit of some elements, of which hydrogen is the most important, to enter into combination with no more than one other element at the same time, while others may combine with two or more.

Thus one atom of oxygen may be combined with two other atoms, an atom of nitrogen with three or five others, an atom of carbon at the most with four other atoms. Why atoms are thus limited in their power of combination is one of the fundamental problems of chemistry.

"Constitutional" formulæ based on notions of valency began to be used soon after 1860, but these formulæ had no pretension to representing the relative positions of atoms in space. They served merely to show in what order the atoms were supposed to be linked one to another in a molecule, and thus served to some extent to epitomise the chief chemical changes to which the compound would be liable. Thus if acetic acid was represented as $CH_3.CO.OH$ or as

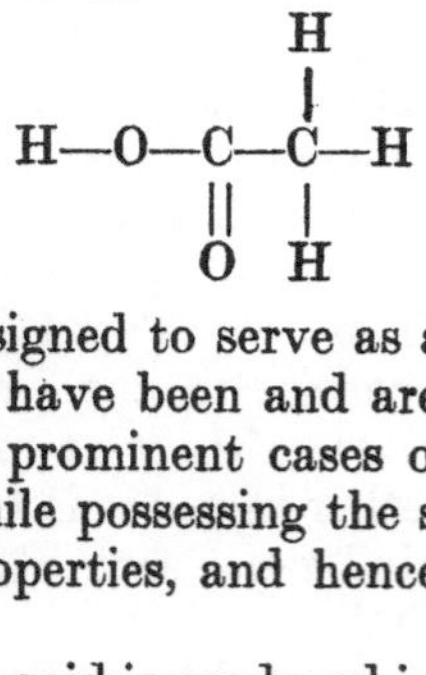

the latter was not designed to serve as a picture of a molecule, though such formulæ have been and are very valuable for distinguishing the more prominent cases of isomerism, that is of compounds which, while possessing the same composition, have different chemical properties, and hence, presumably, different atomic structure.

An acid called lactic acid is produced in sour milk, and another acid having the same composition occurs in flesh, and hence is found in Liebig's meat extract. These acids have the formula $CH_3.CHOH.COOH$. They are very much alike, but the latter of these acids is optically active—that is it causes the rotation of a plane polarised ray—while the other is inactive. The study of these acids by Wislicenus in 1872–3 led to the idea that the differences observed could only be accounted for by supposing different relative positions to be assumed by their constituent atoms in space of three dimensions. But it was not till 1875 that a complete theory was conceived by the late Dutch professor Van 't Hoff, and set forth in his treatise *La Chimie dans l'Espace.* Almost simultaneously the connection between optical activity and asymmetry was discovered by the French chemist J. A. Le Bel.

The theory is based on the following recognised facts :—

1. The four units of valency of carbon are equal in every respect. In the mono-substitution derivatives of marsh-gas CH_4 and ethane C_2H_6 no isomeric modifications have been discovered.

2. All compounds of carbon which in the liquid state rotate a polarised ray, or when crystallised produce hemihedral forms which are mirror-images of each other, are found to contain at least one atom of carbon which is united directly to four dissimilar atoms or groups of atoms, and which is therefore said to be asymmetric.

3. Compounds which are known to contain asymmetric carbon, and which, nevertheless, do not exhibit optical activity, are generally resolvable by one or other of several known processes into two compounds, each of which possesses rotatory power equal and opposite in direction to the rotatory power of the other.

Succinic acid, for example,

```
  H
  HC—CO·OH
   |
  HC—CO·OH
  H
```

is optically inactive ; but when one of the hydrogen atoms is replaced by hydroxyl, so that the C to which it is attached becomes " asymmetric," the result is the production of malic acid,

```
     H
  HO·C—CO·OH
     |
    HC—CO·OH
```

which exists in two isomeric forms, one of which rotates the polarised ray to the right, the other to the left.

An apparent exception is represented by mesotartaric acid, which has the same composition as (1) ordinary, dextro-, tartaric acid, (2) racemic acid which is found in the grapes of certain districts, and (3) lævo-tartaric acid which is obtainable along with the dextro-acid from racemic acid. Mesotartaric acid is not resolvable into two acids, like racemic acid, and therefore cannot be regarded as a compound of the other two. But it contains within the molecule two asymmetric carbon atoms indicated by black type in the formula,

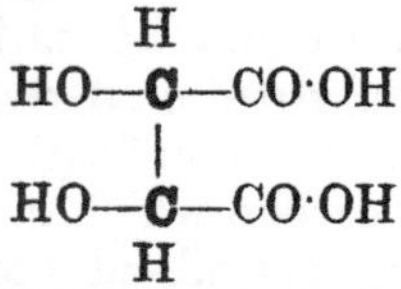

and the action of one of these on the polarised ray may be supposed to be equal and opposite to the action of the other, so that the effect is the same as if they existed in separate molecules, mixed together in exactly equal numbers, as in the case of racemic acid.

Van 't Hoff's hypothesis, which serves to explain these facts, supposes the carbon atom to be situated at the centre of a regular tetrahedron, while the four other atoms united with it are situated at the solid angles; so that the four valencies of the carbon atom are supposed to operate in the directions of four radii of a sphere included in the tetrahedron or which includes it.

Suppose the atoms united with a carbon atom to be represented by letters, then when one atom of carbon is united with 4A, with 3A+1B, with 2 A + 2 B or with 2A+1B+1C isomerism is impossible, that is there can exist only one compound of this constitution. But when all four of the attached atoms or groups of atoms, A, B, C, D, are different two cases occur. These may be represented either by a tetrahedron, the angles of which are lettered as in figure 59; or by a conventional symbol, which is more easily printed thus—

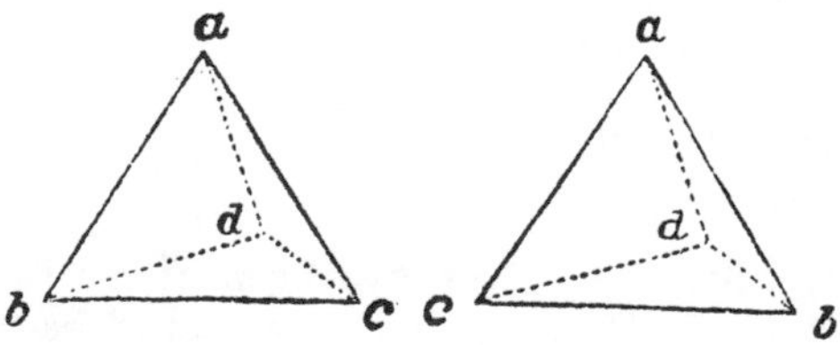

Fig. 59.

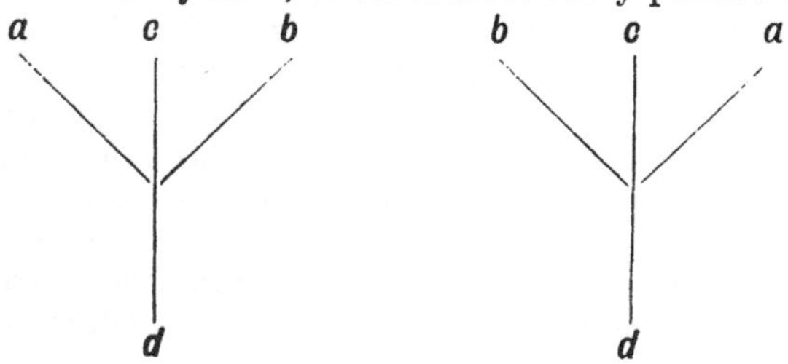

One of these is obviously a reflection of the other, and is not superposable upon it in such a way that the letters coincide.

The use of models assists materially in the consideration of the problems arising out of this hypothesis. One of the first questions which arise relates to the direction in which the several valencies of an atom of carbon may be supposed to be exerted. If the direction of these be supposed to be absolutely fixed, then it can be shown that :—

1. Two carbon atoms cannot unite by two bonds, nor by three, because that would involve the distortion of the atom.
2. Three or more carbon atoms cannot unite to form a ring for the same reason.

But inasmuch as carbon atoms do certainly combine together to form rings or closed chains, and therefore the direction of the valency must be drawn from the normal, there must be something analogous to the action at the pole of a magnet, that is there is a certain *field*. It appears, however, that though two carbon atoms may be apparently united by two or more units of valency, in all such cases the combination is not only not more secure but is decidedly more easily broken up than when one valency of each atom is employed.

A modification of the hypothesis of the tetrahedral carbon is based on the idea that the relative force of attraction between two units of valency depends on the distance through which they have to act. By assuming that the combination between two carbon atoms is not in the direction of the solid angles of the tetrahedron, but that the attraction between the two is in the direction of the normals to the *faces* of that figure, it is obvious that the most intimate union is that in which two of these faces are placed parallel to, and probably very near, each other. A less intimate union occurs when the centres of gravity of two faces of one atom attract two faces of another. The two tetrahedra have then a common edge, the two pairs of faces forming equal angles with each other. And, lastly, three faces of one may attract equally three faces of the other, and so cause the two tetrahedra to be applied to each other by one of their solid angles. These three positions correspond to combination by single, double, and triple bonds. According to this assumption it is only possible for the atoms to touch each other when united by the single bond, as shown at *a* in the following diagram (Fig. 60).

Van 't Hoff long ago indicated that when two carbon atoms are united together by a single bond they may be supposed to be free to rotate about an axis which is in the line representing the

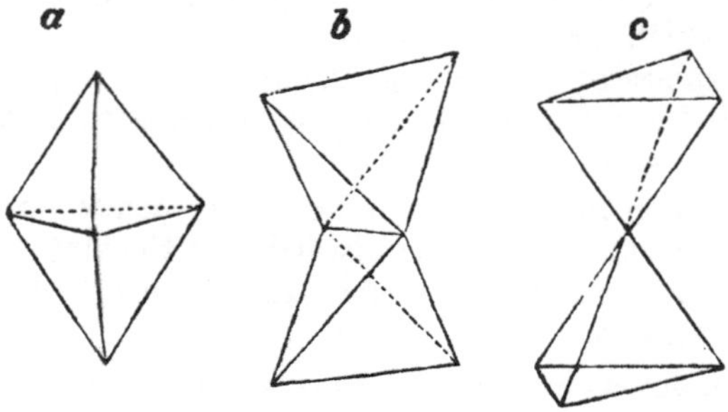

FIG. 60.

direction of the uniting valencies, and that if two carbons are joined by two or more bonds rotation becomes impossible. This hypothesis was first made use of by Wislicenus in 1886, and has been the subject of a good deal of discussion since.

It may be assumed that the radicles united to two adjacent atoms of carbon will be likely to influence each other, and, according as they attract or repel each other, rotation may or may not occur. Thus it may be supposed that in ethylene dichloride, $C_2H_4Cl_2$, chlorine and hydrogen atoms probably attract each other. Hence there can be only one stable form of this compound, viz. :—

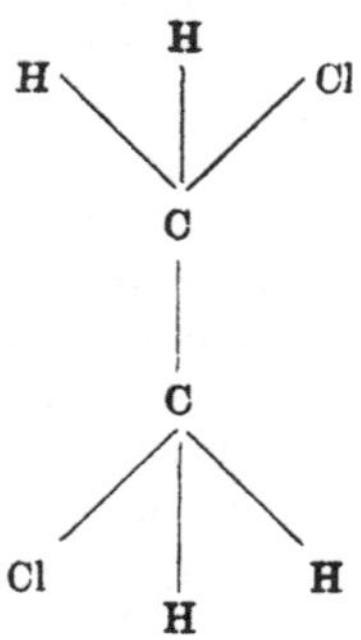

other arrangements passing spontaneously by rotation into this. Succinic acid also is known in only one form, which probably has the following structure :—

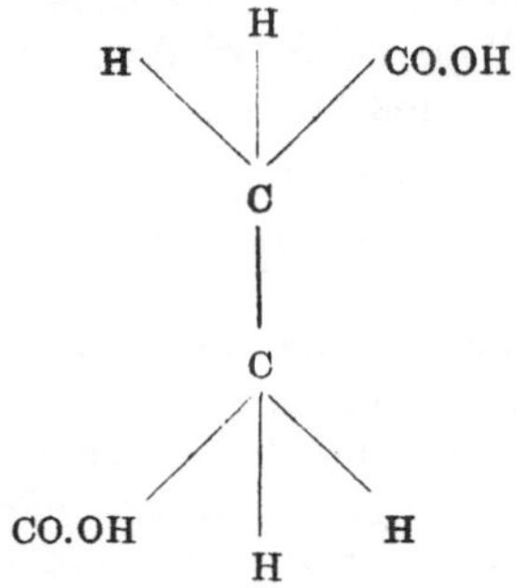

If this is so it follows that when it is heated and is resolved into water and the anhydride, of which the following is the formula, rotation must occur :—

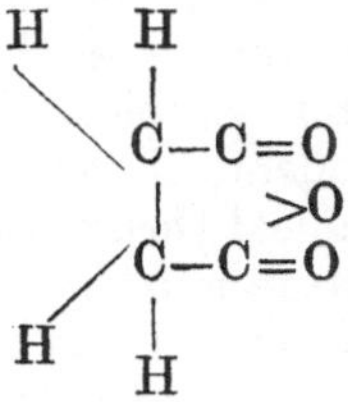

Many similar cases are known. It may be supposed that the atoms attached to the central framework of carbon have the power of attracting or repelling one another, as in the case of ethylene dichloride (Dutch liquid) mentioned above, and it has been thought that the relative masses, that is the atomic weights of such atoms or groups, may have something to do with this. The law connecting mass with optical activity or with rotation of two semi-molecules round one valency has, however, not been discovered.

Within the last few years many cases have become known in which apparently the existence of an asymmetric atom of carbon within the molecule is not essential to the development of optical activity. It appears, therefore, that a modification of the original theory is necessary so as to include those compounds which are optically active, and whose activity can only be attributed to a want of symmetry in the molecule considered as a whole.

The first compounds of this type were obtained in 1909 by Professors Perkin, Pope, and Wallach, and one of them is represented by the formula below :—

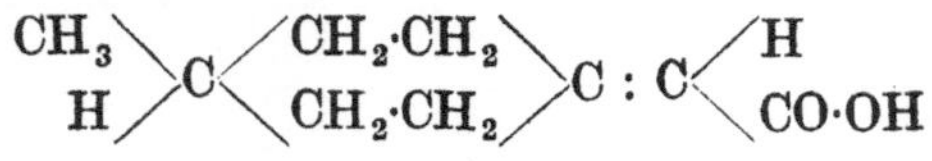

Here there is no carbon atom which is asymmetric according to Van 't Hoff's definition ; nevertheless it is obtainable in a right-handed and left-handed form, which may be accounted for by supposing the two groups at one extremity to lie in the plane of the paper, while the other two stand the one above and the other below the paper.

But at this point the difficulties of the problems involved multiply to such an extent that without a considerable acquaintance with the complicated, and in some cases imperfectly known, compounds of carbon it would be impossible to communicate further information on this most interesting department of chemistry. Moreover for complete demonstration models are necessary. For the student, therefore, who desires to pursue the study of stereo-chemical theory the larger treatises and the current periodical literature must be appealed to.

It may, however, be interesting to the general reader to learn what progress has been made in the application of space chemistry to elements other than carbon, which so far has alone been referred to.

Nitrogen presents itself first, and a case of nitrogen asymmetry, corresponding to carbon asymmetry, has been observed by Le Bel in the compound methyl-ethyl-propyl-isobutyl ammonium chloride :—

$$CH_3.C_2H_5.C_3H_7.C_4H_9NCl$$

Here the four radicles are attached in the same kind of way to the nitrogen, and in accordance with the principles already explained there may be two arrangements, one of which is the mirror image of the other.

A few years later, in 1899, a compound of similar constitution was obtained by Professor Pope and Mr. Peachey and resolved into optically active dextro- and lævo- compounds.

It is uncertain whether the centre of mass of the nitrogen atom should be supposed to occupy one solid angle of a regular tetrahedron, three of the valencies acting along the edges of the figure, or whether the directions of these three should be supposed to lie in one plane, the two others being disposed one above and the other below that plane.

Since the time referred to discoveries have followed one another in rapid succession. Pope and Peachey, and at nearly the same time S. Smiles, succeeded in isolating optically active compounds, of which the atom of sulphur was the directing nucleus. These were followed by compounds of tin, silicon, and phosphorus, which have been shown to be endowed with the same property. And there appears to be no doubt that all the quadrivalent elements to be found in Groups IV and VI of the periodic scheme, as well as the quinquevalent associates or allies of nitrogen and phosphorus, may gather round them groups of other atoms in tridimensional space and can thus act as centres of optical activity.

Perhaps still more remarkable was the announcement in 1911 from the laboratory of Professor Werner of Zürich, whose views were briefly referred to in a former chapter (Electrolysis). From an examination of the complex ammonia compounds, called ammines, produced by several metals, especially platinum, chromium and cobalt, he has been able to prove that metals can act as the central nucleus of stable asymmetric molecules, which may be resolved into optical isomerides exhibiting the optical activity corresponding to that of carbon compounds. The formulæ contrived by Werner, it will be remembered, involve the assumption of two kinds of valency, namely that which applies to the ionisable salt-forming constituents and that which is concerned in holding together the constituents included in the undissociable zone.

A large number of substances when passing from the liquid or gaseous to the solid state produce crystals. Thus common salt, the alums, fluorspar, and many other compounds form cubes or regular octahedrons, while quartz is familiar in six-sided prisms, calcspar in rhombohedrons, and green vitriol is at once distinguished from blue vitriol not only by colour but by the form of the prism assumed by their crystals respectively. Now a crystal is not only characterised by a definite external form, but the material of which it is composed gives evidence of definite internal structure. If for example a rhombohedral crystal of calcite is examined it is found to exhibit double refraction, a phenomenon which is familiar in Iceland spar. And if such crystal is broken into fragments each portion has the same external form and the same action on light. From such facts it is obvious that a crystal consists of a number of exactly similar parts which are repeated over and over indefinitely throughout the mass. And if the subdivision is supposed to be carried so far that the mass consists no longer of a number of molecules joined together but of one molecule alone, the space required for the accommodation of that one molecule within the crystal structure would have the same proportions as the crystal itself.

The atoms which enter into the composition of a molecule must be assumed to act as centres under the operation of two

Fig. 61.

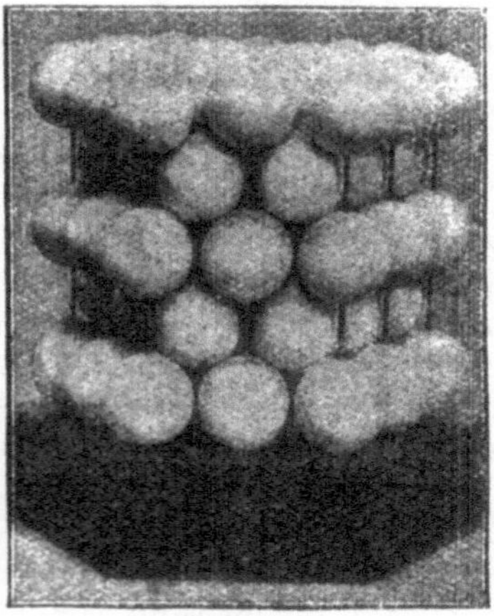

Fig. 62.

opposing forces, namely, a repellent force due to the kinetic energy of the atom and an attractive force which is the result of " chemical affinity," both forces being governed by some unknown law of inverse distance.

If now a crystalline element be considered, the molecules of which consist of one atom only, and every atom is like every other atom, it is found that two homogeneous arrangements of atoms are possible. These correspond respectively to the symmetry of the cubic and hexagonal crystalline systems.

In the two figures above which show the appearance of models representing these two arrangements the atoms are supposed to reside at the centres of the spheres which are seen to be in contact. Each sphere represents the range of the influence or attractive force of an atom, but it is equally convenient to suppose the whole space filled as it would be if the spheres were elastic and compressed till there were no interstices between them.

It is interesting to note that some 85 per cent of the crystalline elements seem to be in general agreement with these assumptions, 50 per cent being cubic and 35 per cent hexagonal, while the remaining 15 per cent which present a divergence still await explanation.

If now a case is considered in which the atoms are not alike but, as in common salt, are of two kinds, it is evident that throughout the crystal structure they must be ranged in regular order alternately, so that each sodium atom is related directly to six chlorine atoms, and each chlorine atom is similarly situated with respect to six sodium atoms. The whole mass might be partitioned into equal and similar spaces, each containing an atom of sodium and an atom of chlorine or the conventional molecule of common salt. Such division would not necessarily correspond to any physical division into molecules, for the selection of partners by the sodium and chlorine atoms respectively does not necessarily occur till the compound is melted or dissolved, when the sodium may become associated with any one of the six chlorine atoms adjacent to it, and the chlorine similarly may pair with any one of the sodium atoms. Each component atom then has a separate existence in space, but a further question arises as to the relative spaces occupied by atoms of different chemical characters and their relations to the crystalline forms assumed by the solid compounds in which they exist. This has been answered by the entirely new conception of valency introduced by Mr. W. Barlow and Professor W. J. Pope in 1906.

In the development of the ideas briefly described in the foregoing paragraphs they found that in a crystalline substance each component atom appropriates a portion of space which is approximately *proportional to its fundamental valency*. Thus the volume of the sphere of influence of carbon is nearly four times that of the hydrogen sphere. This is illustrated by the models shown in figures 63, 64, and 65, which represent the molecules of methane and benzene respectively.

The volumes of other univalent elements are not exactly, though very nearly, the same as that of hydrogen.

The fundamental valency of an element must also be regarded as of a different nature from other valencies exhibited by the same element. This has already been indicated by the researches and hypotheses of Werner referred to in a former chapter. It has not been found necessary by Barlow and Pope to suppose that multivalent elements affect spheres of influence of different sizes corresponding to the several extra valencies thrown out when in other conditions of combination. The sphere of influence of nitrogen, for example, is approximately three times the volume of that of hydrogen, both in ammonia and ammonium chloride, and that of sulphur is approximately twice as large as that of

Fig. 63.

Fig. 64.

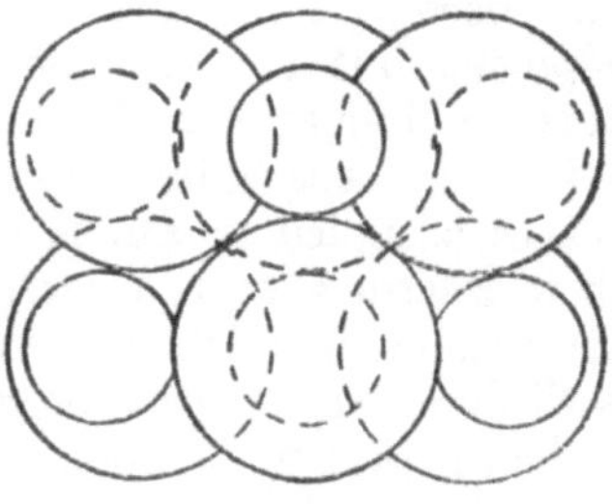

Fig. 65.

hydrogen, even in compounds in which the sulphur is commonly assumed on good evidence to be sexvalent.

Remembering that it is assumed that the volume of the sphere of influence of an atom in a closely packed assemblage is approximately proportional to its valency, it is possible to replace one atom by two or more atoms of a different kind without more than local disturbance which does not require a remarshalling of the whole, provided the total valency of the introduced atoms does not exceed that of the atom replaced. Just as in ordinary constitutional formulæ, therefore, we may find an atom of chlorine replacing an atom of hydrogen without disturbance of the general character of the compound, but if hydrogen is to be replaced by oxygen two atoms of the former are replaced by one atom of the latter element.

The elements which stand in the same vertical column in the periodic scheme are generally very similar in chemical character, and the magnitude of the sphere of influence of the atoms of elements in such a case increases but slowly. Consequently the difference between two such elements as potassium and rubidium would be but small, and the one may replace the other without affecting the general arrangement or marshalling of the other atoms present in a compound. In this way the relation of valency and isomorphism becomes apparent. Measurements of the crystals of some isomorphous groups, comparing potassium sulphate, K_2SO_4, with cæsium sulphate, Cs_2SO_4, for instance, show that while the volume of the entire molecule is changed, the spheres of influence of the atoms present preserve the same ratios, and the same crystal form with but slightly changed axial ratios is retained.

Professor Richards of Harvard has adduced a considerable amount of experimental evidence in favour of the idea that atoms are compressible. How far the compressibility of an atom may affect the volume of its sphere of influence is uncertain, but the independence of the atom, the retention of its mass and, approximately, its volume are facts which up to the present have been very generally accepted.

The theory of Barlow and Pope, which has just been very superficially reviewed, grew out of the study of crystals, their form, structure, optical properties, and relation to chemical composition. It has now been under consideration by the chemical world during ten years, and though much yet remains to be learned, partly owing to the scarcity of crystallographers, the theory in its broad aspect may be regarded as having established its position among accepted chemical doctrines. The theory in no way conflicts with the tetrahedral hypothesis as to the atom of carbon and elements related to carbon, and the assumption as to the volume of the carbon atom being four times that of a hydrogen atom may be regarded as consistent with all that is known of carbon compounds.

Help has come from an entirely new direction, and has arisen out of discoveries which have been made as to the action of crystalline substances on X-rays. The origin of the work is an experiment due to Professor von Laue of Munich, who showed that the ordered array of atoms in a crystal can behave to X-rays in a manner analogous to the action of a diffraction grating on a ray of light. The results were not at first completely understood, but an interpretation by Mr. W. L. Bragg, followed by further experimental researches in association with his father, Professor W. H. Bragg, has led to extremely interesting and important conclusions. Their work has, indeed, been considered of such fundamental importance that the Nobel Prize for last year was awarded to them. Unfortunately it is impossible in a few words to convey a sufficient account of the method. This has been described with some detail in a lecture delivered before the Chemical Society of London on February 3rd, 1916, and printed with illustrations in the Transactions of the Society for 1916, p. 252. The experiments show that in a crystal the atoms are independent of each other, and are arranged in regular order at distances apart, which can be measured. Thus in a crystal of rock-salt, which is cubic, the atoms of sodium and chlorine are stationed alternately in planes which lie at right angles to each other, so that each atom of sodium is not associated with only one atom of chlorine but stands in the same relation of distance to six atoms of chlorine. Each atom of chlorine is similarly placed with regard to six atoms of sodium, and apparently without preference for any one of them. Hence in the solid crystal the idea of molecule disappears, for the entire crystal, small or large, is one molecule. It is only when melted or dissolved that the atoms of sodium and chlorine select partners and move off. These experiments have also revealed the fact that the diffracting power of an atom for X-rays is directly proportional to its atomic weight, so that one result is obtained with common salt in which the atoms of sodium (atomic weight 23) and of chlorine (atomic weight 35·5) are very different in mass, while another result is obtained with potassium chloride in which the atomic weights 39 and 35·5 are more nearly equal. Similarly the differences between potassium chloride, bromide (Br=80), and iodide (I=127) correspond to different X-ray spectra. One of the most obvious and important results is to confirm the accuracy of the conclusion involved in the theory of Barlow and Pope which has already been mentioned, namely that a crystalline substance is an assemblage of atoms in which the molecular unit does not necessarily exist until the crystal is dissolved or melted. But perhaps the most important result of all is the proof of the physical existence of the atom as an independent entity.

The atomic Theory of Dalton formulated about 1808 was for half a century or more employed by the chemical world with reservations, and so late as 1856 it was referred to in a prominent English textbook as "at the best but a graceful, ingenious, and, in its place, useful hypothesis." In 1869 Professor Williamson, then President of the Chemical Society, thought it necessary to lecture the Society on the position of the atomic theory at that time. He pointed out "that on the one hand all chemists use the atomic theory, and that, on the other hand, a considerable number of them view it with distrust, some with positive dislike." And so lately as 1904 Professor Ostwald in the Faraday lecture endeavoured to show that the theory of chemistry was independent of the atomic theory.

The attitude of chemists since the time of Williamson's lecture has certainly been very different from that which he described. Since 1872 the facts and theories connected with stereo-chemical ideas have become so consolidated by experience that the revolutionary notions introduced by Ostwald in 1904 came as a surprise, and it may be confidently stated were never acceptable to anyone.

All wavering, uncertainty, and distrust must now disappear, for, though it cannot be said that the separate atoms in a crystal have been seen by human eye, their effects have been recorded on a photographic plate, and their presence and separate physical existence are as well established as the existence of the crystal itself.

Everything then proves that matter is made up of discrete particles which cling together under the action of the various forces—cohesion, adhesion, and chemical attraction. The separate particles are the atoms, which in a substance called an element are all alike, while in a compound there are two or more different kinds joined together in a certain order. As to the atoms a hypothesis which has already been explained (Chapter X) has been formed concerning their constitution. They are supposed to consist of a large number of electrons embedded in or surrounding a nucleus of positive electricity. But outside this structure there appear to be attached and detachable 1, 2, 3, 4, 5, or 6 (rarely 7 or 8 ?) electrons which correspond to the valency of the atom, and hence its sphere of influence in forming chemical compounds. These external electrons differ from the constitutional electrons in the fact that their removal does not disturb the essential structure of the atom as a whole. When they are restored therefore there is no change of properties and the case is different from that of the radio-active elements of which, when they lose an electron, that is a β particle, the residue is fundamentally changed in properties.

On the assumption that these external electrons which determine valency move round the atoms to which they are attached in small circles, Sir William Ramsay has very recently (Proceedings of the Royal Society, July, 1916) been considering the manner in which atoms, represented as spheres, would place themselves in the formation of some of the simpler molecules, hydrogen, oxygen, hydrogen chloride, water, ammonia. So far as they go the results are consistent with previous ideas, but the difficulties increase seriously when any attempt is made to represent any but the very simplest cases of molecular configuration.

CHAPTER XIII

THE MICROSCOPIC AND ULTRAMICROSCOPIC COLLOIDS

Human existence hangs between two great worlds, the infinitely great and the infinitely little, and into both the chemist can penetrate. Though the mind fails altogether to grasp any clear idea of the space which separates us from any of the stars we can at least realise that it is enormous by a very simple reflection. As the earth travels round her orbit she sweeps out a circle more than 180 millions of miles in diameter. Nevertheless an observer who looks out into the night in summer and again in winter, or in spring and again in autumn, will be able to distinguish no change in the relative positions of the stars forming the constellations or patterns which they have for so many ages written on the background of the sky. And when we remember that though modern instruments are capable of appreciating differences far more minute than any which the unaided eye can trace, the vast majority of stars remain apparently unaltered in position when viewed from opposite sides of the earth's orbit. Into these vast spaces the spectroscope, brought into use as a practical instrument less than seventy years ago, has enabled the astronomer and the chemist to penetrate and to investigate the composition as well as the motions of these distant sources of light. The result has been to show that the elements which enter into the composition of the earth, the seas, and the atmosphere of our planet, are to be found everywhere in the most distant regions of space. They are not equally distributed, for in one star hydrogen, for example, may be found to be present, while in another the lines indicating that element are absent, and similar concentrations of calcium and magnesium, iron, sodium, and the rest may be recognised in stars and nebulæ.

This, however, is not the direction in which attention must be concentrated in these pages. The reader who is interested in astro-physics or astro-chemistry must consult one of the numerous treatises on the subject, from which he may learn something of the progress which has been made in spectroscopic investigation concerning the composition of our sun and the stars, which are believed to be constituted like our sun, as well as the nature of comets and nebulæ. It is sufficient to say that the results of late years have been to some considerable extent employed in connection with theories concerning the origin and evolution of the elements, a subject which has already been dealt with in Chapter VIII.

Systematic and quantitative chemistry is based on the assumption that matter is not divisible *ad infinitum*. It is known to the senses in the form of masses, the most minute of which, discernible by sight, contain many millions of the ultimate particles—molecules or atoms—which are subjects of study by the physicist and chemist. This is true even of the granules, which, produced by the subdivision of ordinary massive matter, are large enough to be just visible under the highest powers of the ordinary microscope. Particles which are too small to be seen by the eye alone or assisted by lenses may be perceived by other senses, especially by smell. Everyone has heard of the grain of musk which, lying for years on the pan of a balance, continues to emit its characteristic odour without betraying any loss of substance by appreciable loss of weight. And yet there seems no reason to doubt that the effect on the olfactory surface is produced by contact with the minute particles of musk substance which enter the organ. The sense of smell, like that of taste, depends on actual contact of the sensory surface with the exciting substance, and it does not appear to depend like the faculties of hearing and of sight on the entrance of impulses in the nature of waves of air or of " ether " into the receiving nervous surface. The extremely minute division of matters capable of exciting smell may be inferred from common observations in everyday life. One of the commonest experiences of a walk in the country is the smell of burning weeds. When the vegetable matter thus disposed of is heated it is partly consumed by the aid of the oxygen of the air and is converted into vapours of water and carbon dioxide which are inodorous. At the same time a portion, but only a very small proportion, of the stuff is by the heat alone, without combustion, made to yield substances such as acetic acid and phenols, which in a state of vapour are acrid and disagreeable to the nose and eyes. They are perceptible at great distances when diffused through the air. They can be smelt when the accompanying smoke is no longer visible and at a distance of half a mile or more. The matter thus diffused would probably not exceed an ounce or two, but it can be recognised when spread in this way through millions of cubic yards of air. One is tempted almost to believe that under such circumstances the sense may be awakened by separate molecules arriving singly or a few at a time.

The subdivision of matter in its extreme forms may, however, be rendered perceptible by the eye if only the illumination is sufficient. The passage of a sunbeam through a room in which nothing could previously be perceived is enough to show that the apparently clear air is filled with myriads of particles which by reason of their small size remain suspended. These little particles are, however, relatively monstrous, for they consist of minute hairs, pollen, yeast cells, animal and vegetable débris of all kinds, each of which possesses an organic structure, mixed with tiny grains of sand or earthy matters blown up by the wind. This process of illumination will help to reveal other effects in which the actual masses of the particles can be calculated. If one of the artificial colouring matters such as magenta, fluorescein, or eosine, of which the composition and molecular weight are known, be dissolved in water the solution may be diluted till it contains no more than 1 or 2 parts of the solid in 100 millions of the liquid, and the colour will still be perceptible. When the solution possesses the property of fluorescence which is beautifully shown by fluorescein and by eosine,[1] this effect is still perceptible when the dilution is 20 million times greater, provided the liquid is examined by the aid of a converging beam from an electric arc directed into the body of the liquid. It can be shown that the particles thus made recognisable are still not resolved into separate molecules, for the dilution would still require to be increased some seven or eight thousandfold before the dissolved substance is reduced to one molecule per cubic millimetre.

Another illustration of minute division, even into separate atoms, has been described in connection with radio-active substances, the escaping atoms making themselves perceptible by their electrical effects and by exciting phosphorescence on certain surfaces. Each atom as it strikes the screen of the spinthariscope produces a separate vivid spark.

The observation of minute particles, approaching molecular magnitudes in some cases, has been accomplished within recent years by the use of the instrument known as the "ultra-microscope." It has long been recognised that however intense a beam of light from the sun or electric arc may be, it remains invisible until it strikes some surface. It is then scattered more or less, and if any of the scattered rays reach the eye they produce the sensation of vision. Hence the course of a beam of sunlight traversing a room containing dust-laden air is visible, but if the same beam be transmitted through a glass tube or other vessel in which the air has been purified from suspended particles the course of such a beam will be imperceptible. In the ultra-microscope[2] advantage is taken of these facts. The medium holding minute particles diffused through it is examined through a high-power microscope, while a strong beam of light, admitted through a very fine slit, is cast through the space in front of the object glass. As the axis of the microscope is vertical while the beam is horizontal it is obvious that no portion of it enters the microscope directly. Hence if anything is seen it is the little rays reflected sideways from the surface of the particles under examination. The image observed therefore depends on the power of the microscope and the intensity of the

[1] Everyone has seen the peculiar blue light (fluorescence) exhibited by a watery solution of quinine sulphate.

[2] The ultra-microscope was invented by H. Siedentopf and R. Zsigmondy early in the present century.

illumination. The size of the particles inspected cannot be observed directly, but only inferred from a knowledge of the amount of solid contained in a known bulk of the liquid and the number of particles which can be counted in a measured space.

Ordinary microscopes may give a magnification amounting, under favourable circumstances, to as much as 3000 diameters, and so render visible objects which have a size represented by about one ten-thousandth of a millimetre across. The ultra-microscope is said to be capable of distinguishing objects having a diameter one-tenth of this with light from an arc lamp, and considerably smaller when illuminated with the brightest summer sun. These minute measures are usually indicated by the symbols μ and $\mu\mu$ which stand respectively for one-thousandth and one-millionth of a millimetre.

It is estimated that the diameters of gaseous molecules must lie somewhere between 0·1 to 0·5 $\mu\mu$. Hence there appears little probability of ever rendering these visible, for the finest particles discernible in suspension have a diameter equal to 0·001 μ or 1 $\mu\mu$. But though separate individual molecules may never be revealed to mortal eye, observations on some of these very small particles afford a vision which gives a lively stimulus to the imagination.

It has long been known that very small particles of any solid when suspended in water are seen under the microscope to be in a state of movement. This has no relation to the composition of the solid, but is dependent solely on the size of the particles which must not exceed 3 μ in diameter. The movement was observed originally by Dr. Robert Brown, the botanist, in 1827–8, and hence is commonly referred to as the Brownian movement. The cause of it was long doubtful, but there seems to be now a reasonable explanation which is generally accepted.

For the study of these movements it has been found convenient to prepare certain emulsions,[1] by which are to be understood milky liquids, which are formed when alcoholic solutions of resin or similar substances are mixed with a large volume of water. The resin being insoluble, or but slightly soluble in water, is thrown out in the form of minute spherical drops, of which the greater part remain for a long time in suspension. Gamboge is a resin familiar as a water-colour pigment which lends itself to

[1] Milk is the most familiar example of an emulsion.

this process very suitably, though other resins such as mastic or even common resin will also answer the purpose. A number of these emulsions and other imperfect solutions have been the subject of numerous experiments during the last twenty years or more. These liquids contain granules in suspension so small that they commonly pass through ordinary filter paper, and the liquid can only be clarified by resort to special processes. When allowed to rest undisturbed for a long time the upper layers of the liquid gradually become clear, while the floating particles accumulate in regularly increasing proportion in the lower layers. When such a liquid is examined under the microscope the granules are seen to be all spheroidal in form and to vary considerably in size, though after the deposition of the coarser particles no solid can be discovered by the unaided eye. They can be most conveniently sorted by submitting the liquid to the action of a centrifuge, the larger masses passing first toward the periphery. This method was introduced by Professor Jean Perrin of the Sorbonne, who has given much attention to the size of particles which exhibit the Brownian movement. A method which has been used for several similar purposes is based on the formula given by the late Professor Sir George Stokes for the velocity of a small sphere of known density falling through a medium of which the viscosity is known. With an emulsion holding in suspension grains of uniform size the rate of fall can be easily measured by observing the time occupied while the top layer of known depth becomes clear.

Another process is to count under the microscope the number of granules in a very small but known volume of an emulsion containing grains of uniform size, whereof the total mass is known.

In the case of gamboge the diameter of the granules was found, by the first method, to range from $\cdot 90\ \mu$ to $\cdot 42\ \mu$, and by the second from $\cdot 92\ \mu$ to $\cdot 42\ \mu$. The last result involved the counting of about 11,000 grains.

The main result arrived at by Perrin is the conclusion that the Brownian movement of colloidal particles in suspension and the *movement of molecules* are of the same type and are due to the action of the same forces. These forces appear to be the translatory forces of the moving molecules of surrounding water; in other words the granules in suspension are pushed about by the molecules surrounding them. In order that movement may result from contact with a molecule the size of the particle must be small, that is of the same order as the size of the molecule. The movements executed by the Brownian particles may therefore be supposed to be very similar to those of the molecules themselves. An idea of their character may be gained from the adjoining figure which was obtained by observing the positions

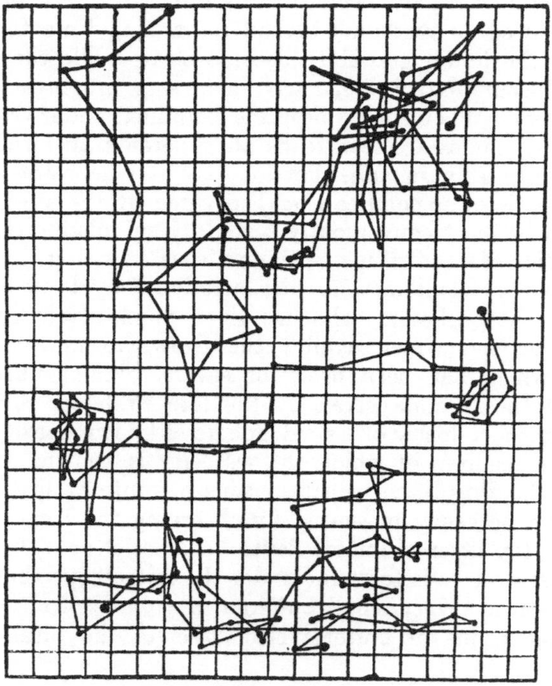

FIG. 66. BROWNIAN MOVEMENT.
Paths of three separate granules.

of three granules every 30 seconds. These positions are then joined, in the diagram, by straight lines.

Investigations of the kind just described have assumed a position of great interest and importance within the last few years.

Something is at last beginning to be understood regarding the constitution of that peculiar state of matter to which the name "colloid" has been given. To Thomas Graham (at one time Professor of Chemistry at University College, London, and later

Master of the Mint) we owe the first and very important researches concerning the diffusion of dissolved substances. The first memoir on the subject is the Bakerian Lecture to the Royal Society for 1849. In the course of these experiments Graham discovered not only that dissolved salts and other compounds *diffuse*, that is they move from place to place in the liquid, with very various degrees of rapidity, but that the same compounds are divisible into two main classes according as they possess or do not possess the power of passing when in solution through animal membranes or parchment paper. These two classes Graham called " crystalloids " and " colloids." The former includes not only substances like salt and sugar which are capable of crystallising, but compounds such as hydrogen chloride (hydrochloric acid) which are not known in the crystalline state. These substances diffuse through membranes with various degrees of rapidity. The other class, colloids, are typified by gelatine or glue (*κόλλα* glue). They are substances which swell up indefinitely when soaked in water, they show no sign of crystallisation, but when dry break with a conchoidal fracture like resin or glass. They are known to have a large molecular weight, and they pass, when in quasi-solution, very slowly through membranes. Many colloids, however, are not of organic origin like glue, but are obtained by a great variety of processes from salts, oxides, and even metals, which under ordinary conditions are practically insoluble in water. An example or two will make the matter clear. Silica SiO_2 is known in the form of rock-crystal, quartz rock, agate, carnelian, opal, cairngorm, flint, and other stones, sometimes nearly pure as in rock crystal, more usually coloured by the presence of small quantities of ferric and other oxides. These agree in being practically insoluble in water, but if melted with caustic soda or merely boiled for some hours with a solution of the same the silica is rendered soluble, and a moderately strong solution of the product in water constitutes the " water-glass " of the shops. If a diluted solution of water-glass is poured into an excess of dilute hydrochloric acid no precipitate is formed and the solution remains clear. It contains silicic acid, a colloid, together with the sodium chloride which has been formed and the excess of hydrochloric acid, both crystalloids. By pouring the solution into a bag of parchment paper and suspending it in water the latter substances diffuse away, and if fresh water is supplied several times in place of the diffusate a liquid is ultimately left within the bag which contains the whole of the silicic acid with mere traces of the chlorides. Such a liquid is clear as water, it may be concentrated by evaporation, provided no solid crust is allowed to be formed at the sides of the dish, till it flows like syrup, and ultimately it will dry up into a glass-like mass. The dissolved substance exhibits extremely small osmotic pressure and neither reduces the freezing point of water nor raises the boiling point. The liquid is, however, very sensitive to the presence of small quantities of salts and acids and speedily sets into a gelatinous mass when any electrolyte is added. The pseudo-solution of silicic acid thus obtained, called by Graham a "colloidal solution," is now simply called a "sol," while the gelatinous mass resulting from slow change or the addition of saline substances is called a "gel."

By similar processes which involved the preparation of the colloid form and its purification from the attendant crystalloids by the method of dialysis through a membrane, Graham succeeded in preparing sols of the hydroxides of iron, chromium, and aluminium, tungstic acid and other substances. Some of these preparations may be made by other methods. Thus aluminium hydroxide (hydrated alumina) may be obtained in the form of a sol by making first a solution of aluminium acetate and then boiling the solution in an open dish till the acetic acid produced has been completely driven off. Water must be added from time to time to replace that which is lost by evaporation. The residual hydrous alumina is the basis of the aluminous mordant used in calico printing.

A very interesting sol is produced by boiling white arsenic (arsenious oxide As_4O_6) in water and adding hydrogen sulphide to the solution. A bright yellow colour is immediately developed but no precipitate is formed, although arsenious sulphide, As_4S_6, is insoluble in water. If now a few drops of hydrochloric acid are added to the clear liquid a yellow precipitate of the sulphide is immediately formed, and after it has had time to subside the liquid is seen to be colourless.

The most remarkable sols, however, are those which are producible from certain metals, especially those which have been so long called the "noble metals," namely gold, silver, and platinum.

Faraday discovered in the earlier half of the nineteenth century that a solution of gold chloride on which is floated a solution of phosphorus in ether will yield a liquid of various colours, blue, violet or rose according to circumstances. A mixture of tin salts added to a solution of gold also yields the purple of Cassius which is long retained in suspension. The most beautiful ruby glass also owes its colour to gold which is certainly in the metallic state, not only on account of the fact that the temperature of melting glass is far above that at which all known compounds of gold are decomposed, but the particles have been examined by the ultra-microscope and are known to be for the most part spherical. Some of the coloured liquids just mentioned must have been known to the alchemists, and it is not improbable that one of them was an ingredient in the " elixir," the composition of which was their chief subject of study. " Soluble gold," which was actually used in medicine down to the end of the seventeenth or beginning of the eighteenth century, probably consisted of one of them.

To prepare a coloured solution or pseudo-solution of gold it is only necessary to add to a weak solution of the chloride in water any one of many easily oxidisable substances or as they are called " reducing agents." Phosphorus has already been mentioned, but phosphorous and sulphurous acids, essential oils of various kinds, formaldehyde, sugars, hydrazine, hydroxylamine and many other substances have been used. The colour produced depends on the agent used. Thus when a weak solution of pure chloride of gold is exposed to contact with carbonic oxide gas bubbled through it, a *red* colour is produced which is pretty stable, and the hydrochloric acid which is left in the liquid as a consequence of the deposition of the gold can be removed by dialysis.

If the same gold chloride solution is neutralised very carefully by means of a weak solution of sodium carbonate and a very dilute solution of hydrazine hydrate is added drop by drop a *blue* liquid is formed. And further, according to A. Gutbier (*Zeit. Anorgan. Chem.*, 1904, pp. 112–114), the same reducing agent added to the same gold chloride solution in successive portions is capable of producing several colours successively. If a very weak solution of gold chloride is mixed with a few drops of a very weak phenylhydrazine hydrochloride solution a red colour is first produced, a little more of the reagent produces a violet colour, which with the addition of further quantities becomes bluish and ultimately deep blue.

These coloured liquids are producible in a very curious manner by electrical dispersal of the metal itself below the surface of water or other liquid. By making a small electric arc between gold wires under water purple-red clouds and a coloured liquid result from dispersal of the metal which comes from the cathode only. It would not be surprising to find that the fine particles of metal thus held in pseudo-solution are electrically charged, but it is very significant of the character of colloids generally to find that the particles of such substances as ferric hydroxide and arsenious sulphide produced by the processes already mentioned are in this condition. The apparent attraction or repulsion of such substances in the colloidal state has been examined by Messrs. Linder and Picton, who began their researches so far back as 1892, and who must be regarded as among the chief pioneers in this difficult field of enquiry. They describe the electrical convection of arsenious sulphide as follows, using the yellow liquid in which no particles are visible under the microscope, while the liquid is filterable:—

"The resistance of such a solution is extremely high and the current passing through it in one case amounted to only 0·000007 ampère. The conductivity is probably due to the presence of small traces of arsenious oxide; but, however that may be, the passage of this small amount of current is accompanied by the repulsion of the colloidal sulphide as a whole from the negative electrode." (*Trans. Chem. Soc.*, 1897, p. 569.) They also observed that of different colloids examined some are positive and some negative: thus arsenious sulphide is negative in a solution faintly acid to litmus, while ferric hydroxide is electropositive under the same conditions.

All these hydrosols or colloidal solutions are coagulated when mixed with an electrolyte. It has already been mentioned that the addition of a little hydrochloric acid to the arsenious sulphide liquid causes immediate precipitation of the solid sulphide, but an acid is by no means necessary, as in most cases any soluble neutral salt will bring about the same effect. It is interesting to notice that the coagulating effect is delayed or prevented altogether in some cases by the protective effect of mixing the hydrosol with one of the more stable colloids, such as a solution of gelatine. Faraday discovered, for example, that his red gold solution when mixed with jelly "is rendered much more permanent than before; and then it may by a little warmth be had in the fluid state, or by cooling as a tremulous jelly, or by desiccation as a hard ruby solid, presenting all the transitions between the gold fluid and the ruby glass." (Bakerian Lecture, 1857.)

Some of these metallic sols are now stated to have a germicidal action which may result in valuable medical results. The late Mr. Henry Crookes has introduced certain preparations of this kind containing silver and mercury which he called "collosols." They are stated to be fatal within a few minutes to all the pathogenic organisms examined, at the same time that they have no injurious effects on animal tissues. These liquids contain the metallic particles in a state of suspension and so small as to exhibit the Brownian movement actively. How they attack bacteria is not certain, but it appears probable that the particles in any one preparation being all in the same electrical state, positive or negative, do not touch one another and do not lose their charge. When a neutral foreign body such as a microbe is introduced into the liquid it probably receives the charges of many thousands or millions of particles of metal, and this may account for its death. Favourable notices concerning the germicidal power of these collosols have been published both in the *Lancet* (12th December, 1914) and the *British Medical Journal* (16th January, 1915).

We may now turn again to the consideration of the question of the sizes of particles which are recognisable by the ultramicroscope. So far as molecules of gases are concerned it has already been pointed out that the probability of their being ever seen is remote. But since something is now known of the molecular size of the molecules of some albuminoid matters, all of which are undoubtedly very large, it seems possible that in such cases the molecule may be seen. Experiments on the diffusion of various colloids by Herzog and Kasarnowski in 1908 led to the following figures, which are of course only approximations and are to be compared with the molecular weight of hydrogen as 2.

Substance.	Molecular weight.
Egg albumin	17,000
Pepsin	13,000
Invertin	54,000
Emulsin	45,000

Calculating the molecular diameter of such bodies the largest molecules reach the size represented by 6 $\mu\mu$, which brings them within the range of the ultra-microscope. It is obvious, however, that in view of the complicated chemical constitution of such substances it cannot be regarded as certain whether these are chemical individuals or aggregates of two or more chemical units, especially in view of the commonly recognised tendency to association exhibited in the liquid state by many chemical compounds. There is in fact a gradual mergence of granules into molecules among these minute particles. The subject is comparatively new, and the further study of colloids may lead to still more wonderful results.

The molecule, however, is not the ultimate limit of subdivision of matter. Molecules are made up of atoms, and as the whole is always greater than the part, atoms are smaller than molecules. Spectrum analysis proves that an atom is itself a complex structure, and what is the nature of the attraction which binds atoms closely together into molecules and the law of this attraction is at present unknown. The spaces between the atoms in a molecule are doubtless small as compared with the volume of the molecule as a whole. Nevertheless there are strong reasons for believing in the independence and entity of each atom. The capacity for heat of each atom, that is its "specific heat," is, for example, nearly the same, whether the element is in the "free" state or in chemical combination. The facts of stereo-chemistry (see Chapter XII) also afford strong evidence.

At the end of the nineteenth century every chemist and physicist believed in the permanence of the atom as the fundamental unit of mass. Since that time it has been shown not only that certain special types of atoms break up spontaneously, but that all atoms under cathodic discharge yield smaller bodies, electrons, which have only about $\frac{1}{1800}$th part of the mass of an atom of hydrogen. Attempts to conceive or to express such dimensions are vain and ineffectual. It is calculated that the absolute mass of an electron is 6×10^{-28} gram and its radius 10^{-14} of a millimetre, but such figures convey no idea to the mind, since dimensions of such an order correspond to nothing within human experience.

PART III

MODERN APPLICATIONS OF CHEMISTRY

CHAPTER XIV

HYDROGEN

Up to the middle of the eighteenth century several gases were known which were confused together under the general name "inflammable air." But in 1766 Henry Cavendish, in a paper on "Factitious Air," showed that the gas which is procured by the action of diluted sulphuric or muriatic acid on zinc and iron is to be distinguished from the inflammable air known as marsh-gas, and the other variety obtained by passing air over or through red-hot charcoal. He afterwards proved that this kind of "inflammable air" unites with half its volume of "dephlogisticated air" (oxygen) to form water.

The inflammability of the gas coupled with its extraordinary lightness led the discoverer and others to suppose that it was actually "phlogiston" itself, the then assumed hypothetical principle of combustibility. The name hydrogen, which means water producer, was given to the gas by Lavoisier, in accordance with the system of nomenclature contrived by him some years later. The lightness of the gas always excited curiosity, and after the phlogistic theory had been abandoned and phlogiston a thing of the past, the idea arose that hydrogen was in fact a realisation of the notion which had come down from ancient times as to the existence of a *πρώτη ὕλη* or protyl, the primal matter out of which all else was composed.

These speculations are now merely matters for antiquarian curiosity, but the fact still remains that of all forms of matter known to the chemist hydrogen is the lightest. Standing next to it is the rare gas helium, which is just twice as heavy bulk for bulk, and the next in order is methane or marsh-gas, CH_4, which is eight times as heavy as hydrogen.

The development of the balloon into the airship is one and perhaps the chief reason for enquiry into the methods by which hydrogen may be manufactured in quantity. Formerly aeronauts were content with charging their balloons with an impure kind of methane, obtained by distilling coal at a relatively high temperature, and collecting the last portions of the gas escaping from the retort.

The difference of density, however, is so great that gas of this quality does not satisfy the modern requirements. Not that it is by any means necessary to secure chemically pure hydrogen, as the proportions of the ordinary impurities, such as sulphuretted hydrogen, arsenetted hydrogen, and the vapours of volatile hydrocarbons, or a little carbon monoxide are so small as to exercise but little influence on the density, and hence the lifting power of the gas. The chief objection to the presence of such impurities as sulphuretted hydrogen is that they slowly oxidise into acids which are apt to destroy the fabric of the balloon.

The largest Zeppelins are said to be 600 feet long by 65 feet in diameter. Suppose we assume an envelope 500 feet long by 60 feet in diameter has to be filled with hydrogen gas. The volume of gas required is approximately 40 million litres, or nearly $1\frac{1}{2}$ million cubic feet, which, at freezing-point and under the pressure of 1 atmosphere (or 15 pounds per square inch of surface), would weigh 3584 kilograms or about $3\frac{1}{2}$ tons. Air is nearly $14\frac{1}{2}$ times heavier than hydrogen, consequently the same bulk of air, under the same conditions, weighs just over 50 tons. The lifting power of such a volume of hydrogen gas would therefore be about 46–47 tons. When the frame of the balloon, the car, the engines, the fuel, and the guns, if any, are allowed for, it is obvious that the machine is still capable of carrying a considerable number of men and bombs or other cargo. But it must be remembered that when the airship leaves the ground it rapidly rises into regions in which the pressure is very much less than 1 atmosphere, and the gas expands, and if closely confined would probably burst the envelope. This effect is at high elevations counteracted to some extent by the reduction of temperature. Some of the gas would, however, escape through a relief valve, and on descending would probably have to be replaced by means of a reserve of compressed gas carried in the car.

The airships, however, are commonly provided with a second envelope containing air into which the exhaust gases from the engine are delivered for the purpose of regulating the temperature of the hydrogen, warming it when the ship has risen to such a height as to be exposed to considerable cold. It is also important to observe that the materials of the envelope, which consist of a strong double cotton cloth enclosing a thin layer of rubber, are not impervious to hydrogen gas, and that in course of hours an appreciable escape occurs by the process of diffusion. All these facts make the available carrying power of any sort of balloon less than the amount calculated from the volume of hydrogen with which it is charged at starting.

Another application of hydrogen to an entirely different chemical purpose, namely the hardening of oils and fats, does require the hydrogen employed to be free from impurities such as sulphur, phosphorus, and arsenic, even in minute quantities, as they so readily attach themselves to metals. Reasons for this condition of purity in the hydrogen employed in this new industry are given in the chapter relating to "Catalysis."

Hydrogen, then, is now required on a scale never contemplated until quite recent times. The first patents relating to the hydrogenation of oils date only from 1903, and it was not till later that the high standard of purity required for this purpose was recognised.

Another process in which hydrogen may before long be required in large quantity is the production of ammonia synthetically by Haber's method, as described in the chapter on "Catalysis." In this case, as in the hydrogenation of oils, the gas employed must be free from sulphurous and other impurities.

Other uses of hydrogen have developed or expanded of late years, such as the melting of platinum, the working of fused quartz in making apparatus of silica, the autogenous soldering of lead sheet, and the making of joints in lead pipes employed in chemical works. These are found to resist corrosion by acids much more successfully than when solder is used.

We may now review the principal methods by which hydrogen may be obtained, and from a consideration of the materials and conditions of each some conclusion may be arrived at as to their practicability from the industrial point of view.

We may at once exclude those processes which are too costly or otherwise objectionable, such as the action of metallic sodium on water or the dissolution of metallic aluminium in caustic soda. These are frequently resorted to in the laboratory, but are unsuitable for use on a large scale.

The available methods may be ranged under several heads as follows :—

1. The action of metals on acids.
2. The action of metals on steam at a red heat or on water.

3. The action of carbon on steam at a red heat.
4. Electrolytic processes.
5. Miscellaneous methods.

1. *Metals and acids.* The traditional laboratory process for making hydrogen by the action of diluted sulphuric acid on zinc cannot be considered, as the cost of the zinc would be far too great. The only metal by which it could be replaced is iron, but inasmuch as 28 parts by weight of iron are required to produce 1 part by weight of hydrogen, the mere mass of material would be an objection. One ton of iron would require 3920 pounds or $1\frac{3}{4}$ tons of sulphuric acid, and would produce only 13,917 cubic feet of gas, a quantity which would be about one-twentieth of the capacity of a small balloon. The action of metals on acids may therefore be at once ruled out of the practicable processes for generating hydrogen for industrial purposes.

2. *Decomposition of water by metals.* The action of red-hot iron on steam results in the production of hydrogen gas and a residue of magnetic oxide of iron.

$$3Fe+4H_2O=Fe_3O_4+4H_2.$$

In this case iron theoretically yields one-third more hydrogen than when it is made to act on sulphuric acid. By associating with this process another for the restoration of the oxide of iron to the metallic state, and working the two alternately, a plan for the production of hydrogen on a practical scale results. The reduction is most advantageously effected by means of water gas, the product of the action of steam on red-hot coke :—

$$C+H_2O=CO+H_2.$$

The equation indicates that theoretically the interaction should result in the production of equal measures of carbonic oxide and hydrogen gases. This, however, involves the assumption that coke is pure carbon, and that the steam employed is free from air, while in practice neither of these conditions is fulfilled. Consequently the amount of carbonic oxide is somewhat below the theoretical amount, a small quantity of carbon dioxide being formed by the intrusion of a little air, the oxygen of which is of course accompanied by four times its bulk of nitrogen. These operations have been the subject of numerous patents.

Another very remarkable process for the decomposition of

water by iron has been patented by a German chemical engineer named Bergius, and it is stated that hydrogen, containing a very minute percentage of impurity, can be produced at a cost of about ¾d. per cubic metre (about 35 cubic feet). Water is heated to between 300° and 340° C. in a strong steel cylinder containing iron turnings in contact with copper, and a little common salt. The cylinder has a long neck fitted with a screw cock by which the hydrogen is allowed to escape under the pressure generated, which may amount to 300 atmospheres. The gas can therefore be stored under pressure in gas cylinders and is thus rendered portable. The oxide of iron produced is left in the form of a fine powder which is easily reduced to the metallic state by carbonic oxide.

3. *Decomposition of water by carbon.* The hydrogen contained in water gas to the extent of about half its volume may be secured by the comparatively simple process of freezing out the attendant impurities. These consist of a nearly equal volume of carbonic oxide, together with 2–5 per cent of carbon dioxide, and about the same bulk of nitrogen and traces of hydrocarbons, also sulphuretted hydrogen. The boiling-point of hydrogen being about –253° C., it boils at some sixty degrees below the boiling-point of nitrogen (–195°–196°), oxygen (–183°), or carbonic oxide (–190°), and consequently when cooled by liquid air under a moderate pressure these impurities are liquefied and removed, while the hydrogen retains the gaseous state.

4. *Electrolysis.* Hydrogen is liberated in several operations as a by-product which till recently has had but little value. In the Castner-Kellner process for obtaining caustic soda by electrolysis of brine, the sodium ions in contact with mercury dissolve, but the amalgam formed is in the presence of water, and consequently hydrogen is set free, while sodium hydroxide is produced. In the manufacturing method for production of metallic sodium, by electrolysis of fused caustic soda, hydrogen equivalent in quantity to the sodium is liberated. But these two methods are necessarily associated with the caustic soda and sodium which are of greater value, and to collect the gas a compressing plant would be required.

With cheap electric energy available, as in Germany, hydrogen can be produced, it is said, at a cost of about three farthings per cubic metre. The electrolyte is a solution of potassium carbonate which yields hydrogen at the cathode, oxygen and potassium bicarbonate appearing at the anode. A solution of potassium hydroxide at a temperature of 60° to 70° is also employed; in this case gaseous hydrogen and oxygen are the only products liberated.

The hydrogen obtained by any electrolytic process is apt to be accompanied by small quantities of oxygen, while the oxygen simultaneously set free is liable to contain a little hydrogen which in certain cases would be objectionable or even dangerous.

5. *Miscellaneous methods.* Some of these are designed to furnish the means of generating hydrogen in moderate quantities in the field by requiring only simple and portable apparatus. One substance proposed for this purpose is impure calcium hydride, CaH_2, a white powder to which the name *hydrolith* has been given. It is produced under a French patent. When mixed with water about one cubic metre of hydrogen is evolved from one kilo of the material, and this costs five francs.

Hydrogenite is the name given to another material in which ferrosilicon is mixed with a relatively large quantity of caustic soda and some lime. When heated this mixture evolves hydrogen and leaves a mass of silicates of sodium and calcium. Ferrosilicon in fine powder also dissolves in solution of caustic soda.

Acetylene is producible by igniting carbon in hydrogen at the temperature of the electric arc, and the process is endothermic, that is, a large amount of heat is absorbed and the energy is stored up in the gas. In common with other endothermic compounds it is therefore somewhat unstable and it can be decomposed by heat alone, being resolved into hydrogen gas and finely divided carbon. As acetylene is produced pretty cheaply from calcium carbide this principle has been made the subject of a patent by the German Carbonium Company. The gas contained in steel cylinders is decomposed by electric sparks, and the very fine carbon deposited is valued for making printer's ink. The process has been used at the Zeppelin factory at Friedrichshafen, but it appears not to be entirely free from danger, as it is said that explosions have occurred.

Another process of Dutch origin yields a gas, which though reported to be very light is certainly far from pure hydrogen. The process consists in heating coke to bright incandescence by means of a blast of air, and then blowing in hydrocarbon oils as long as the temperature is high enough. The blast of air is then again applied. The gas requires to be scrubbed with oil of vitriol and caustic soda.

CHAPTER XV

OXYGEN AND NITROGEN

EVERY schoolboy is acquainted with the process, common a few years ago and still used on a small scale for producing oxygen gas, by heating potassium chlorate, either alone or mixed with a small quantity of manganese dioxide. Priestley's original method of heating mercuric oxide (red precipitate) is described in most chemical textbooks, as well as the decomposition by heat of a considerable number of peroxides and other metallic oxides, and highly oxidised substances such as potassium permanganate, bleaching powder, sulphuric acid. But for industrial purposes and manufacture on a fairly large scale, a process was introduced, by patent in 1880, by MM. Brin frères which soon took the place of all the others and became established as a successful commercial undertaking. This was based on the fact that at a low red heat barium oxide in a stream of air, deprived of carbon dioxide, is converted into barium dioxide, $BaO+O=BaO_2$. The latter compound heated to a higher temperature gives off again the absorbed oxygen, while barium monoxide, baryta, BaO, is reproduced.

The inconvenience of alternately raising and lowering the temperature of the retorts in which the baryta was heated, and the wear and tear involved in these operations, led to the substitution of change of pressure for change of temperature with great advantage.

An apparatus was devised in which, by means of automatic reversing gear, the air could be alternately introduced into the retorts under slight pressure, and after absorption of the oxygen and escape of the nitrogen, the oxygen could be pumped off by reducing the pressure below atmospheric, but without altering the temperature of the retorts and their contents. Oxygen made by this process had a purity of 93 to 96 per cent. Some of the plants erected a few years ago may be still working, but new conditions have led to the practical abandonment of the Brin process.

Electrical power having become of late years more readily available and cheaper, and simultaneous demands having arisen for hydrogen gas, the process of submitting water to electrolysis, or rather a solution of caustic potash or soda, has found some considerable application. A variety of apparatus has been devised with the object not only of carrying the current through the liquid, but preventing local action or divergence of the current, with the risk of intermixture of the oxygen and hydrogen. As a fact, in the normal process small quantities of the one are generally found intermixed with the other. Obviously if more than a very small amount of such intermixture took place the use of the gas might lead to serious explosions.

The use of the oxyacetylene blowpipe flame for welding and for cutting through metal plates has recently extended so much as to lead to a greatly increased consumption of oxygen independently of hydrogen. The readiness with which air is now reduced to the liquid state and from the liquid, both nitrogen and oxygen can be separated in a condition of approximate purity, are circumstances which again have led to a new position of affairs. Since also nitrogen is required in such vast quantities in the manufacture of cyanamide the liquefaction process has taken the place of all the others in the production of oxygen for industrial purposes.

Before proceeding to describe the production of the gases from air it will be worth while to glance at the information now available as to the use of acetylene and oxygen for the purposes referred to. Acetylene is a gas familiar enough for lighting purposes and produced by the action of water on calcium carbide. The gas is supplied for the use of engineers dissolved in acetone and contained under pressure in steel bottles. When burnt acetylene gives out more heat than is represented by the carbon and hydrogen it contains, for it is an endothermic compound, that is to say, in the union of carbon with hydrogen in the production of acetylene, heat is absorbed.

Hence when the carbon and hydrogen are again separated the same amount of heat is evolved, and if this occurs while they are both uniting with oxygen a greater amount of heat is produced and a flame of higher temperature results.

By means of an appropriate blowpipe the two gases, acetylene and oxygen, delivered from their respective cylinders meet, and when ignited produce a pointed flame of which the temperature is considerably higher than that of a mixture of oxygen and hydrogen. A flame of this kind can be used for welding together iron surfaces of all kinds.

The use of the blowpipe, however, is still more wonderful in the feats which it is now capable of performing in the direction of cutting thick sheets of metal. In this case the oxyacetylene flame is produced at the mouth of the blowpipe, and through the middle of this a pointed oxygen flame is directed on the surface of the iron or steel to be cut. The following quotation from Thorpe's Dictionary will serve as a sufficient illustration of the capacity of the method.

"A plate 12 inches thick of nickel chrome steel armour plate was cut through at the rate of 1 foot in 4½ minutes with a consumption of 50 feet of oxygen per foot run."

With such an instrument at hand the older shearing, sawing, and boring methods of the engineering workshop are likely soon to disappear.

The illustrations Figs. 67, 68 convey an idea of the apparatus required for the production of liquid air and the separation of the oxygen and nitrogen from it. The principles made use of in the cooling of a gas below its critical point, and the system of intensive or cumulative cooling have already been explained sufficiently in the chapter on apparatus (p. 85). It is therefore unnecessary to do more than indicate with the aid of the plan the relative positions of the several parts of the machinery. The description which follows is taken from an article in *Engineering*, vol. 87 (1909). The plant at Odda on the Sondre Fiord, Norway, for the production of the nitrogen required in the manufacture of cyanamide, was "constructed by the Linde Eismaschinen Gesellschaft, Munich; the English patents are held by the British Oxygen Company Limited. The process is the invention of Professor Linde. It is based on the fact that at atmospheric pressure nitrogen boils at –196 deg. Cent.; liquid air boiling at –194 deg. Cent.; and liquid oxygen at –183 deg. Cent.

"By using the well-known rectification process followed in the manufacture of alcohol, the nitrogen can be completely separated from the oxygen. The Linde Company guarantees that the nitrogen does not contain more than 0·4 per cent of oxygen, a percentage which has neither an unfavourable influence on the chemical reaction,[1] nor leads to the burning of the electrodes. The plant . . . was put down for a production of 375 cubic metres (about 13,000 cubic feet)[2] per hour, and is run by a 200 horse-power electric motor.[3] All parts are in duplicate in order to prevent any long interruptions in the working, the second half of the plant serving as a stand-by. The diagram shows how the separation of oxygen and nitrogen from the air takes place, and will better explain the process than would sections through the various apparatus. The right half of the diagram represents the part of the plant in actual working; the left half shows the part held as a stand-by. We shall deal only with the former part. The air to be treated is drawn from the atmosphere by the largest of the four cylinders of the compressor, and through two towers, through which a soda liquor is made to trickle, the object being to free the air as much as possible from carbonic acid. The air is compressed to approximately 4 atmospheres (57 lb. per square inch); it is then cooled, first in a water tower, down to the temperature of the cooling water, and further by being passed in pipes through a reversing air cooler. Around these pipes flow cold oxygen or nitrogen from other portions of the apparatus. The water condensed is drained off at the bottom. In the rising leg of the apparatus ice is apt to form, as the temperature is below freezing-point. It is got rid of by periodically reversing the direction of the flow by the valves indicated. The air next passes to an ammonia cooler, where its temperature is further reduced to about –20 deg. or –25 deg. Cent. In this almost the whole of the remaining moisture is abstracted. On leaving this, almost absolutely pure and dry, the air passes next to the separator or still, which it enters through a counter-current interchanger consisting of the usual system of concentric pipes. Flowing itself through the inner of these pipes it is cooled by an oppositely flowing current of nitrogen or oxygen evaporating from the liquid state. The incoming air thus passes from the interchanger at a very low temperature, and being led next into a coil immersed in a tank of liquid oxygen liquefies there, since it is at a pressure of 4 atmospheres. From this coil it expands through a throttle valve with result that a large portion of it is obtained in the liquid state, and at about atmospheric pressure.

[1] This refers to the use of the nitrogen in the production of cyanamide.
[2] Of nitrogen.
[3] The power is derived from falling water. See later "Cyanamide."

Fig. 67.—LINDE PLANT FOR SEPARATING ATMOSPHERIC NITROGEN AND OXYGEN. NORTH-WESTERN CYANAMIDE COMPANY, LTD., AT ODDA, NORWAY.

To face page 252.

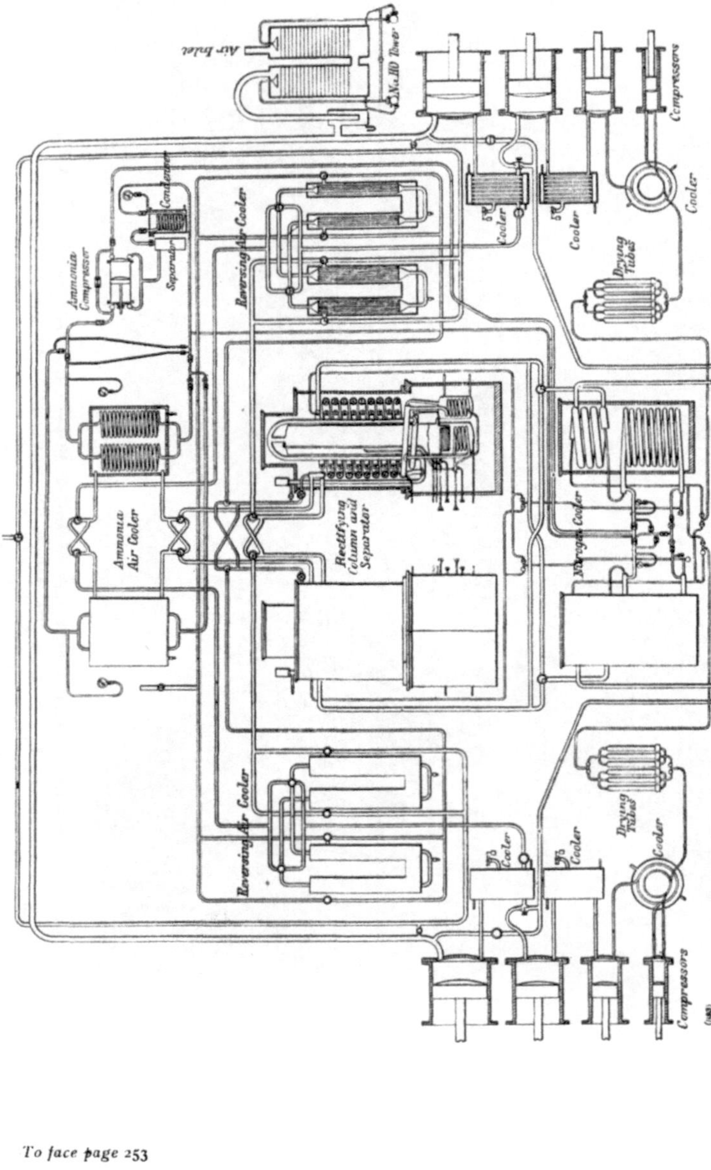

FIG. 68.—DIAGRAM OF THE LINDE PLANT AT THE WORKS OF THE NORTH-WESTERN CYANAMIDE COMPANY, LTD., AT ODDA, NORWAY.

To face page 253

"The liquid thus produced is led to a point near the top of the rectifying column, filled with glass marbles, over which it trickles to the bottom. On its way down it meets with an ascending current of gas from liquid below. This gas is rich in oxygen, and this oxygen having a higher temperature of liquefaction than the liquid air it meets in the rectifier, is condensed by an equivalent proportion of nitrogen being distilled off from the descending liquid. The latter therefore enters the tank at the bottom enriched in oxygen, whilst the gases passing off above are nearly pure nitrogen. To remove the remaining traces of oxygen, the rectifying column is extended at the top above the point at which the liquid air is introduced. The gases ascending through this extension meet a downward flow of pure liquid nitrogen which robs them completely of oxygen, so that pure gaseous nitrogen alone escapes from the top of the column." In order to obtain the liquid nitrogen needed in the upper part of the rectifying column the gas is led into the second cylinder of the compressor and thence, through intermediate coolers, to the other cylinders, from the last of which it is delivered at a pressure of 120 atmospheres.

The compressed gas is next passed through a coil contained in a tank of liquid oxygen derived from the separator, and after expansion through a throttle-valve the supply of liquid nitrogen at low pressure needed in the rectification is obtained.

The oxygen separated from the air is evaporated in a special receiver, and is used in cooling the compressed air as already explained. It may then be compressed into cylinders or used in any other way.

As the moisture and carbonic acid contained in the air drawn from the atmosphere cannot be completely removed at the beginning of the cycle of operations the separator gradually gets choked up with ice and solid carbon dioxide. The working period lasts from six to ten days, and the second half of the plant is then brought into operation a few hours before the first half is stopped, so that when this occurs the second half is already yielding nitrogen.

Chemical processes for the isolation of nitrogen, or at least those which can be used on a large scale are always based on the withdrawal of the oxygen from atmospheric air.

Copper may be used for this purpose, as at a red heat it rapidly unites with oxygen forming solid copper oxides. The metal is readily regenerated by passing over it a stream of producer gas or any gas containing hydrogen or hydrocarbon vapours. A simple arrangement would therefore be to provide two cylindrical vessels filled with scrap copper maintained at a red heat, the one being supplied with air, while the other is fed with producer gas alternately.

The remarkable chemically active form of nitrogen gas discovered by Professor Strutt has been described (p. 121). It seems not improbable that its properties may be hereafter turned to account in connection with the fixation of nitrogen from air.

CHAPTER XVI

WATER AND ITS PURIFICATION

The provision of a sufficient safe and suitable supply of water has always been a subject of great public importance. But it is only within the memory of the present generation that the character of the impurities occurring in water used for drinking has been completely understood, and that due precautions have been taken in the selection and treatment of water to be supplied to towns for all purposes. Water as it falls from the skies in the form of rain, snow, and hail may be said to be, from the dietetic point of view, pure, that is, it contains in solution only a small quantity of the gases of the atmosphere. This is true of rain water falling in the country, but as is well known the rain in towns is always contaminated with soot and with acids, which are the result of burning coal containing sulphurous and arsenical minerals, to say nothing of acid impurities emitted from works where chemical operations, such as alkali, glass, or cement making are carried on. The water supplies are, however, always drawn from districts as remote as possible from influences of this kind, and are subject only to sources of contamination provided by nature, and dependent chiefly on the geological character of the strata through which the water rises or over which it flows. The impurities thus naturally introduced are of two kinds, namely, the inorganic and the organic. With regard to the former we have to remember that some saline or earthy matters are soluble in water without any addition or assistance, while others are dissolved only by water holding carbonic acid derived from the air.

Thus while silica, alumina, and minerals in which these substances predominate as constituents are practically insoluble in water, common salt, Epsom salt (magnesium sulphate), and gypsum (calcium sulphate) are more or less soluble in water, the two last giving rise to the quality commonly called *permanent* hardness in many natural waters. On the other hand, the carbonates of lime and magnesia are practically insoluble in pure water, but they are found as constituents of many natural waters derived from springs, lakes, and rivers, owing to the presence of carbonic acid, by which they are taken up, forming unstable bicarbonates which are decomposed by boiling the water. The presence of these compounds gives rise to *temporary* hardness. Both these forms of hardness are the cause of some discomfort in washing and destroy soap, but, at least in moderate amount, they are certainly not injurious to health. They are, of course, mischievous when used in steam boilers, as they give rise to calcareous deposits and incrustations.

With regard to the organic substances found in water, much of this material is derived from the decay of vegetable matter, and it is commonly the cause of the various shades of green, yellow, or brown which are observable in the waters of streams and lakes. It is especially noticeable in water which has flowed or soaked through beds of peat, and may occur in water which, having passed only over hard silicious rocks, is comparatively free from saline or earthy impurity, and is therefore soft. Organic matter of this kind in moderate amount, as it occurs in the majority of waters supplied to communities, is not known to be definitely harmful as a constituent of drinking water. Down to comparatively recent times much ingenuity was expended in devising processes for estimating the amount of such substances in drinking water, but it is now recognised that the dangerous constituents in water are living organisms, and that the presence of much decomposing nitrogenous organic matter is significant chiefly as pointing to the probable contamination of the water with animal excreta. The last may be derived from influx of sewage, surface drainage from land supplied with manure or other similar sources. The examination of water with the object of determining whether or not it is fit for drinking by human beings is therefore now dependent less on chemical analysis than on bacteriological processes, in which the number of pathogenic organisms in measured quantities of the water can be counted and their character determined.

The softening of hard waters containing carbonates of lime and magnesia can be effected by boiling the water, when the bicarbonates are decomposed and carbon dioxide escapes. This, however, is impracticable on a large scale, and in practice such water is dealt with by the addition of slaked lime in quantity which must be accurately estimated from a knowledge, obtained by analysis, of the composition of the water to be treated. The chemical change which occurs is represented in the following equation :—

$$CaH_2(CO_3)_2 + Ca(HO)_2 = 2CaCO_3 + 2H_2O$$

calcium bicarbonate	calcium hydroxide (slaked lime)	calcium carbonate	water

A precipitate is formed which consists of the lime which has been added together with the lime previously held in solution, both in the form of carbonate. The water is, therefore, deprived of its hardness to this extent, and any hardness remaining is due to the presence of lime or magnesia in the form of chloride or sulphate. The latter can only be removed by the addition of washing soda which consists of sodium carbonate. The process of liming has the additional great advantage that the formation and deposition of the fine particles of the precipitated carbonate leads to removal of nearly the whole of the suspended organisms, and thus reduces in a great degree the probability of the spread of disease, even when the water was known to be previously infected. The results recently obtained in experiments on the use of lime for the purpose of water purification are described with full detail in the Eleventh Report by Dr. A. C. Houston to the Metropolitan Water Board, published in July, 1915. There it is stated that "the credit belongs to Aberdeen of being the first town in this country to show that what was proven to be feasible in the laboratory could be achieved under practical conditions in the case of a water supply to about 165,000 inhabitants. Since writing this report the Aberdeen Town Council on February 15th, 1915, decided by an almost unanimous vote to adopt a threefold system of purification by (A) liming, (B) storage, and (C) filtration, and to apply to Parliament in April for a Provisional Order."

Experiments on Thames River Water, which forms a large part of the supply to London, have led to similar encouraging conclusions. These are expressed by Dr. Houston in the Report (p. 19) in the following words: "No hesitation is felt in expressing the opinion that river water *no matter how impure*, may be brought into a condition of *absolute safety bacteriologically*, and of *great relative purity chemically* by means of *lime*."

The process of water purification sketched in the foregoing lines is applicable on a small scale to the water derived from wells or pools or streams in isolated country districts unprovided with a common supply. But it necessarily requires the use of several rather large tanks, even when the service of one house only has to be provided for, and a little trouble is involved in the regular operations of the preparation of the lime, its intermixture under proper conditions with the water, and the disposal of the sludge which gradually accumulates. In such cases it is very convenient to have a means of softening the very hard waters derived from wells in chalky or limestone districts by an operation which is simple and requires only the use of common and familiar materials. The use of the material known as "permutit" affords a very efficient way out of the difficulty which besets the householder in many country houses whether the water supply is used raw, without treatment, or if recourse is had to the lime process.

Permutit is an interesting case of the application of a mere laboratory product to practical purposes. The permutits are complex silicates, artificially produced, which have the property of exchanging their basic constituents when immersed in appropriate solutions. By melting together china clay (an aluminium silicate) and soda, a compound is formed which after being crushed and washed with water contains the constituents soda, alumina, silica, and water in proportions represented approximately by the formula $Na_2O.Al_2O_3.2SiO_2.6H_2O$. Its use is for the softening of waters which owe their "hardness" to the presence of lime and magnesia in the form not only of carbonate, but of sulphate, or chloride. The presence of these compounds in any considerable proportion is the cause of the formation of scale in steam boilers, and the destruction of much soap with formation of an insoluble curd when used for washing. If a permutit mineral is immersed in such water an exchange takes place between the soda of the mineral and the lime and magnesia of the water, resulting in the formation of a solid compound containing the latter bases, in quantity chemically equivalent to the soda removed, while the water retains the harmless sodium carbonate, sulphate, and chloride. After a certain amount of the hard water has passed through the material the latter will naturally cease to act for the obvious reason that it is fully charged with lime or magnesia, and has nothing further to exchange for the earthy constituents of the water. The chemical law of mass action may then be put into operation, by shutting off the hard water and passing slowly a moderately strong solution of common salt in amount considerably in excess of the quantity of soda required to restore the permutit to its original composition. Under these circumstances the calcium and magnesium pass away in the form of chlorides into the solution.

After the action is over the filter bed is washed free from the excess of salt and earthy chlorides, and is then ready for its renewed activity as water softener. In practice the softener consists of a cylindrical tank containing a bed of permutit of proper depth (according to the degree of hardness exhibited by the water) arranged between two layers of fine gravel. All the water supply required during the day passes through this bed and issues completely deprived of the hardening constituents. At night the water is turned off and the regenerative salt solution flows slowly through. In the morning the permutit is washed by passing a little water through it, and running the solution of chlorides to the drain. The ordinary water supply can then be resumed.

The same principle is applied when the purpose is to remove iron which would be objectionable in manufacturing operations. In this case a manganese compound is prepared, and the regenerative liquid is a solution of permanganate.

CHAPTER XVII

METALS AND SOME OF THEIR COMPOUNDS

The term *metal* is still in use without the possibility of a strict definition. Seven metals were distinguished by the ancients and were in alchemical times associated in a fantastic manner with the names of the seven "planets." Of these names Sol (gold), Luna (silver), Mercury (quicksilver), Venus (copper), Mars (iron), Jupiter (tin), Saturn (lead), only one, namely Mercury, has been retained in common use. The crude practices in the laboratories of the alchemists led to a few useful discoveries, among them probably the metal zinc, which was first mentioned by Paracelsus in the sixteenth century. In the latter half of the eighteenth century, at the time of Lavoisier, the number of recognised metals was seventeen. The popular idea at that time was, and is down to the present day, that a metal is a hard, shining, and heavy substance, which can be melted only in a hot fire. When therefore Davy in 1808 discovered potassium and sodium, which are both lighter than water, some perplexity was caused by their anomalous qualities and for a time it was proposed to designate them merely *metalloids*. This term, however, with the authority of Berzelius, soon received a different application, and with further knowledge of the physical and chemical properties of these elements they were included in the category of metals. Somewhere about fifty substances are now called metals, but it would be difficult to secure complete unanimity among chemists as to whether particular elements should be included. There is, however, one test which would probably be accepted generally. In the process of electrolysis the metals are always deposited at the cathode and are therefore spoken of as positive elements, notwithstanding differences in other physical characters, such as density, fusibility, ductility, or brittleness.

Gold, which occurs in nature almost always in the native or metallic state, was probably the first known to primeval man. The others occur chiefly in the form of sulphides or oxides, and the common useful metals are for the most part obtained by reduction of their oxides. Even those which, like copper, lead, and zinc, are found in combination with sulphur are usually submitted to a preliminary process of roasting in contact with air, so that much of the sulphur is burnt off and an oxide of the metal remains which is subjected to further treatment.

The art of metallurgy has, however, undergone great developments and many modifications within recent years, owing especially to the introduction of the electric arc, which gives temperatures far above ordinary furnace heat, and the electric current by which the method of electrolysis can now be applied economically on a large scale. By the application of these modern agents some metals, calcium for example, are now obtained on a fairly large scale which a few years ago would have been found only in the form of small specimens, the product of a troublesome laboratory operation.

Space will not allow of the description of many of the processes which are now applied to the production of metals for industrial or practical purposes, but a few of the more important may be briefly mentioned.

SODIUM

In September, 1807, Humphry Davy began those experiments on the action of an electric current on caustic potash and caustic soda which resulted in the isolation of the two strange metals, potassium and sodium. On the 19th November he gave his second Bakerian Lecture to the Royal Society in which he announced his discovery. Very shortly after this Gay Lussac and Thénard succeeded in obtaining potassium by heating caustic potash to redness in contact with iron turnings. These metals were afterwards made by distilling at a red heat a mixture of the carbonate with charcoal, and by this process these metals were made for upwards of fifty years. A modification of these methods was then introduced by Castner about 1887, but this has long been superseded by a process, also invented by Castner, which is identical with that of the discoverer but adapted to operations on a large scale. Electric current is now obtainable at moderate expense, and sodium is made in large quantity by the electrolysis of caustic soda fused and kept at a temperature about 20° C. above its melting point. Sodium is at the present time of much greater importance than potassium, as it is used in considerable quantities in the manufacture of various chemical compounds, among them indigo, several of the synthetic drugs, and the cyanides. These metals are not familiar to the public and cannot be handled safely by the inexperienced. Both potassium and sodium are silvery white, almost as soft as cheese, and melt easily. They cannot be exposed unprotected to air, as they absorb oxygen and instantly become covered with a coating of oxide. Thrown into water they decompose it explosively with evolution of hydrogen gas and formation of a solution of the caustic alkali.

A few years ago sodium was consumed in rather large quantity in the manufacture of aluminium by heating it with the anhydrous chloride of that metal, but the extension of facilities for electrolytic methods, together with the dangers and uncertainties of the sodium process led to its abandonment.

ALUMINIUM

Aluminium is probably the most abundant metallic element in the earth, as in the form of the oxide, alumina Al_2O_3, it is the chief constituent of many crystalline rocks and of all clays. The metal was first isolated in the form of powder by Wöhler, but it was not until about 1845 that it was obtained in a compact state and on a manufacturing scale by Deville. Aluminium is distinguished by its low density, which is only about 2·7 times that of water, and therefore about one-third the weight of iron. It is a good conductor of electricity, though inferior to copper. It forms a very valuable alloy with copper, which is known as aluminium bronze. This was manufactured by the Cowles process before the difficulties in the reduction of pure aluminium had been overcome. To obtain the bronze a mixture of corundum (alumina) with charcoal and granulated copper is heated in an electric furnace. The carbon takes the oxygen of the alumina, while the copper unites with the aluminium and forms a fusible alloy to which larger quantities of copper can afterwards be added if required. This alloy has nearly the colour of gold, while it has great strength and elasticity.

Aluminium has been manufactured for many years by submitting to electrolysis alumina (prepared bauxite) dissolved in fused cryolite, the double fluoride of aluminium and sodium. The operation is carried out in an iron pot lined with carbon which forms the cathode. The current is introduced by means of thick carbon rods forming the anode which dips into the mixture. The metal sinks to the bottom and is tapped off at intervals, while carbonic oxide gas escapes. In proportion as the metal is removed the supply of alumina is kept up by adding it to the molten mixture.

In the production of aluminium, as in so many other cases, the source of power is the energy of falling water, and factories have been established in connection with many of the great waterfalls of the world, such as Niagara, the Falls of the Rhine at Schaffhausen, and in our own country at Kinlochleven in Argyllshire. The bauxite consumed in the production of the metal is obtained chiefly from the South of France, and in 1907 amounted to 260,000 tons. The total world output of metal in 1909 was estimated at 30,000 tons.

Aluminium is valuable not only for its lightness but on account of its peculiar behaviour toward acids and alkalis. It dissolves rapidly in diluted hydrochloric acid, but is very slowly attacked by sulphuric or nitric acids, and still less by vegetable acids. On the other hand it is dissolved by alkaline solutions with evolution of hydrogen. Hence aluminium is used in a variety of ways for making cooking utensils and in the storage of a great variety of foodstuffs, but it is important to remember that saucepans or pots of aluminium must not be cleaned with the assistance of soda.

Aluminium melts below a red heat, and when heated quickly in the air it becomes covered with a white film of oxide which prevents rapid oxidation. A thin piece of aluminium foil in a bottleful of oxygen gas if touched with a red-hot wire disappears instantly with an extremely brilliant flash, leaving the white oxide behind. The combination of aluminium with oxygen is attended by the evolution of a larger amount of heat than is disengaged by the combustion of an equivalent quantity of any other metal. The consequence is that a mixture of aluminium powder with the oxide of another metal when heated at a single point enters into a violent chemical reaction, at the end of which the aluminium is converted into oxide while the other metal is found in the metallic state. The action is so violent in some cases, copper oxide for example, that a kind of explosion occurs and part of the metal is volatilised. A mixture of aluminium with various oxides has been turned to account for the isolation of some metals not previously known or obtainable with difficulty. The metal chromium, for example, is obtainable in this way, in a state of purity, also manganese, which had previously been known only in combination with carbon or with iron.

An ingenious application of this property of aluminium is found in the "thermit" process. A mixture of ferric oxide with aluminium powder is placed in a crucible with a removable bottom, and a fuse placed in the top of the mass being ignited the whole mass becomes incandescent, and in a few minutes a layer of molten iron sinks to the bottom of the pot and can be run off into a mould. The method is applied to the repair of broken castings, or to joining the ends of tramway rails without removal. The mould being placed round the rail end receives the melted metal, and after solidification the excess of iron can be cut or ground away to the level of the rail. The temperature produced in the mixture is said to be about 3500° C.; it is sufficiently high to melt every known metal.

The reaction in thermit being once started cannot be stopped, and this material has been found in many of the incendiary bombs used in the war.

STEEL

Man has been described as a tool-using animal, but the materials accessible in prehistoric times were very different from those which are available now. The use of stone and bone certainly preceded that of any metal, and naturally the metals which were obtainable either in the native state, like gold and copper, or by very simple operations, would come into use before those which were more difficult to procure. The Stone Age therefore preceded the Bronze Age, and this came before the Iron Age, though doubtless these periods overlapped.

Modern metallurgy is the result of constant experiment and research. It has given the world modern steel, which means greater security on railway and steamship, greater capacity in foundry and forge, and consequently the monster ocean-going passenger ships as well as ships of war and big guns.

One of the difficulties which surround any attempt to give an account of some of these developments in a small space is to find a definition of steel. Everyone knows that it is a sort of iron but with qualities of its own. Pure elemental iron is a product which is extraordinarily difficult to obtain and is not found among commercial metals. The nearest approach to it is the finest malleable iron of which wire is made. This is distinguished by its fibrous texture, toughness, and capability of welding. Wrought-iron is fusible only at a white heat, but at any temperature above redness it is soft and can be hammered or drawn into any desired shape, and if at this temperature two pieces are hammered together they become completely united. This is, of course, the basis of the blacksmith's art.

The cast-iron from which wrought-iron is made is the product of the blast furnace. Iron ore, coal or coke, and limestone being heated together, the materials melt and settle to the bottom of the furnace in two liquid layers. The upper is slag, the lower is iron in union with 2 to 5 per cent of carbon, and small quantities of sulphur, phosphorus, and silicon. Iron of this kind is brittle, though hard and much more easily fusible than wrought-iron.

Steel is made by several processes, all of which have for their object the production of a carbide or mixture of carbides of iron containing an amount of carbon which may range from ·1 per cent in mild steel up to about 2·0 per cent or a little more in hard tool steel. The presence of sulphur and phosphorus in steel is detrimental. Steel is in all varieties less brittle than cast-iron and has a greater tensile strength than wrought-iron. The milder varieties, that is those which contain the smallest percentage of carbon, can be forged and welded. The character which formerly was considered distinctive of steel is its property of becoming hardened by quenching in water or oil when at a high temperature. The degree of hardness to be given can be regulated by the temperature to which it is heated and the rate at which it is cooled down. This is called *tempering* and is sometimes regulated by observing the colour of the film of oxide which is formed on the surface when the metal is heated. The blue colour of a watch-spring is familiar and is indicative of great elasticity: heated to redness and then plunged into cold water it becomes brittle. But during the last thirty years great advances have been accomplished in the knowledge of the internal structure of the metal by the aid, not only of chemical analysis, but the use of the microscope and a study of the peculiar phenomena which iron exhibits in changing temperatures. The introduction of electrical resistance thermometers now enables the manufacturer to test and regulate the temperature of his furnaces, a point of great importance which was beyond control only a few years ago. The constituents of steel have been the subject of very numerous researches, and even now authorities differ in some points of detail.

On heating a mass of steel to redness and then allowing it to cool slowly it is observed that the temperature does not drop regularly but at certain points the cooling seems to hesitate and proceed more slowly. There are three of these critical points or points of recalescence, namely at about 825°, 735°, and 660°, observed in very mild steel. In hard steel the critical temperatures are both relatively and absolutely somewhat different, but similar phenomena are noticed. On gradually heating up a mass of steel the rise of temperature is retarded, showing an absorption of heat at about, but not exactly, the same points. These and other observations have led to the hypothesis that iron is capable of existing in two or more *allotropic* states, that is, conditions in which a molecule of the metal is composed of different members of atoms. In the harder steels it must be assumed that the carbon plays a very important part. It seems to be capable of dissolving in molten iron and on cooling it enters into chemical combination with the metal. One, and perhaps the most important compound formed, is *cementite*, to which is attributed the formula Fe_3C. The different varieties of steel when in the solid state may be supposed to be mixtures in various proportions of this compound with one or other of the allotropic forms of the metal, or of a solidified solution of carbon in the metal. Many of the steels introduced into modern practice for special purposes contain other ingredients.

Manganese has been recognised as a necessary ingredient in steel ever since the introduction of the Bessemer and open-hearth processes for the manufacture of the metal. The amount present does not usually exceed 1 per cent, but for special purposes manganese steels are made containing much larger quantities.

Nickel is a familiar white metal, which is about as difficult to melt as wrought-iron. Some thirty years ago, when it began to be available on a large scale, various alloys of nickel with iron were tried and since that time have rapidly extended in use. Nickel added to iron has a toughening effect, and when added in proportions from 12 to 20 per cent it increases greatly both the tensile strength and elastic limit. Nickel steel has been largely used for armour plate. The magnetic properties of this alloy are remarkable ; when about 25 per cent of nickel is present in steel it is almost non-magnetic unless exposed to a temperature of –40° C. After cooling to this low temperature it remains magnetisable at ordinary temperatures, but if heated to 600° C. it recovers its original non-magnetisable condition.[1]

Chromium is a metal which was almost unknown till Moissan's introduction of the electric furnace. A ferro-chromium alloy was formerly made in the blast-furnace, but chromium and other alloys for use in steel making are now manufactured in the electric furnace. The metal can also be obtained by the thermit process already explained. Pure chromium is hard enough to

[1] *Harbord's Steel*, 2nd ed., 1905, p. 628.

scratch glass, it is somewhat similar to iron when polished but is not magnetic and is unaltered by moist air. It combines with carbon in several proportions, forming the carbide Cr_4C and at the higher temperature of the electric furnace the compound Cr_3C_2 (Moissan). Its hardening effect on steel appears to be closely connected with the amount of carbon present. It is now used in fairly large quantity for the manufacture of steel tires, springs, and axles, and for armour plate.

Tungsten is a very infusible metal found in the form of the mineral wolfram which is a tungstate containing the oxide WO_3. This compound can be reduced to the metallic state by heating with coke in the presence of cast-iron. The product is known as ferro-tungsten, and is used as an alloy in steel for the production of so-called "self-hardening" steels. It is associated with manganese in the well-known Mushet steel, and is sometimes introduced together with chromium in steel required for machine tools working at high speed. *Molybdenum* which is very similar to tungsten has also been used.

Vanadium and *tantalum* are other elements which have been tried as ingredients in steel. Vanadium has a curious history, for the substance which during forty years had passed as the metal itself was shown by Roscoe to be a compound of that substance with nitrogen. Vanadium though not rare is far from abundant, and the cost will necessarily have the effect of limiting its application in steel-making to special purposes. Fortunately the addition of very small quantities of vanadium is sufficient to modify the properties of steel substantially. The addition of 0·6 per cent of vanadium to a pure iron and carbon steel (containing 1·1 per cent of carbon) raised its tensile strength from about 30 tons to 85 tons per square inch. There is need for much further research in this direction.

Titaniferous iron ores exist in immense quantities in Sweden and in the form of sands in the United States, Canada, and New Zealand. In the blast furnace only a portion of the titanic oxide, TiO_2, is reduced and passes into the iron. There has been difficulty in introducing titanium into steel, but the presence of a small quantity is said to assist in the production of a sound ingot. It has a tendency to combine with nitrogen and may in this way prevent the formation of blowholes.

NICKEL

This metal is very familiar in the form of nickel plating, and has long been used as an ingredient in the white alloy with copper of which electro-plated dishes, spoons, forks, and other table furniture are made. But the very interesting and remarkable process by which a large proportion of the pure metal is made is based on a modern discovery which dates back no further than 1890. In that year a paper in the *Transactions of the Chemical Society*, by the late Dr. Ludwig Mond, associated with Dr. C. Langer and Dr. F. Quincke, announced the discovery of the fact that when metallic nickel, especially in a finely divided state, is heated gently in a stream of carbon monoxide gas a volatile compound is formed which consists of the metal in union with carbon monoxide. The escaping gas burns with a brightly luminous flame, and when heated to a temperature about 180° it is resolved completely into the gas and the metal, the latter being deposited in the form of a lustrous mirror-like solid. When the mixture of gases is passed through a glass tube surrounded by a freezing mixture of ice and salt the compound is condensed to a colourless, mobile liquid, a little heavier than water, and boiling at 43° C. The liquid has the formula $Ni(CO)_4$; it is not acted on by acids or alkalis. It precipitates copper and silver from ammoniacal solutions of the chlorides of those metals, but in general it behaves as a neutral compound. Attempts have been made to obtain compounds of the same order from the metals nearly allied to nickel, but no success has been met with in the case of cobalt. Iron and platinum have been found to yield carbonyl compounds, but with greater difficulty, and the compounds formed are much less volatile than nickel carbonyl.

It is easy to see how these observations may be turned to account in the extraction of nickel from the mixed ores from which so much of this metal has been obtained. The ores which contain a number of metals, iron, copper, cobalt, nickel, etc., in the form of sulphide and arsenide are first roasted, by which the greater part of the sulphur and arsenic is expelled and the metals converted into oxides. These are then heated moderately in a stream of producer gas whereby the oxides are reduced to the metallic state, and the temperature being duly regulated, the carbonic oxide in the gas unites with the metallic nickel and carries it off, while the other metals which form no volatile compound are left behind in the residue. The nickel is recovered by causing the mixed gases to pass through a heated pipe before being returned to the furnace to play the same part over again.

Very large quantities of nickel are also made from the mineral called garnierite, which consists of a hydrated silicate of nickel and magnesium and is found in New Caledonia. This mineral is practically free from other metals, and the nickel is obtained from it by a furnace process which consists in first converting the metal into sulphide, and then reducing it by a series of operations similar in principle to those by which copper is obtained from its sulphide ores.

Nickel is a white metal a little heavier than iron but having the advantage of practical permanency in the air whether dry or moist. It is somewhat magnetic. The addition of nickel to steel increases the toughness, and on this account nickel is employed in armour plate, as already mentioned in connection with steel.

LAMP FILAMENTS

There must be a considerable number of persons still living who remember the dim light with which the public of sixty years ago had to be contented. Up to the time of the great Exhibition in 1851, and for many years later, the interiors of houses had been lighted by candles, and the snuffer tray was a necessary article of daily domestic use. The streets of London and of most towns were at the same time provided with lamps for coal gas which was burnt at flat flame burners, giving a degree of illumination which would be regarded as intolerable at the present time.

The discovery of large quantities of petroleum in Pennsylvania about 1860 provided a new and cheap source of light, and there was soon great activity among the lamp makers. The use of gas as an internal illuminant for houses was still, previously to 1860, though common, far from universal.

With the development of various forms of magneto-machine and ultimately of the dynamo which occupied many years, the electric arc gradually became available for use in lighthouses, and here and there for large spaces such as the Thames embankment and in a few large workshops. But the arc was never suitable for domestic use, and it was only when the incandescent lamp, with a carbon filament enclosed in a vacuous glass globe, was invented that electric light became a practical source of light for internal illumination. Several conditions were necessary; first electric current was wanted at a moderate price, and next the means of producing a high vacuum pretty easily was also indispensable, especially as the filaments to be made luminous by the current were all at that time made of carbon. Sprengel's mercury pump, invented in 1864, provided the means of getting a vacuum, but there were great difficulties about the production of threads of carbon. The late Sir Joseph Wilson Swan after experiments made so long ago as 1860 exhibited the first electric glow lamp in February, 1879, at a meeting of the Newcastle Chemical Society, and in November, 1880, gave a demonstration at the Institution of Electrical Engineers in London. His carbon filaments were first made by heating parchmentised thread, and later by squirting collodion through a die. These were exhibited at the Inventions Exhibition in 1885.[1] Mr. T. A. Edison, the well-known American inventor, had in the meantime begun working on the subject, but rival claims disappeared under a prudent and amicable arrangement, and in the end the Edi-Swan lamp became familiar to everyone.

In more recent years, however, the filament to be heated, and so made luminous by the current, is more usually made of some metal of low conducting power. Platinum was the first in which the phenomenon of luminosity produced by the current was studied, but platinum, though its melting point is high, is too fusible for use in the lamp. It has one property which has made it useful in the lamps and that is its low coefficient of expansion. As it expands and contracts with change of temperature to nearly the same extent as glass a wire of this metal can be melted into glass and on cooling the glass does not crack. Hence platinum may be used for making the connections between the fittings outside the lamp and the filament which gives the light within.

Many metals and alloys have been tried for the production of lamp filaments, the object being to obtain the maximum of light with the minimum expenditure of current. Tungsten and tantalum have found the most success, but the details of the processes by which these metals are obtained in the form of sufficiently thin but yet strong wires have so far been kept secret.

Tungsten (p. 266) when quite pure is tough and ductile,

[1] A collection of apparatus used in Swan's early experiments is now exhibited in the Science Museum, South Kensington.

though the presence of a small quantity of carbon renders it brittle. The Osram lamp contains a filament of practically pure tungsten, which is said to be produced by mixing the powder of the metal with some binding material and then squirting it into threads by forcing it through a fine hole in a steel plate. The threads are subsequently treated by a secret process to remove the binding matter and produce a continuous thread.

Tantalum was discovered by Ekeberg in 1803. It was found in two Swedish minerals, tantalite and yttro-tantalite, but until the electric furnace provided the high temperature necessary it had not been melted and was known only in the form of a black powder. Tantalum is a white metal with a specific gravity 16·8. It melts at a very high temperature, said to be 2798° C., according to tests made in 1912 at the University of Wisconsin. It is malleable and ductile at a red heat, but like many other metals the presence of impurities, especially carbon, renders it brittle. It resists the action of most acids, and, considering the present very high price of platinum, its employment for making crucibles, dishes, and other chemical vessels seems likely to extend.

PYROPHORIC ALLOYS

A return to the flint and steel with the tinder-box of our forefathers is improbable for everyday purposes, but the discovery of the property exhibited by cerium and some of the other metals of the same group of so-called rare earths (see periodic scheme of the elements, p. 125) has led to the invention of contrivances for striking fire which are occasionally useful and are certainly curious.

The metal cerium, obtained by the electrolysis of its chloride or fluoride, resembles iron in appearance but is far more fusible as it melts at 623° C. It is also much more easily oxidisable, decomposing water slowly, and in moist air it soon becomes coated with a film of oxide. The metal burns when ignited even more brightly than magnesium, and when scratched with a steel edge or struck by a flint it emits brilliant sparks. This property has been turned to account in the production of gas lighters and cigarette lighters for the pocket. Pure cerium is, however, less suitable for this purpose than certain mixtures of the cerium metals with iron, which are harder and yield sparks much more readily. An alloy of cerium with magnesium and aluminium when heated below the melting point in a stream of hydrogen gas absorbs hydrogen in consequence of the formation of solid hydrides of the cerium metals. This solid alloy has the property of sparking in a remarkable degree, and when rubbed against a rough steel surface produces a shower of sparks which will ignite a jet of coal-gas or the wick of a small lamp.

Very large numbers of these pyrophoric metals have been sold, especially in France, where matches are expensive, but these contrivances can at present be regarded only as toys.

COMPOUNDS OF METALS

It is scarcely necessary to say that the metals by combining with other elements produce a very large number of definite compounds. Some of them yield several distinct oxides by combination with different proportions of oxygen, and all metals produce salts, which may be said to be their characteristic compounds. Many of these, such as common salt, alum, green and blue vitriol, have been known from the most ancient times, but scores of new compounds of this kind are added to the list every year, as the result of operations in the course of chemical research or manufacture. Many metallic compounds are applied to useful purposes as chemical agents, in medical practice, or as pigments and otherwise. These, however, are the commonplaces of practical chemistry, and information concerning them is stored up in the larger textbooks and dictionaries of chemistry, and to these sources of information the reader who desires it must be referred.

A few words may be added here as to the artificial production of certain gems. This is the outcome of conveniences for the production of high temperatures, such as the oxyhydrogen flame and the electric furnace. It need scarcely be repeated that the diamond is a form of carbon and that the diamonds produced by Moissan's process (p. 83) so far are minute and are of no use as gems. The ornamental stones, often very beautiful, which are sold in imitation of diamonds are merely artificial silicates containing lead and other heavy metals. Other stones such as emerald, garnet, topaz, etc., are also imitated by glasses, but the most interesting of artificial stones are the ruby and the sapphire which are now produced commercially by the fusion of pure alumina, of which both are essentially composed. The artificial stones are not merely imitations, they are identical in hardness, density, and crystalline form with the natural gems. Ruby owes its colour to the presence of a minute quantity of chromium in a peculiar condition of oxidation, and according to the late Professor Frémy, sapphire contains the same element in a different state. In this case the blue colour is more like that which is imparted to fused substances by cobalt, though it has also been attributed to titanium. As to the rest of the precious stones the cause of the characteristic colours is not in all cases known. Thus the emerald and aquamarine are varieties of beryl which probably owe their green colour to the presence of iron in the ferrous state; that of emerald has, however, been attributed to chromium and even to organic matter. Iron in the ferric state and manganese are capable of producing various shades of yellow, red, or purple, according to the states of combination in which they occur, and to one or both of them together the amethyst, the garnet, and other stones probably owe their colour.

CHAPTER XVIII

LUMINOSITY OF FLAMES

THE INCANDESCENT MANTLE INDUSTRY

EVERYONE is familiar with the facts that the flame of a candle, an oil lamp, or ordinary coal-gas is more or less luminous, while burning hydrogen, spirit of wine, or gas mixed with air, as in the Bunsen burner, gives so little light that such a flame in broad daylight is scarcely perceptible and in sunlight is actually invisible. What is the cause of this difference? The attempts to answer this question have occupied experimental chemists for upwards of a hundred years, and though many hypotheses have been put forward it cannot be said even now that the complete solution of the problem has been discovered.

Sir Humphry Davy was the first to enquire systematically into the source of light in flame, and in 1816 he put forward the opinion that the production and ignition of *solid* particles within the flame itself is the cause of the light. With regard to the luminosity of coal-gas he attributed the effect to the decomposition of a part of the gas towards the interior of the flame, where the air is in smallest quantity, and the deposition of solid charcoal, which by its ignition and afterwards by its combustion increased to a high degree the intensity of the light. "A few experiments," he says, "convinced me that this was the true solution of the problem." Nearly half a century later, however, Frankland drew attention to the fact that flames may be produced which are brilliantly luminous but contain no solid matter, and further that a gas like hydrogen, which burns under ordinary conditions without emission of light, may be made to give out light if burned under increased pressure. It appeared, therefore, that the light of a flame might be due only to the presence in it of dense gases or vapours.

Notwithstanding these results it appears certain for reasons which cannot be discussed that ordinary hydrocarbon flames, such as those of coal-gas, do contain solid particles, and a reason has to be sought for the deposition of carbon from the burning gas. The hypothesis brought forward a few years ago by the late Professor Vivian Lewes, which involved the production and immediate decomposition of acetylene within such flames, was at one time much discussed, but seems to be no longer tenable, and we are still waiting for "the true solution of the problem" which Davy thought he had got hold of a hundred years ago.

That the introduction of solid matter into a non-luminous flame causes the emission of light is a matter of common knowledge, and Davy himself showed that it is of no consequence whether the solid is combustible or not, for he showed that not only did dust of charcoal but fine powder of silica or magnesia thrown into a flame produces light. From the time when coal-gas in the earliest years of last century became a common source of light, efforts have continuously been made to increase its illuminating power, first by the improvements in the jets or burners at which the gas was burned, later by the introduction of gases or vapours rich in heavy carbonaceous compounds.

The last is a method still employed in our own day (see cracking of Petroleum, Chap. XIX). But it has long been known that the introduction into a flame of different solid substances is attended by the emission of very different amounts of light.

Berzelius in 1829 noticed that thoria, zirconia, and other of the rare earths in a non-luminous flame give out a very brilliant light, and similar observations were made later by Bunsen and other chemists. Lime is one of the substances which when strongly heated gives a bright light, but the temperature required in this case to give a satisfactory effect is higher than that of an ordinary flame. The well-known limelight, in fact, requires the use of oxygen with the gas to produce the necessary temperature. It was only toward the latter end of last century that serious attempts began to be made to utilise the earlier observations of Berzelius and Bunsen on the peculiar incandescence produced by several of the rare earths. It is unnecessary to trace the various attempts to utilise the incandescence of magnesia, zirconia, and other substances, for the discoveries made by Dr. Carl Auer,[1] as a consequence of his studies of the rare earths begun about 1885, require all the space which can be spared for this subject. The "mantle," which is familiar in almost every household where gas is the illuminant, consists of a mixture of thorium oxide with about 1 per cent of cerium oxide. The use of this mixture was the result of a long series of trials, and was protected by patent in 1893. Since this discovery minerals which contain thorium have become very important and valuable.

At one time they were known chiefly as of Swedish origin and were even called the Swedish earths. But the immense quantities now required are supplied from *monazite* sands found in extensive deposits in N. and S. Carolina and especially on the coast of Brazil. Other important minerals are *thorianite*, an oxide very rich in thoria, found in Ceylon, and *thorite*, a silicate which occurs in various localities in Scandinavia. Monazite, which is essentially a phosphate of cerium containing relatively small quantities of thorium, is the chief material now used in connection with the mantle industry.

Its composition is, however, very complicated, and the extraction of the small percentage of thorium present is a matter of difficulty.

In dealing with Brazilian monazite sand the first operations are mechanical, and advantage is taken of the high specific gravity of the mineral (about five times heavier than water) to remove by streams of water much of the lighter material. Electro magnets are also used for extracting ferruginous particles, and a concentrated material is ultimately arrived at which retains only 2 or 3 per cent of impurity. The mineral is then treated with sulphuric acid for the extraction of the earths.

Here, however, we may halt, for an exposition of the details of the process to be followed would be extremely tedious

[1] Now known as Baron Auer von Welsbach.

to the general reader, and would be useless to the technical reader unless furnished with minute particulars.

For those who desire to pursue the subject there exist several recent works which contain full information. Of these Böhm's *Fabrication der Glühkörper für Gasglühlicht* (1910) gives a fairly complete account of the extraction of the earths from monazite, and the application of these substances to the manufacture of "incandescent mantles." Mr. S. I. Levy's recent work on the *Rare Earths* contains a very full account of the minerals, followed by a condensed description of the mantle industry (1915).

For the accompanying views of the chief operations by which incandescent mantles are manufactured in this country the author is indebted to Mr. C. S. Garland, Managing Director of the Volker Lighting Corporation of Wandsworth, London.

The process of mantle making consists in first knitting from a fine ramie, cotton, or artificial silk thread, a continuous cylindrical hose on a rotary knitting machine, the length of the stitch and the tension of the thread being regulated according to the width and depth of mantle required. The hose is then made up into loose bundles for washing. The yarn contains about 1 per cent of non-cellulose matter, chiefly fatty material introduced in spinning, and a certain amount of siliceous and calcareous matter from the original bast fibres of the plant "Rhea Elastica," from which the ramie fibre is prepared. (See Fig. 69 facing page 276.)

The earlier mantles were very liable to shrink and suffer contortion whereby they were often withdrawn from the hot part of the flame and so produced less light. The luminosity was also reduced in course of use by the presence of the small quantities of mineral impurities left after the burning of the vegetable fibre. Ramie appears to be very superior to cotton as a basis for the mantle-maker, and artificial silk is still better, being practically free from mineral matter, and therefore leaves no ash. But the use of artificial silk has not become general owing chiefly to the cost.

The very careful and elaborate washing of the fabric before impregnation is, therefore, a part of the manufacturing process, and is of great importance.

The washing process varies in different factories, but its object is, first of all, the removal of the grease and silica by digestion of the stocking with a weak solution of caustic alkali, followed by thorough rinsing in distilled water. The excess of alkali and the lime are afterwards removed by steeping the knitted fabric in a weak solution—1 to 2 per cent—of hydrochloric acid or acetic acid. The acid is then removed by agitation of the fabric in distilled water, and, if the hose is required to be kept any length of time, this is followed by a bath of dilute ammonia. The stocking is dried at a low temperature by hanging over poles or drawing continuously through a hot chamber through which a current of hot air is blown. It is then cut up, usually on a cutting machine similar to the one illustrated (Fig. 70), which cuts the fabric into pieces of equal length and width, stacks them, and automatically counts them as cut.

We must now treat the upright and inverted mantles separately.

The heads of the upright mantles require to be reinforced by stitching on to them a piece of cotton or ramie tulle, with the object of strengthening the head and keeping it to a uniform size. They are then dipped in the previously prepared strong solution of thorium nitrate with about 1 per cent of cerium nitrate and ·5 to 1 per cent of other hardening materials, such as aluminium, zirconium, or calcium nitrate, put upon the elevator illustrated in figure 71, and carried through gutta-percha rollers, carefully adjusted to leave exactly the right quantity of thorium nitrate in the mantle. The stocking is then dried upon glass forms shown in figure 72, and when dry the head is reinforced with a further supply of solution containing a higher proportion of the hardening materials. Such "fixing" fluid consists often of a solution of aluminium, magnesium, and calcium nitrates in water. The mantles are again dried and then sewn with asbestos thread to form the head.

The next operation is that of burning off the organic matter of the mantles, and at the same time decomposing the nitrates with which they are impregnated. The stocking is first shaped over a wooden form and hung up by means of an iron hook (Fig 73), fired, and allowed to burn until nothing but the white ash of thorium and cerium oxides remains. This ash is now in a loose condition and is extremely fragile. With the help of a burner supplied with either gas or air at a pressure of from 5 to 15 lbs. per square inch, the mass of oxides is blown out to its correct shape, and, by an up and down movement of the burner inside the mantle, the latter is brought to its final state of hardness, the change being due to the partial "fritting" of the more fusible oxides

FIG. 69.—VOLKER LIGHTING CORPORATION, LTD. KNITTING THE HOSE

FIG. 70.—VOLKER LIGHTING CORPORATION, LTD. CUTTING THE HOSE.

To face page 276.

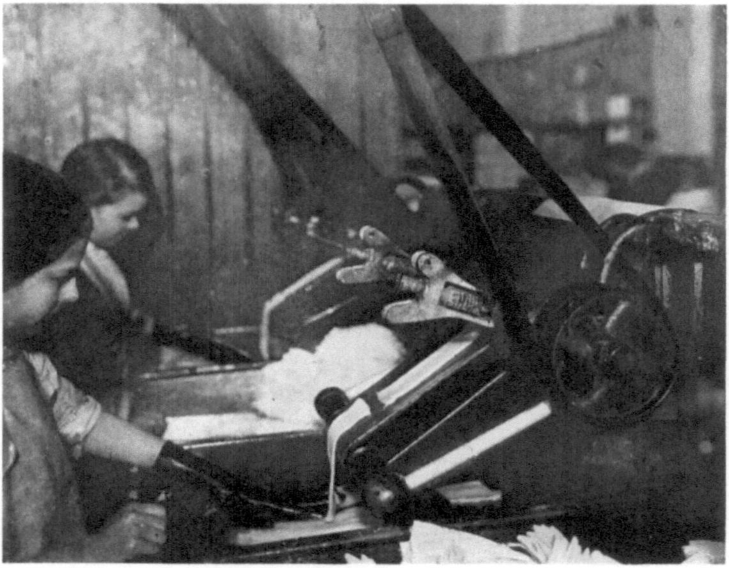

Fig. 71.—VOLKER LIGHTING CORPORATION, LTD. IMPREGNATING MANTLES.

Fig. 72.—VOLKER LIGHTING CORPORATION, LTD.
DRYING IMPREGNATED MANTLES.

To face page 277.

in the mantle. In this state the mantle is in the same resistant condition as when it is finally used upon the burners, but for purposes of transport it is coated with a varnish consisting of nitro-cellulose and various oils, to modify the too rapid character of the combustion of the nitro-cellulose. The mantles are dipped, usually from 40–60 at a time, in this "collodion" solution, allowed to drain and dried in ovens heated by high pressure steam (Fig. 74). They are then able to withstand the handling necessary for examination, cutting to length, and testing to size and for minute holes, due to traces of silica adhering to the original yarn.

The perfect mantles are packed each one in a box having a pad of cotton wool at either end. A dozen such mantles are packed into an "outer" for the shopkeeper.

In the case of the inverted mantles, these are impregnated after being cut into suitable lengths and are dried upon glass forms. They are then mounted upon rings with an asbestos thread, and the other end of the stocking is drawn together by hand or by machines, and darned with a thread impregnated with the same solution as that in the body of the mantle (Fig. 75). The fixing fluid is then sprayed on to that portion of the mantle which is adjacent to the magnesia ring in order to give it extra strength at this point. It is then ready for burning, which is conducted in the same way as with the upright mantles, except that the magnesia ring is supported during burning in an iron ring. Ten or twelve of these make up a tray, which goes into the burning and seasoning machines. In the burning off operations the eyes of the workers require to be protected from the glare by the use of shades or blue spectacles.

The inverted mantles are dipped in collodion, dried, tested to see that they are perfect and that they fit the burners properly, and packed into units and dozens for sale.

The machines for the testing of the yarn consist of three parts.

The first, upon which 500 metres exactly of the thread is wound off from the cones upon which it is delivered into a skein, which is then weighed upon the second part. This records directly the "count" of the yarn. The other machine is a stretching and strain-testing machine, which determines the breaking strain of the yarn measured against a weighted bob, which moves over the graduated arc. The breaking strain of 2-25 ramie thread should be from 7 to 9 lbs.

In modern factories a certain proportion, from 1 to 2 per cent, of the mantles are tested for resistance to shock by being burnt off and mounted on the burners illustrated with the shocking machines. At the foot of these burners is fixed a bar at right angles to the burner stem, and the machine is arranged so that a little stamp battery (Fig. 76), having weights of 2 to 5 oz. on each side, can be run at different speeds upon this bar, and submit the mantle to the same sort of vibration as it might experience when mounted in a lamp-post adjacent to a road gully over which a heavy lorry is passing.

The rings for support of the inverted mantles are composed of a mixture of china clay and silica with a small proportion of magnesia. For rings which have to resist a very high temperature, as when high-pressure gas is used, from 10 to 40 per cent of carborundum is added. The mixture of powders, moistened with a special oil to make it cohere when pressed, is squeezed into a metal die, made of three or four pieces which move together in a machine contrived for the purpose. After pressing into shape the rings are weathered for a period from two days to a week, and are then scraped to remove rough edges and give them their finished shape. They are then packed in fire-clay boxes, called saggars, which are stacked together in a pottery furnace or kiln where they are baked. The baking which requires slow heating up and cooling down occupies about two days. The rings are then examined, brushed to remove dust, and the perfect ones packed in layers to be sent to the mantle factories.

The manufacture of the rings has been chiefly in the hands of one firm having branches in Germany, France, England, and America. It is estimated that in the year 1914 about 25 to 30 million rings were manufactured in this country, and probably another 15 millions were imported from Germany.

The works of the Volker Corporation in Wandsworth were established in 1895 to work a patent of Dr. Voelker's, who seasoned the mantles in an electric furnace. Very beautiful mantles appear to have been made, but they were far too expensive, and ultimately the Auer von Welsbach patents, in which the thorium-cerium process already described were put into operation, were adopted. At the present time the Company employs about seven hundred hands and turns out from 18 to 20 million mantles per annum.

With regard to the total world production the best way to

Fig. 73.—VOLKER LIGHTING CORPORATION, LTD.
SEASONING UPRIGHT MANTLES.

Fig. 74.—VOLKER LIGHTING CORPORATION, LTD. DIPPING AND DRYING.

To face page 278.

Fig. 75.—VOLKER LIGHTING CORPORATION, LTD.
SEWING INVERTED MANTLES.

Fig. 76.—VOLKER LIGHTING CORPORATION, LTD. STAMP BATTERIES.

To face page 279.

arrive at an estimate appears to be to calculate from the amount of monazite sand annually employed in the manufacture of thorium nitrate. From these figures it has been estimated that approximately 250 million mantles are made per annum in the whole world. More than half of these were produced by the German Mantle Trust (the Deutsche Auer-Gesellschaft of Berlin) in its various branches. This country probably manufactures from 50 to 60 millions per annum, and has imported about the same number from Germany. The figures of imports prior to the war are as follows :—

Year.	Value.
1908	£227,486
1909	£295,950
1910	£292,467
1911	£277,357
1912	£322,631
1913	£303,576

Such figures supply occasion for remarking on the rivalry which has existed for many years between the several systems of lighting our streets and houses. When the incandescent electric lighting began to make way, the gas industry appeared almost doomed to extinction so far as illumination was concerned, and great efforts were made to increase the candle-power of the gas manufactured, and to improve the burners in use. Then came the Welsbach mantle, and all seemed well for a time. But a new difficulty arose when the supply of the Swedish earths began to be exhausted, and the earliest Welsbach Company came to an end. The search for thorium-bearing minerals in other parts of the world, however, was soon rewarded by the discovery of inexhaustible deposits of monazite sands on the other side of the Atlantic. These deposits are now the basis of the vast industry which has been described in this chapter. Simultaneously the character of the coal-gas produced at the gas works has gradually undergone considerable modification, for, with the assistance to the illuminating power afforded by the mantle, it is no longer necessary to furnish gas of the relatively high candle-power formerly demanded. The effect of this is that a larger quantity of gas can be extracted from a ton of coal than was formerly possible when an illuminating power equal to 16 candles per 5 cubic feet was required. In fact whereas less than 10,000 cubic feet of gas were obtained per ton of coal heated, the yield is now commonly 13,000 cubic feet. A cheap gas produced by the addition of water gas charged with vapours from petroleum (see petrol) also helps to reduce the cost. The testing of gas, for statutory purposes, now relates more to its power as a source of heat than as a source of light.

At the same time this improvement in gas illumination has not been without its effect on the system of electric lighting. The improved efficiency of the incandescent electric lamp by the substitution of various metal filaments for the carbon filament, exclusively used a few years ago, may not have been due entirely to the increasing success and economy of the gas mantle light, but invention has undoubtedly been stimulated by these results.

The mantle industry is only one of many examples which could be quoted of the ultimate practical application of the results of purely scientific research to common industrial purposes. A generation ago the salts of thorium and cerium, lanthanum and didymium, and the rest were interesting only to a few enthusiasts. The place of these elements was and continues to be among the problems perplexing to the scientific chemist, and they were only known to exist in a few comparatively scarce minerals found chiefly in Scandinavia, and all this was implied in the name which for so long a time they bore, namely "the rare earths." Now these elements are known to be widely diffused and available in any required amount from mineral deposits which are actually handled to the extent of thousands of tons annually.

CHAPTER XIX

PETROL

Less than twenty years ago the arrival of any sort of motor vehicle in a country place would have been sufficient to bring together a crowd of wondering folk, and less than ten years ago any street in London exhibited a collection of omnibuses, carriages, wagons, carts, and other vehicles of which the majority were drawn by horses. The proportions of motor to horse-drawn carriages are now so much changed that even in England, where the movement has not been so rapid as in America and some other countries, there is a prospect that in a few years the horse as an agent of traction will become as great a curiosity as the motor was a generation ago.

This state of things has been brought about by the development of various forms of internal combustion engines, of which the gas-engine, introduced about 1876, is one form from which engines capable of working with the vapour of a volatile liquid have been developed. The perfection of engines of this type applicable to every kind of moving vehicle, including the aeroplane, has occupied the attention of the engineer for many years past, and the supply of suitable "spirit" for the motor is a business of importance second only to that of the supply of coal. Motor spirit is derived almost entirely from the lighter and more volatile portions of natural petroleum, which is now known to exist in vast quantities in the earth, and distributed very widely. There are, in fact, few countries in which it is not found in greater or less amount. The value of petroleum as a fuel has become greatly enhanced of late years, since it has become more commonly used as a substitute for coal in locomotives, and in ships, especially in ships of war. Apart, however, from its utility as a fuel petroleum has been the subject of innumerable researches in the hands of the chemist, and in addition to the use of the constituent hydrocarbons actually existent in it which receive a great variety of applications, there is some reason for believing that certain of them may hereafter be transformed by chemical processes so as to yield valuable compounds of a totally different nature, such, for example, as synthetic rubber.

The history of petroleum in its practical applications belongs to modern times. For while the Fire-worshippers from India and the East resorted centuries ago to the district near Baku on the Caspian Sea, where inflammable gas issued from the ground and where remains of some of their temples existed down to our own times, the use of the oil which was associated with the gas was of little practical importance.

The great petroleum industry of the United States began in 1859. Up to this time the oil which was used almost exclusively as a medicinal agent, both internally and externally, had been collected in a crude way from the water of the Seneca Lake in Allegheny County, New York, and other places. The production of burning and lubricating oils by distilling shales and other low-grade coal-like minerals was introduced in 1850 by James Young of Kelly, and it is probable that the recognition of the similarity between these oils and the natural petroleum led to experiments on the latter. It was soon found that on distillation a number of useful products could be obtained, including oil suitable for burning in lamps and a denser oil applicable as a lubricant and preservative to machinery.

The extraction of petroleum from the earth is accomplished by operations which are very simple in principle, and which have not changed in fundamental character during the sixty to seventy years since the commencement of the industry in the United States. Petroleum is usually associated with more or less salt water, and the strata which contain it are commonly charged with inflammable gas often existing there under enormous pressures. It may, however, happen that one of these products may exist without the others, and either salt water alone, or gas alone may be obtained when oil was expected.

The first business in commencing to bore for oil is the erection of a wooden structure, called the " derrick," about 70 feet high, tapering upwards from a base of about 20 feet square to about 4 feet square at the top (Fig. 78).

The derrick has a wheel at the top over which passes a rope which hangs vertically and carries the steel boring tools at one end, while the other end, by means of an arrangement worked by a steam-engine, placed at a little distance from the derrick, is alternately pulled and let go, so that the tools are alternately raised and dropped in an iron tube previously inserted upright in the ground. The tools have various forms according to the character of the rock which has to be pierced. From time to time the tools are lifted out of the pipe, and a " sand pump " is introduced in order to withdraw the pulverised rock and sediment at the bottom of the hole.

The wells thus sunk vary in depth in different oil fields, those in Pennsylvania ranging from 300 feet to 3700 feet (Redwood). In a district which has been found to be productive the borings of wells soon multiply and the numerous derricks form a curious and striking feature in the landscape. This is shown in the accompanying pictures. An early stage is shown in the erection of a derrick in a new field in Southern Russia (Fig. 78), and when compared with the other pictures it will be seen that in

Fig. 77.—PETROLIA, CANADA.
THREE-POLE DERRICK USED FOR CLEANING AND REPAIRING PUMPING WELLS

Fig. 78.—GURIEFF NEAR MOUTH OF URAL RIVER. S. RUSSIA.

To face page 282.

Fig. 79.—TRIUMPH HILL ON THE ALLEGHANY RIVER. PENNSYLVANIA (1885).

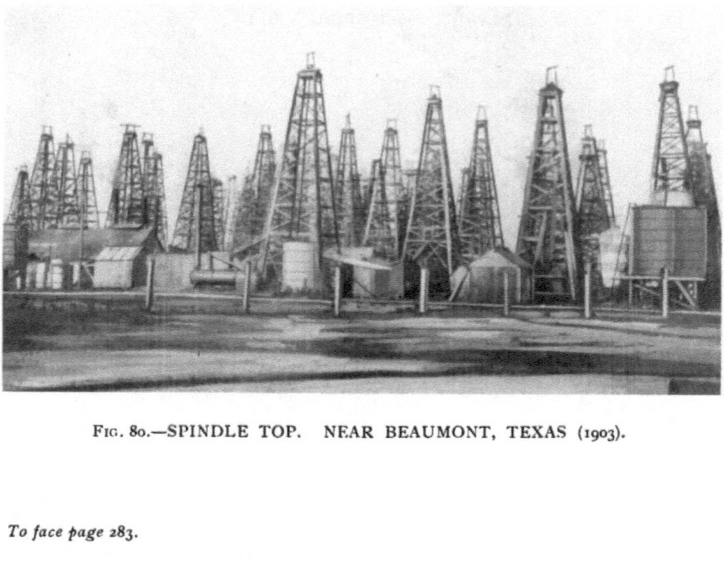

Fig. 80.—SPINDLE TOP. NEAR BEAUMONT, TEXAS (1903).

To face page 283.

widely separated countries the operations are essentially similar.

Another picture (Fig. 79) shows the appearance of an oil-field in the Pennsylvanian regions, and for comparison with it a view in Texas (Fig. 80) and another in California (Fig. 81) nearly twenty years later. From these it will be seen that the aspect of a petroleum field with its clustering derricks and associated tanks for oil remains much the same as in the earlier periods of development in the older fields of Pennsylvania. The country, however, is very different superficially, as the Pennsylvanian oil-fields are situated among well-wooded hills, while the Californian wells are sunk in a district which is largely desert. The geological character of the rocks from which the oil is extracted is also different in these three regions. In Pennsylvania the oil-bearing sandstone rocks belong to the Devonian system below the carboniferous series, while the Californian oil is derived from Eocene and Miocene beds, and that of Texas also from other formations more recent geologically than those of Pennsylvania. In no case is the oil deposit found in contiguity with beds of coal or shale, but in cavities generally, but not always, not in communication with one another. This is important to remember in connection with the question as to the origin of petroleum, which will be briefly discussed on a later page.

Turning now from the western hemisphere toward the east, the great Russian oil-field, extending through the Caucasian region chiefly along the shores of the Caspian Sea, is of the utmost commercial importance.

From prehistoric times this district has been the resort of all the East for the sake of the oil exuding from the ground. The inflammable gas which escapes in so many places was naturally the wonder not only of the natives, but of numerous pilgrims from afar.

Jonas Hanway, an English merchant in the reign of George II, visited the Caspian, and on his return, published in 1754, *An Account of British Trade over the Caspian Sea.* In this book he gave an interesting account of the phenomena which had attracted the Fire-worshippers from Persia and India for many centuries. The worshippers in Hanway's time all came from Bombay, the home of so many of the Parsees, the last disciples of Zoroaster. The ruined remains of the temples in which the priests tended the eternal flames which were the object of devotion existed in the neighbourhood of Baku down to our own times. These were described by Mr. Arthur Arnold (afterwards Sir Arthur Arnold), M.P. for Salford, who visited the district in 1875,[1] but it is almost unnecessary to add that the pilgrims who now visit this region are attracted by considerations altogether different from those which brought the followers of the Magi.

The history of the commercial development of the Russian oil-fields is comparatively simple, and the chief steps in its progress can be readily traced. When Baku became Russian territory, in the early part of the nineteenth century, having been taken from Persia, the extraction of oil was made a crown monopoly. The result of this was that the trade grew very slowly, and in the meantime American oil found its way into all the markets of the world. The monopoly at Baku was maintained down to 1872, but though the restriction was then removed an excise duty was imposed, which for five years longer served as an impediment to free production. Since that time all restrictions have been removed, and the number of individuals and companies engaged in boring for oil and in the business of refining is very large. But the rapid expansion of the Russian oil industry owes almost everything to the influence of the two brothers Robert and Ludwig Nobel.[2]

Up to their time oil had been carried from the wells to the refineries in barrels, and the refined oil to the Russian consumer also in barrels. In place of this slow and costly system the use of pipe lines for transmitting the crude oil, the introduction of tank steamers on the Caspian in 1879, tank barges on the Volga, the subsequent establishment of tank cars on the Russian railways, the provision of storage tanks at convenient points on the railways all over the country are due to the initiative of the Nobel Brothers.

In the accompanying illustrations a few scenes are shown characteristic of the district (Figs. 82, 83, 84, 85, 86).

"After that of Russia the petroleum industry of the districts of the Carpathian range next claims attention by its importance and antiquity. On the northern slopes will be found the oil-fields of Galicia ; while on the south-eastern and southern slopes of the southern Carpathians or Transylvanian Alps lie the im-

[1] *Through Persia by Karavan.* London, 1875.

[2] Alfred Nobel, a third brother, was the inventor of dynamite. See "Explosives."

Fig. 81.—N. END McKITTRICK, NEAR BAKERSFIELD, CALIFORNIA (1903).

Fig. 82.—BINAGADI NEAR BAKU, CASPIAN.
AMBAR OR EARTH STORAGE RESERVOIR CONTAINING CRUDE OIL IN FOREGROUND.

To face page 284.

FIG. 83.—BALAKHANG NEAR BAKU, CASPIAN.

To face page 285.

portant deposits of Rumania and the less known fields of Bukowina and Hungary. . . . The petroleum industry in this country is of considerable antiquity. The earliest historical records show that oil was collected in a primitive fashion and used as a cart grease from very early times, and old timbered oil-wells still existing in Galicia and Rumania indicate that this practice prevailed to a considerable extent" (Redwood's *Petroleum*, Vol. I). The older wells were dug out, but modern methods of boring were introduced about 1881, and since then the development has gone on steadily till in 1910 the total production in Galicia approached two million tons. The petroleum deposits of Rumania are continuous with those of Galicia; they are also supposed to be of about the same age as the petroleum bearing beds of the Caucasus with which they are said to be continuous (Redwood). The total Rumanian production was estimated in 1910 at upwards of one million and a quarter tons.

The Figures 87, 88, and 89, facing pages 288 and 289, show some of the modern wells in these regions. Figure 90, as well as Figure 78, facing page 282, are views taken in the new oil-field in the Uralsk Province of Southern Russia. It will be noticed that throughout the eastern oil regions, whether in Central or Eastern Europe, it is customary to enclose the derricks by boarding them over.

In consequence of the general association of the oil with gas confined in the oil-bearing rock or sand under pressure it frequently happens that when the rock is pierced the oil is forced up the bore-hole with great violence, producing a fountain of oil.

Many of these fountains which are very frequent in the Baku district have been described by the late Charles Marvin in his *Region of the Eternal Fire*, published in 1888, and we cannot do better than quote his account of the impressive spectacle exhibited by the famous Droojba fountain in 1883. He says, p. 211, "The oil was flying twice the height of the great Geyser in Iceland, with a roar that could be heard several miles round. When the first outburst took place the oil had knocked off the roof and part of the sides of the derrick, but there was a beam left at the top against which the oil broke with a roar in its upward course, and which served in a measure to check its velocity. The derrick itself was 70 feet high, and the oil and the sand, after bursting through the roof and sides, flowed fully three times higher, forming a greyish black fountain, the column clearly defined on the southern side, but merging into a cloud of spray thirty yards broad on the other. A strong southerly wind enabled us to approach within a few yards of the crater on the former side, and to look down into the sandy basin formed round about the bottom of the derrick where the oil was bubbling and seething round the stalk of the oil-shoot like a geyser. The diameter of the tube up which the oil was rushing was ten inches. On issuing from this the fountain formed a clearly-defined stem about eighteen inches thick, and shot up to the top of the derrick, where in striking against the beam, which was already worn half through by the friction, it got broadened out a little. Thence, continuing its course more than 200 feet high, it curled over and fell in a dense cloud to the ground on the north side, forming a sand bank, over which the olive-coloured oil ran in innumerable channels towards the lakes of petroleum which had been formed on the surrounding estates. Now and again the sand flowing up with the oil would obstruct the pipe, or a stone would clog the course; then the column would sink for a few seconds lower than 200 feet, to rise directly afterwards with a burst and a roar to 300. . . . Some idea of the mass of matter thrown up from the well could be formed by a glance at the damage done on the south side in twenty-four hours,—a vast shoal of sand having been formed which had buried to the roof some magazines and shops, and had blocked to the height of six or seven feet all the neighbouring derricks within a distance of fifty yards." The fountain belonged to a small Armenian Company, the Droojba, having ground enough to establish the well, but nothing to spare for reservoirs. Consequently all the oil flowed away on other people's property and the owners of the well were ruined. This oil volcano was estimated to have thrown up from 1,600,000 to 2,000,000 gallons of oil every day from the first outburst which occurred on the 1st of September. In the middle of November it was still spouting at the rate of 240,000 gallons a day.

The pictures 91 and 92 give an idea of the sort of scene Marvin describes. Grosnic is situated north of the Caucasian mountains in country occupied by the Terek Cossacks. In the first picture the fountain is shown, having pierced the top of the derrick. The second picture shows the destruction wrought by the jet of oil which, after the fountain has subsided, continued to flow

Fig. 84.—TANK CARS ON RAILWAY SIDINGS NEAR BLACK TOWN, BAKU.

To face page 286.

Fig. 85.—ASTRACHAN. TANK BARGES ON THE VOLGA.

Fig. 86.—ASTRACHAN. TANK STORAGE NEAR THE VOLGA.

To face page 287.

down a channel across which one of the workmen is seen astride.

In such cases there is usually an immense loss of oil and gas. Attempts are usually made, not always with success, to control the outflow by means of an iron cap with appropriate valves by which the stream of oil can be directed into a reservoir or tank.

In cases where the pressure of gas is not sufficient to bring the oil to the surface it has to be forced up by the application of compressed air.

One important oil-field further east exists in Burma, and is interesting to English readers as it exists in British territory. A very large proportion of the oil comes from the rich district of Yenangyoung, where for many generations hand-dug wells were worked by a class of hereditary oil winners under the Burmese kings. The rights of these people were recognised by the British Government after the annexation of Upper Burma, and a certain area was reserved for them in which they were annually allotted oil-well sites. In most cases they sold or leased these sites to the Oil Companies which introduced the American methods of drilling. Most of the oil existed under high gas pressure, so that when a well was bored into the oil enormous quantities gushed out (Figs. 93 and 94).

The upper sand has now been exhausted and with it the gas pressure. Wells have now to be driven down to much greater depths and pumping is necessary, but the local experts are of opinion that there are numerous sands, one below another, in the district. Drilling has hitherto been done by the American percussion system, but a rotary drill is being tried. The oil from the field used to be carried to the refineries, which are situated in the immediate neighbourhood of Rangoon, solely by barges towed down the Irawaddy River. The Burma Oil Company some years ago constructed a pipe line underground the whole distance.

Burma petroleum contains a very high percentage of wax, and this has been the cause of much difficulty in pumping the oil through the pipe lines, by reason of its viscosity.

The following figures, extracted from the *Rangoon Gazette* for April 29th, 1915, show that a fairly steady increase in the production of this field has been going on for some years past.

PETROLEUM IN BURMA

Year ending December 31st, at 260 gallons per ton.

Year.		Gallons.		Tons.
1905	..	142 millions	..	547,000
1906	..	138 ,,	..	528,000
1907	..	149 ,,	..	568,000
1908	..	173 ,,	..	665,000
1909	..	230 ,,	..	885,000
1910	..	212 ,,	..	815,000
1911	..	222 ,,	..	854,000
1912	..	245 ,,	..	943,000
1913	..	273 ,,	..	1,050,000
1914	..	255 ,,	..	975,000

The petroleum obtained from the Koetei district in Borneo has become specially important in consequence of being found to contain a large quantity of the hydrocarbons of the benzene series, including benzene, toluene, *m*xylene, and mesitylene, beside members of the naphthalene series. The less volatile portions, like some of the Russian oils, are optically active. (H. O. Jones and H. A. Wootton. *Trans. Chem. Soc.*, 1907.)

Crude petroleum is an unattractive brown or nearly black liquid which floats on water. Though inflammable and when blown into spray mixed with a sufficient supply of air it may be employed in locomotive engines or under boilers for raising steam, it is not well fitted for burning in lamps, as a source of light, owing to its viscidity and the presence of impurities. In order to obtain from it various useful products it requires to be submitted to distillation. The refineries in which petroleum is thus dealt with are usually situated at a considerable distance from the oil-field. Consequently the question of conveyance becomes one of great practical importance.

During the first ten years, or thereabout, of the American oil industry the crude oil was carried chiefly in barrels from the wells to the refineries, but as the business grew this method was not only expensive, but quite inadequate to deal with the very large quantities of oil obtained. The greater part of the oil has long been conveyed by means of lines of iron pipes which commonly run alongside the railway track. These pipes vary in diameter from four inches up to as much as eight inches, but owing

Fig. 87.—WELLS AT MORENI, ROUMANIA.

Fig. 88.—MORENI (ROUMANIA) VIEW ON THE AMERICAN PROPERTY.

To face page 288.

FIG. 89.—GALICIA. DERRICK AND ENGINE SHED.

FIG. 90.—GURIEFF NEAR MOUTH OF URAL RIVER. S. RUSSIA.

To face page 289.

to friction and differences of level it is not practicable to force the oil beyond a distance of forty or fifty miles. At intervals, therefore, the oil is delivered into a tank at each station where an auxiliary pump passes it along to the next stage, and so on, through the entire distance which commonly reaches several hundred miles. The illustration (Fig. 95 facing page 293) shows a view of the storage tanks near to an oil field, together with one of the pumping stations referred to.

The conveyance of oil in the Baku district has been already sufficiently described. Fig. 96 facing page 293 gives a view of a steamer of the American river type, used for towing tank barges.

Arrived at the refineries the oil has to undergo the process of distillation. As everyone knows, this process consists in heating the liquid in some kind of boiler or "still" with a head which confines the vapour given off and conducts it into a pipe or series of pipes, cooled, if necessary by water, where the vapour is condensed into the liquid state, and is run into a receiver. In operating on such a mixture as petroleum the distilled liquid differs in properties and composition from the original, for it consists of those ingredients the boiling points of which are lower than the boiling points of those portions which are converted into vapour at a later stage.

The stills actually in use differ in different countries both as to form and capacity. One form commonly used in the United States is shown in the Figs. 97 and 98 facing page 294.

It consists of a cylindrical iron vessel, capable of holding between 50,000 and 60,000 gallons, connected by a wide pipe which delivers the vapour into a box furnished with a large number of pipes surrounded by water for the purpose of condensation. Many modifications have been introduced with a variety of objects, such as making the process continuous by causing crude petroleum to flow into the first of a series of connected stills, the temperature of each successive still being higher in proportion as the more volatile portions pass off. In more recent years, with the object of "cracking" the oil and obtaining a larger yield of certain light portions, the upper part of the still is kept comparatively cool, so that the less volatile portions condense and drop back into the hot boiler.

In order to explain the process of cracking it is necessary to give a brief account of the nature and constitution of natural petroleum, and the chief products obtained from it.

Petroleum is a mixture of compounds which consist of the elements carbon and hydrogen only. The chemist speaks of them as hydrocarbons. With these compounds are associated in the natural oil minute quantities of compounds containing nitrogen and sulphur, which, however, may be dismissed from further discussion at this point.

The American oils consist chiefly of hydrocarbons called "paraffins" by the chemist. These are characterised by the comparative indifference which they exhibit toward chemical agents, such as strong sulphuric acid and bromine or chlorine.

By the process of distillation these are separated into successive fractions, the first portions collected in the condenser being accompanied by a certain quantity of gas which has been held in solution by the oil. These first and lightest portions of the distillate constitute the liquids which are sold under the names gasoline, rhigolene (petroleum ether), naphtha, and benzine (petroleum spirit). Following these in order of volatility come kerosene burning oil or mineral colza, and heavier oils used for lubrication, and from which vaseline and solid paraffin are extracted. After everything has been distilled off the residue consists of coke.

Some forty years ago it was found out that when solid paraffin is heated pretty strongly it is converted into a mixture of *liquid* products, while only a very little gas is given off. The liquid contains one or more paraffins of simpler constitution than the solid, together with a liquid having the composition of a paraffin less two atoms of hydrogen, or, in chemical language, it belongs to the series of olefines. The action of heat on the heavy paraffins, therefore, is to break them up in such a way that each molecule yields two smaller molecules, one being a paraffin, while the other is an olefine. This is fundamentally the change which is effected by the process of cracking, minor changes being brought about at the same time.

Russian oil yields a smaller proportion of the lighter and more volatile portions of spirit, but a larger proportion of heavy oils and solid paraffin. It is characterised by the presence of a considerable quantity of compounds called *naphthenes*, which present most of the characters of the paraffins, especially in their insolubility in strong sulphuric acid. They are believed to possess a constitution like that of benzene (see coal-tar, p. 308)

Fig. 91.—GROSNIC (S. RUSSIA) OIL FOUNTAIN.
AKHWERDOFF. No. 7 FOUNTAIN SPOUTING.

To face page 290.

FIG. 92.—GROSNIC (S. RUSSIA) DERRICK AFTER DESTRUCTION BY OIL FOUNTAIN.
AKHWERDOFF. No. 7.

To face page 291.

with additional hydrogen. They are represented by the formulæ C_6H_{12}, C_7H_{14}, C_8H_{16}, etc., and their boiling points range from about 69° C. to near 250° C. Beside the naphthenes Russian petroleum is believed to contain small quantities of benzene and toluene.

The " cracking " of petroleum for the purpose of producing a very volatile or gaseous mixture of hydrocarbons, which when burnt give a luminous flame, has for many years been practised, at first especially in the United States, as an auxiliary to the manufacture of common coal-gas.

The gas with which towns are supplied now consists to a large extent of a mixture containing gas distilled from coal by heating it in retorts, together with what is known as " water-gas." The latter is a mixture of carbonic oxide and hydrogen formed by passing steam through masses of red-hot coke, the temperature being maintained by substituting air for steam when the mass has cooled below a certain point. The introduction of air gives rise to carbonic oxide which remains mixed with the nitrogen of the air. As all these gases, carbonic oxide, hydrogen, and nitrogen together, yield a mixture which burns with a non-luminous flame a small quantity of petroleum is injected which being " cracked," as already explained, by the heat furnishes the light-giving ingredient required. Large quantities of petroleum are consumed in this way.

Fig. 101 facing page 300 gives a view of the carburetted water-gas plant on the works of the " Gas Light and Coke Company " in London.

In the year 1911 the amount of oil used by this Company for carburetting the water-gas amounted to 13,401,101 gallons. As it seems probable that the present high price of coal will be maintained the consumption of petroleum in gas-making will doubtless continue to increase.

At this point a glance may be taken at the following table which displays an estimate of the output of crude petroleum over the whole world. From this it appears that the largest amount by far comes from the United States, while Russia follows with a yield of less than one-third of the American. In the next few years this proportion will probably be disturbed to some extent.

WORLD'S PRODUCTION OF CRUDE PETROLEUM IN 1911

Source.	Metric tons.		Percentage of total.
United States	29,393,252	..	63·80
Russia	9,066,259	..	19·16
Mexico	1,873,522	..	4·07
Dutch East Indies . .	1,670,668	..	3·52
Rumania	1,544,072	..	3·21
Galicia	1,458,275	..	3·04
India	897,184	..	1·87
Japan	222,187	..	0·48
Peru	186,405	..	0·40
Germany	140,000	..	0·29
Canada	38,813	..	0·08
Italy	10,000	..	0·02
Other Countries Estimated .	26,667	..	0·06
	46,526,334	..	100·00

(Day quoted by Redwood.)

COMPOSITION AND USES OF PETROLEUM

Before attempting to enumerate the various uses to which petroleum and products from it are applied, it will be well to take a survey of the general nature of these products. The compounds which have been identified among the products of distillation are very numerous, and in the following table it has been thought sufficient to include only those which are the most abundant and characteristic. They are ranged according to their physical condition at the common temperature of the air, and according to their chemical composition.

PENNSYLVANIAN PETROLEUM

GASEOUS

Paraffins.		*Olefines.*	
Marsh-gas	CH_4	Ethylene	C_2H_4
Ethane	C_2H_6	Propylene	C_3H_6
Propane	C_3H_8	Butylene	C_4H_8
Butane	C_4H_{10}		

Fig. 93.—AUNGBAN YO (BURMA).

Fig. 94.—IN THE YENOUNGYOUNG DISTRICT (BURMA).
KHODARY BRIDGE.

To face page 292.

Fig. 95.—LOUISIANA. STORAGE TANKS.

Fig. 96.—STERN-WHEEL STEAM TOW-BOAT ON PANUCO RIVER, MEXICO. USED FOR TOWING OIL BARGES.

To face page 293.

LIQUID

	Paraffins.	Formula.	Boiling point C.
	Pentane normal	C_5H_{12}	About 38
	iso-	,,	30
A	Hexane normal	C_6H_{14}	69
	iso-	,,	61
B	Heptane normal	C_7H_{16}	97·5
	-iso-	,,	91
C	Octane normal	C_8H_{18}	125
	iso-		118
	Nonane	C_9H_{20}	136
	Decane	$C_{10}H_{22}$	158
	Endecane	$C_{11}H_{21}$	182
	Dodecane	$C_{12}H_{26}$	198
D	Tridecane	$C_{13}H_{28}$	216
	Tetradecane	$C_{14}H_{30}$	238
	Pentadecane	$C_{15}H_{32}$	258
	Hexadecane	$C_{16}H_{34}$	280
	Octadecane	$C_{18}H_{38}$	?
	?	?	

SOLID

Paraffin, $C_{27}H_{56}$, melting about 60°
,, $C_{30}H_{62}$,, ,, 66°
etc. etc.

The first and most important use alike of the crude natural petroleum and of portions separated by distillation is for burning as a source of heat or power.

The fluid residue from the refineries has for many years been employed under the name *ostatki*, as fuel on the locomotives in Southern Russia, and on the steamers on the Black Sea and Caspian. The use of liquid fuel has many obvious advantages coupled with some disadvantages. First of all it is composed almost entirely of the combustible elements carbon and hydrogen, while bituminous coal contains not only oxygen, but a considerable quantity of mineral matter which is left in the form of ash or clinker after burning.

A further advantage possessed by a liquid is that it occupies much less space than an equal weight of a solid in lumps, and may be stowed in bunkers of any shape or in spaces which could not be utilised for coal. The bunkers can also be filled by simple pumping with the expenditure of much less labour, the flame produced by burning is instantly available, and with a properly constructed furnace no smoke need be produced. There are also no ashes to be cleared out from the furnace and removed by the stoker.

In order to burn a thick liquid like natural petroleum it is necessary to distribute it in the form of spray, and a great deal of ingenuity has been expended in contriving burners for this purpose. In many forms of "atomiser" the oil is sprayed by means of a jet of steam into a chamber where it finds the air requisite for combustion, and by regulation of the oil supply a blue smokeless or a smoky flame may be produced at pleasure. In other forms of jet the oil is pulverised by being forced under pressure through a small orifice of peculiar construction.

The number of distinct substances obtained from petroleum is very large, but reference to the table (pp. 292–293) will simplify the classification and enumeration of these products.

The gases associated with petroleum in nature consist chiefly of methane or marsh-gas mixed with small quantities of other hydrocarbons, together with a little nitrogen and traces of hydrogen, carbon dioxide, and sometimes sulphuretted hydrogen. This wonderful natural supply of gaseous fuel was for many years allowed to escape in the neighbourhood of the American oil wells, but this deplorable waste was ultimately put an end to, and for the last thirty years or more natural gas has been applied, not only to the lighting of towns, but in the manufacture of steel and glass, and generally for the production of heat on a large scale.

From petroleum a number of liquid products are obtained which are indicated in the table by the letters A, B, C, and D. Every one of these is, however, a mixture of several compounds, olefines as well as paraffins, and may for special purposes be further subdivided by distillation.

A includes a small quantity of very volatile liquid which can only be retained in the liquid state by pressure or cold. This is sometimes used in surgery as a local anæsthetic for cooling the surface to which it is applied. The greater part of A is used under the names *gasoline*, *light benzine*, or *benzoline* for producing "air-gas" employed in lighting country houses.

B is benzine, used extensively as *motor spirit*, also for cleaning purposes and as a solvent.

FIG. 97.—LOUISIANA. GENERAL VIEW OF STILLS.

FIG. 98.—LOUISIANA. STILLS AND FURNACES.

To face page 294.

C represents a heavier benzine which often goes under the indefinite name *naphtha*.

D is lamp oil, known in different countries by various names, such as *kerosene*, *mineral colza*, *paraffin oil*, etc.

Paraffin wax is a familiar solid used for making candles and for other purposes. Mixtures of solid paraffin with some of the liquid members of the series in different proportions constitute machine oil used for lubrication, and vaseline a well-known semi-solid or jelly-like substance used also for protecting metals from rust and corrosion, and as an ointment or dressing in surgery.

It is unnecessary to dwell on the purposes to which all these and other special fractions of petroleum are applied. Motor spirit alone would occupy many pages in the discussion of the important questions involved in the various suggestions which have been made as to possible substitutes. Of these the only liquid which, in the event of deficiency, appears likely to be available is alcohol.

ORIGIN OF PETROLEUM

Everyone is familiar with the idea that common coal, in all its different varieties, is a product of chemical change, taking place through long geological periods, in masses of vegetable matter accumulated in certain strata of the mineral matters forming the crust of the earth. Concerning the origin of petroleum there is, however, no such unanimity of opinion. Many speculations have been put forward from time to time, and without entering into much detail, these may be at once ranged under two main divisions.

The question is did natural oil result from purely chemical processes taking place in the earth, or did it result from the action of heat on the remains of organic beings, animal or vegetable ?

That rock oil may have been formed without the previous existence of living things is the view which was promoted chiefly by the famous Russian chemist Mendeléeff, and it has much to recommend it. Thus some deposits of oil are found in the most ancient Silurian rocks, where it is difficult to suppose a previous sufficient accumulation of organic matter.

The formation of hydrocarbons can be accounted for by supposing that the interior of the earth consists largely of compounds of carbon with such heavy metals as iron and manganese in a heated state. During the upheaval of mountain chains, and in times of superficial disturbance owing to contraction, water may penetrate through the crust of the earth and thus come into contact with these carbides, the hydrogen of the water uniting with the carbon, while the oxygen of the water forms oxides with the metals. From direct experiment on cast iron and on *Spiegeleisen*, both of which contain iron and manganese united with carbon, it has long been known that this change is produced by contact of the metal with water or acids. This hypothesis harmonises also with what is known of the constitution of our earth, which is probably a metallic mass, having a thin earthy crust on the surface, like so many of the meteorites which fall from the skies. The mean density of the earth as a whole is about 5·5, that is to say it would weigh 5·5 times as much as an equal bulk of water. But the crust of the earth accessible to us consists of minerals which are generally not more than about twice as heavy as water, and therefore the interior must contain a large quantity of something much denser.

An alternative theory has been proposed more recently, about 1904, by Professor Paul Sabatier of Toulouse, whose discovery of the remarkable catalytic actions of certain metals has been referred to elsewhere (p. 202). He assumes that in the depths of the earth are found deposits of the alkaline metals, sodium, potassium, etc., as well as compounds of these metals with carbon. This assumption is less novel than might appear, if it were not remembered that Davy, a century earlier, had supposed that deposits of this kind might occur in the earth's crust, and had attributed volcanic explosions in certain cases to the entrance of water, with which as is well known they react violently.

Sabatier imagines that such deposits of metal may be the cause of the production of hydrogen, while their carbides in contact with water simultaneously produce acetylene. These two gases in variable proportions are then assumed to come into contact with one or other of the metals, nickel, cobalt, or iron, in a finely divided state and at an elevated temperature, the result being that by condensation of the acetylene alone or by union with hydrogen under the influence of the heavy metal a mixture of hydrocarbons results.

These may consist of (1) saturated paraffins as in the Pennsylvanian petroleum, or (2) cyclic paraffins such as occur in the Caucasian oil, or (3) a mixture containing also benzenoid hydrocarbons and unsaturated compounds, such as are found in the petroleum of Galicia and other districts. Sabatier does not pretend that other theories of the formation of these natural hydrocarbons are necessarily excluded, but he claims that no other theory serves to account for the diversity observed in the composition of petroleum from different regions, and especially for the production of the naphthenes.

On the other hand, the suggestion is offered that large masses of vegetable or animal matter may have accumulated in past geological times, that they may have undergone decay, and that subsequently a process corresponding to that which has given rise to coal gradually set in. Later a rise of temperature sufficient to cause complete chemical decomposition must be assumed, and this would cause the oil, and gas which would be formed at the same time, to pass from lower to higher strata, and still be retained under the pressure of superincumbent rock.

Such a view is consistent with the fact that oils closely similar to natural oil are obtained by the action of heat on coal and shales as in the manufacture of paraffin oil from these materials.

It also would explain the existence of petroleum in so many cases in strata in which no animal or vegetable remains are found, and in fact the general distribution of petroleum in the crust of the earth without apparent connection with the geological age or character of the rocks in which it is found.

That petroleum may have originated in beds of animal or vegetable matter is rendered probable by the fact that certain varieties have the power of rotating a ray of polarised light, while nearly all contain an appreciable quantity of nitrogen in the form of nitrogenous bases similar to those which are derived from the distillation of coal or animal matter. A small quantity of sulphur in the form of an organic sulphide is also frequently present.

It is quite possible, in view of the variation in the composition and character of the oil found in different regions, that petroleum may have been produced from both mineral and organic sources.

There is, however, something fascinating in the idea that this valuable natural product may be actually in process of formation now and always from the body of the earth itself without the intervention of living beings.

There are, however, outstanding difficulties in respect to all these theories, and it is probable that it will be some time before the formation of the higher terms of the paraffin series (paraffin wax, ozokerite, etc.), and the differences of constitution between the American and Russian oils are fully explained.

CHAPTER XX

COAL-TAR

It seems a far cry from the black, sticky, and stinking liquid coal-tar to the brilliant or delicate colours adorning a lady's dress, to the perfume of the Tonquin bean, or the little white tabloid of aspirin or phenacetin which affords relief from pain. But the black tar is the chemical ancestor of these and many other valuable products unknown to our grandfathers, and the business of extracting from it the intermediate substances which are more directly the parents of the dyes and other things is a matter of great national importance.

When coal is heated strongly in a closed vessel, so that it does not burn, it yields four chief products, viz. :

Gas.
Watery ammoniacal liquor.
Tar.
Coke.

At the gas works the first of these is the primary object of the manufacture, while the liquor and tar are spoken of as residuals, together with the coke which is left behind in the retorts. But in connection, specially, with the production of iron and steel the purpose in view in heating coal is not so much the production of gas as of the residual coke. The process is carried out in coke "ovens," and down to comparatively recent times the coke alone was preserved, all the gas and other volatile products given off were burned to waste. This, however, is no longer the case, and since the more general recognition of the value of the by-products many of the coke ovens of the present day are constructed so as to provide for collecting and so utilising what was formerly wasted.

The accompanying diagram shows in section the construction of what is known as a beehive oven. A number of these are always placed side by side with the object of economising heat and convenience of charging. Each oven is a brick structure lined with firebrick. There is an opening below by which the coal is charged into the oven and by which the coke when produced is withdrawn. The oven is filled up to the shoulder, giving a layer of coal about five feet thick, and the combustion is started at the top and proceeds downwards till the whole is deprived of volatile matter and the luminosity of the flame at the top practically ceases. During the burning the admission of air through the door is regulated by a damper or a movable brick. When the operation is over the coke is withdrawn and quenched with water. The oven while still hot receives a fresh charge of coal. This process yields a quantity of coke amounting to about

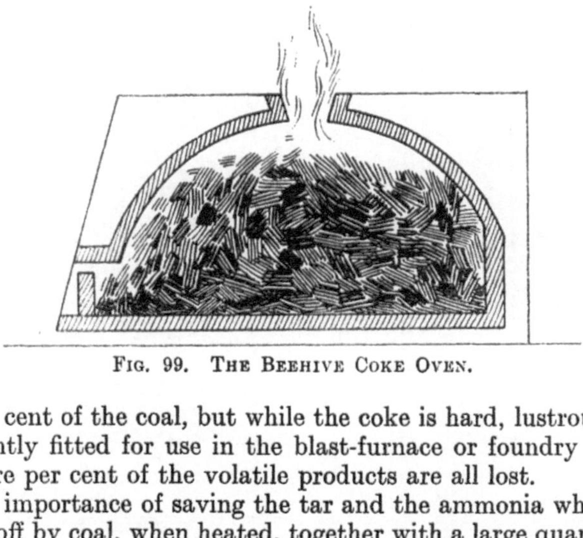

FIG. 99. THE BEEHIVE COKE OVEN.

60 per cent of the coal, but while the coke is hard, lustrous, and eminently fitted for use in the blast-furnace or foundry the 30 or more per cent of the volatile products are all lost.

The importance of saving the tar and the ammonia which are given off by coal, when heated, together with a large quantity of inflammable gas has led to an immense amount of invention, and an almost innumerable variety of ovens have been devised with the object of recovering and utilising these by-products. One of the first contrivances introduced was the coke oven known as Jameson's. This was shaped like a beehive, but the floor sloping forward was perforated with holes which opened into pipes underneath. These communicated with a large horizontal main which ran along the front of a series of ovens and in which tar and ammoniacal water collected. The gas was drawn forward by a pump and delivered into a gas-holder. The more modern ovens generally take the form of some modification of the Coppée oven, in which the principle is quite different. Instead of generating the requisite heat by the combustion of a portion of the coal the charge in these ovens is never brought into contact with air, but is submitted to a process of distillation by heating in closed chambers. The heat required is obtained by

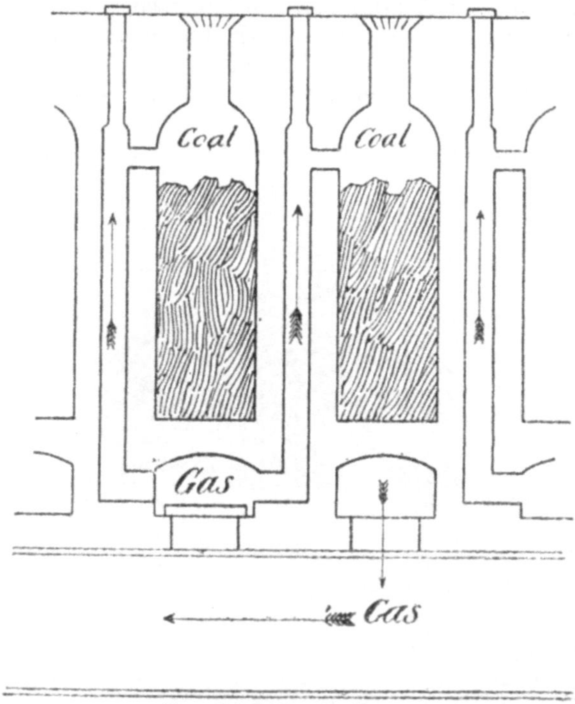

FIG. 100. COPPÉE OVEN.

combustion of the gas given off by the coal within, and further economy is secured by making use of the regenerative principle, that is to say the gas before being burnt and the air required to burn it are both heated by passage through ducts arranged between the retorts or ovens. The principle of the construction of such ovens will be understood by referring to the adjoining diagram. Here is seen a section across two of the ovens which are chambers some 25 feet long, having a door at each end by which the finished coke can be withdrawn. The coal is charged

FIG. 101.—CARBURETTED WATER GAS PLANT. GAS LIGHT AND COKE COMPANY.

To face page 300.

into the ovens from wagons which run on rails over the set of ovens. The gas and other volatile products given off from the coal are conveyed away through pipes to a series of collecting wells and purifiers and a gas-holder. The gas, having then deposited nearly all the tar and ammoniacal liquor, is brought back to a series of burners where, with admixed air, it is burnt in the spaces between the ovens, the walls of which are thus raised to a bright red heat. The products of combustion, chiefly carbon dioxide and water vapour, are then carried into an underground flue, as shown by the arrow, and so to the chimney-stack. By means of apparatus of this kind great economy is secured. There is a yield of about 70 per cent of coke equal in quality to the ordinary coke, together with all the tar and ammonia which the variety of coal used is capable of yielding.

In some districts where the manufactures can be combined together the surplus gas from the coke ovens, of a quality which is up to the present-day standard of luminosity, is delivered into the ordinary coal-gas mains.

For the purpose of generating combustible gas a variety of systems have been adopted by which air in limited quantity is drawn or driven through a mass of red-hot coal contained in a firebrick chamber called a "gas-producer." A gas of this kind contains about one-third of its volume of carbonic oxide mixed with nitrogen from the air and a little carbon dioxide, etc. The ammonia is partly destroyed, but of late years a method introduced by the late Dr. Ludwig Mond has been very extensively adopted whereby a cheap coal slack can be used, at the same time that nearly the whole of the nitrogen of the coal can be recovered in the form of ammonia. To secure this result the temperature must not be allowed to rise too high, and the yield of ammonia is augmented by driving into the coal a considerable quantity of steam. A portion of this is decomposed with formation of hydrogen gas and carbonic oxide, but the chief effect is to keep down the temperature. The increased collection of ammonia is attended by a considerable reduction in the value of the accompanying tar, which is quite different in character from the tar obtained at the gas works and from the coke ovens. The latter is produced at a much higher temperature, and its characteristic ingredients are hydrocarbons, which are related more or less closely to benzene (benzol) and its homologues. The tars produced at lower temperatures contain very little of these valuable compounds, their chief constituents resembling the oils obtained by moderate distillation of shales and consist chiefly of paraffins.

The Mond gas system has been adopted on a large scale in Staffordshire, a company having been formed a few years ago which generates the gas and supplies it through a system of mains to the surrounding district. Here it finds extensive employment in the gas engines which now assume such enormous dimensions, and in metallurgical furnaces as well as for raising steam and other purposes. A view is shown facing page 29 of a small Mond gas plant installed for experimental purposes at the Imperial College at South Kensington.

The tar obtained from gas works is black, heavier than water, and the amount produced per ton of coal varies considerably according to the character of the coal. When the distillation of the coal is so conducted as to yield 10,000 to 11,000 cubic feet of gas, which is the usual practice, the tar produced amounts to 10 to 12 gallons of specific gravity about 1·2. Roughly speaking then coal yields at a red heat about 10 per cent of its weight of tar.

From whatever source it is obtained, the distillation of coal-tar forms a separate industry, some of the features of which, on account of its great practical and scientific importance, it is desirable to survey.

Before proceeding further, however, it will be interesting to glance at the following tables, which will give a rough idea of the enormous quantities of tar available and its connection with other valuable products.

The tendency is to save more and more of the by-products obtained by heating coal, and as the demand for the innumerable dyes, drugs, antiseptics, and other chemical compounds increases, so will the utilisation of the hydrocarbons from coal-tar continue to extend. At the time of writing these lines the chief nations of Europe are engaged in a war upon which it is useless to waste epithets. But on both sides there is an unlimited demand for what are called high explosives, such as picric acid and trinitrotoluene, which are obtained by the chemical treatment of certain distillation products of coal.

It is obvious, therefore, that there is every inducement to practise economy in the treatment of coal whether considered from the domestic or industrial point of view. The national importance of this economy has not as yet attracted the serious attention which it deserves, at any rate so far as this country is concerned. At the same time figures tend to show that the question has not altogether escaped the notice of manufacturers, and the total abolition of wasteful methods will be accomplished, it may be hoped, within a few years.

In an address to the Society of Chemical Industry in 1900, Dr. Beilby supplied the following figures, pointing out that if, at that time, beehive ovens had been entirely replaced by recovery ovens in the manufacture of blast-furnace coke the yield of tar, ammonia, and gas from this source would have been increased more than tenfold, since the proportion of beehives to recovery ovens was about 10 to 1.

Products of Distillation of Coal in United Kingdom, 1899.

Source.	Millions tons Coal distilled.	Millions tons Coke produced.	Tons Tar.	Tons Ammonia Sulphate.	Millions cubic feet of Gas.
Gas Works .	13	7–8	650,000	130,000	130,000
Blast Furnaces	2	—	150,000	18,000	360,000
Coke Ovens .	$1\frac{1}{4}$	9/10	62,000	11,000	12,500
Totals . .	—	—	862,000	159,000	—

A more recent estimate has been given by Professor Bone in his address to the Chemical Section of the British Association at Manchester in September, 1915. From the following passage in this address we learn that the beehives form only about 30 per cent of the coke ovens in the United Kingdom at the present time.

"Of the 189 million tons of coal consumed in the United Kingdom in the year 1913, about 40 million tons, or (say) approximately one-fifth of the whole, were carbonised either in gas works, primarily for the manufacture of towns' gas, or in coke ovens for the manufacture of metallurgical coke—in practically equal proportions. Two-thirds of the latter was carbonised in by-product recovery plants; the remainder in the old wasteful beehive ovens. So that, roughly speaking, we have—

Total Coal Carbonised = 40 Million Tons.

In Gas Works.		In By-Product Coke Ovens.		In Beehive Coke Ovens.
20	...	13·5	...	6·5

"At present there are 8297 by-product coke ovens built in this country, of which 6678 are fitted with benzol recovery arrangements, capable of producing something like 10 million tons of coke per annum.

"The yields of the various by-products obtainable on such coke oven installations naturally vary with the locality and character of the coal seam; but they probably average somewhat as follows—expressed as percentages on dry coal carbonised:—

District.	Ammonium Sulphate.	Tar.	Benzol and Toluol as Finished Products.
Durham	0·9 to 1·45	2·5 to 4·5	0·6 to 1·0
Yorkshire	1·3 to 1·5	3·5 to 5·0	0·9 to 1·1
Derbyshire	1·3 to 1·6	3·5 to 5·0	0·9 to 1·1
Scotland	1·4 to 1·6	3·5 to 5·0	0·9 to 1·1
South Wales	0·9 to 1·1	2·0 to 3·5	0·6 to 0·75

Or, to put the matter a little differently, each ton of dry coal carbonized yields from 20 to 35 lbs. of ammonium sulphate, from 56 to 112 lbs. of tar, and from 2 to 3½ gallons of crude benzol, etc., according to the locality. About 65 to 70 per cent of the crude benzol is obtained as finished products—benzene, toluene, solvent and heavy naphthas.

"How rapid has been the development of the by-product coking industry in this country during recent years may be judged from the following official returns of the quantities of ammonium sulphate annually made by such plants, as compared with the corresponding quantities produced in gas works.

Tons of Ammonium Sulphate Produced in

Year.	By-Product Coke-Oven Plants.	Gas Works.
1903	17,435	149,489
1908	64,227	165,218
1913	133,816	182,180

"In the natural course of events the final disappearance of the wasteful beehive coking oven from this country is now only a matter of a few years; but I venture to suggest that public interest would justify the Government fixing, by law, a reasonable time-limit beyond which no beehive coke oven would be allowed to remain in operation, except by express sanction of the State, and then only on special circumstances being proved."

In a work of this kind the subject cannot be pursued further, but it must be obvious to the most superficial reader that reform in the use of our national coal supplies is urgently necessary. It must be remembered also that though industrial waste is more readily open to supervision and control, the domestic hearth is responsible for a large part of the useless consumption which goes on.

Now resuming the subject of this chapter, which is Coal-Tar, we may notice in passing the estimate which was drawn up in the year 1901 of the amount produced by all the countries of the world. From the nature of things this can only be regarded as approximate, and from what has been said the total amount is probably now much greater.

	Tons.
United Kingdom	908,000
Germany	590,000
United States (including water-gas tar)	272,400
France	190,680
Belgium, Holland, Sweden, Norway and Denmark	272,400
Austria, Russia, Spain and other European countries	199,760
All other countries	227,000
Total	2,660,240

Coal-tar is distilled in large iron stills set in brickwork and heated by a fire below. Each still is a sort of boiler which holds from 20 to 30 tons. It has a slightly domed head with a wide pipe by which the vapour is carried into a spiral continuation of the same called a condenser and in which it is liquefied. The condensed product runs into different receivers. First come some ammoniacal liquor and the light oils which float on water. These are followed, as the temperature rises, by heavy oils which sink in water. The residue left in the still is pitch, which while still hot is run out into large iron vessels where it cools and gradually becomes solid. Ordinary gas works tar is an exceedingly complex mixture of upwards of two hundred different chemical compounds. These may be divided into hydrocarbons, of which the most important are benzene, toluene, xylene, naphthalene, anthracene, etc., and compounds containing oxygen, of which the most important are phenol, commonly called carbolic acid, and other substances of that class. A small proportion of basic substances containing nitrogen, and others containing sulphur are also present, but are of minor importance and at this point may be neglected.

On distillation 1 ton of average gas works tar will produce approximately :—

Ammoniacal Liquor	5 gallons
Crude Naphtha	5·6 ,,
Light Oils	26·0 ,,
Creosote Oils	17·0 ,,
Anthracene Oils	38·0 ,,
Pitch	12 cwt.

The general arrangement of the stills is shown in the accompanying photograph taken in the works of Messrs. Major and Co. of Hull, Fig 102. The importance of coal-tar and of economy in its manipulation has led to invention of many modifications in the form of the stills and the process of distillation. One of the most important of these is the continuous process introduced by Messrs. Hird Chambers and Hammond of Huddersfield.

The principle here applied is very simple, although to the non-expert reader the details may look complicated. If crude tar is run into a heated still it gives off first its most volatile constituents, which may be passed into a condenser and collected. The residue in the still may be allowed to flow into a second still alongside of the first, and being heated to a higher temperature the portion distilling away consists of liquids having a higher boiling point. Similarly a third still may be arranged so as to receive the residue from the second as it parts with its vaporisable constituents. A series of three stills is found to be sufficient, and the final residue runs into the pitch-cooler. The process is made automatic by placing the crude tar tank at a higher level than the stills so that it flows by gravitation, and on the way it is warmed by making it pass through the vessels in which are coiled pipes conveying the hot vapours from the stills to the condensers. All this will become clear by consulting the diagram

Fig. 102.—TAR STILLS.

To face page 306.

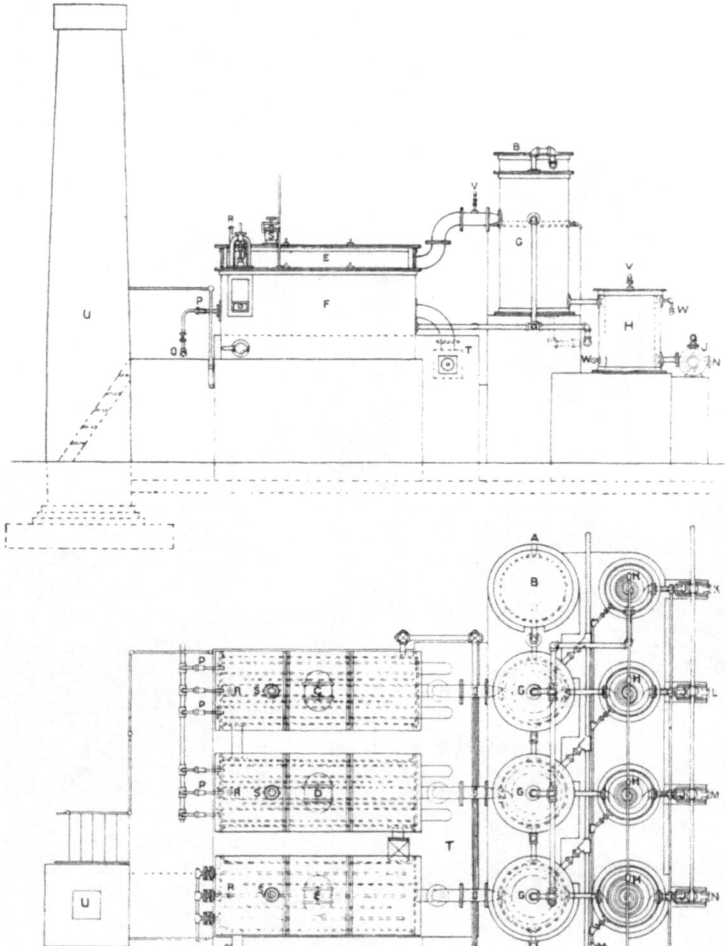

Fig. 103.—CONTINUOUS TAR DISTILLATION PLANT.

To face page 307

Fig. 103 which shows an end view and a plan of the stills, heaters, condensers, etc. The plant shown has a capacity of 10 tons per day. It is only necessary to add that the gas required for heating is derived from a gas producer not shown in the plan, or from coke ovens.

REFERENCES TO THE DIAGRAM

A. Crude Tar Inlet.
B. Tar Regulating Tank.
C. No. 1 Still.
D. No. 2 Still.
E. No. 3 Still.
F. Pitch-Cooler.
G. Heater-Coolers.
H. Water Coil Condensers.
J. Sight Boxes.
K. Water and Naphtha Outlet.
L. Light Oil Outlet.
M. Creosote Oil Outlet.
N. Anthracene Oil Outlet.
O. Pitch Outlet.
P. Bunsen Burners.
Q. Gas Main.
R. Thermometers.
S. Safety Valves.
T. Flue.
U. Chimney.
V. Steam Pipes.
W. Water Pipes.

The picture Fig. 104 provides a view of this continuous plant in operation.

We may now proceed to consider the next step, which consists in dealing with the several products of the distillation in the tar stills. The ammoniacal liquor goes to a different department where it yields sulphate of ammonia. The crude naphtha and light oils are usually redistilled for the separation of benzene[1] and its homologues. But as it contains small quantities of basic substances, including aniline, it is first shaken up with sulphuric acid which removes these compounds, and subse-

[1] These hydrocarbons are called commercially benzol, toluol, etc., but the pure hydrocarbons are in chemical language distinguished as benz*ene*, tolu*ene*, etc.

quently with caustic soda which separates any carbolic acid and the residual sulphuric acid. After washing with water the hydrocarbons boiling between 80° and 150° C. are transferred to the naphtha still. The accompanying photograph, Fig. 105, taken in the works of Messrs. Major of Hull, represents one of these stills with its rectifying column. The scaffolding shows that the whole plant is about to be surrounded by a brick building.

Fig. 106 shows the arrangement of the benzene rectifying plant, of which a view is given in the preceding picture.

The capacity of the boiler is about 5300 gallons, and it is heated by steam. Its purpose is to separate from impure benzol or naphtha the pure hydrocarbon benzene. In order to do this it is necessary to provide for the condensation, from the vapour which passes up the fractionating column, of those constituents toluene (b.p. 110° C.), the xylenes (b.p. 137° to 142° C.), and other hydrocarbons present in the crude benzol which boil at temperatures above the boiling point of benzene (b.p. 80·5° C.). With this object the column itself is divided by horizontal plates, in each of which are placed a number of valves opening upwards, and a tube open at top and bottom and projecting an inch or so above the plate. The latter is thus kept covered with a layer of liquid through which the vapour bubbles in its passage upward. A further effect is produced by the tubes at the top of the column, and the vapour which leaves it then passes into the cooling arrangement furnished with thermometers by which the temperature of the passing vapour is observed. From the cooler any remaining vapour other than that of pure benzene is condensed and returned to the fractionating column by means of the wrought-iron pipes shown in the figure. The vapour of the pure benzene then passes into the coil surrounded by cold water and is thus delivered in the form of liquid into the receiver. Benzene of any lower degree of purity can be obtained by simply cutting off the connection between the cooler and the pipes leading back to the column.

With regard to the remaining fractions of the tar distillates it will be sufficient to follow the course of three of them.

Naphthalene is a white crystalline solid which melts at 79° C. and boils at 218° C. It is contained in the light oils and in the creosote oils. When the latter are stirred up with caustic soda the carbolic and cresylic acids (phenol and cresol) present are dissolved out and on standing the watery liquid separates from

Fig. 104.—HIRD'S CONTINUOUS TAR DISTILLATION PLANT.

To face page 308.

Fig. 105.—NAPHTHA STILL WITH RECTIFYING COLUMN.

To face page 309.

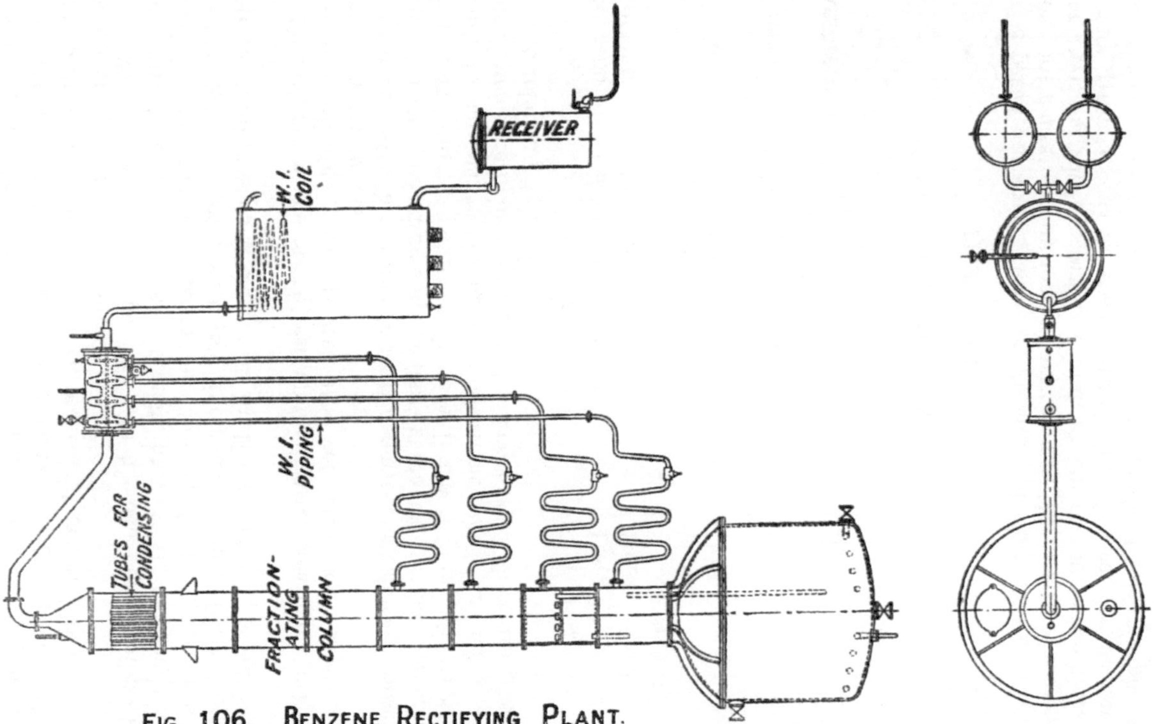

FIG. 106. BENZENE RECTIFYING PLANT.

the undissolved oil which contains much naphthalene. The tar acids are separated from the caustic soda solution by heating the liquid in the presence of waste carbonic acid gas, and by further treatment pure *phenol* is obtained.

Anthracene is a crystalline solid which when pure exhibits a beautiful blue fluorescence. It melts at 213° C., and boils at 351° C. Anthracene separates on standing from the crude oil in the form of a greenish crystalline solid which is filtered off, pressed, and washed free from oil by the use of a little solvent naphtha.

Creosote oil, which is the residuum left after treatment in the manner indicated above, is used on a very large scale as a preservative for timber.

Crude coal-tar finds some applications without being separated by distillation into its components. These are familiar enough. But as the starting point for chemical industry it may be worth while to summarise the chief applications which are made of the chemical compounds isolated from coal-tar. These are given in the following table :—

	Crude.	Pure.
Benzene .	Motor fuel, Solvent.	Dyes.
Toluene .	Solvent.	Dyes. Explosive, T.N.T.
Xylenes .	Solvent Naphtha.	Dyes.
Naphthalene	Insecticide and Mild Antiseptic.	Dyes.
Phenol .	Strong Antiseptic and Disinfectant.	Dyes. Explosive.
Anthracene .		Dyes.

CHAPTER XXI

PRODUCTION OF DYES

It is perhaps scarcely necessary to state the fact that coal-tar does not contain ready-formed anything in the nature of a dye, nor anything which possesses colour, except of course the black highly carbonaceous substances which are left after distillation. But that such an idea still lurks in the minds of some people is indicated by the legend formerly current, that Perkin, the discoverer of mauve, the first of these artificial colouring matters, was attracted to the subject by noticing the beautiful tints displayed by a film of tar floating on water. Such colours are, of course, due to the thinness of the film, and have nothing whatever to do with its composition.

In order to render intelligible the chemical changes which are involved in the successive stages of the production of the numerous synthetic dyes, drugs, perfumes, and other so-called organic compounds, a very brief statement of the meaning of chemical symbols is necessary. How they are arrived at, and the full extent of their meaning, are questions which are beyond the scope of this work, but can be answered by reference to any one of the best textbooks of chemistry.

A chemical symbol such as C or H is usually the initial letter of the name of an element, carbon or hydrogen for example. It is used to signify one *atom* of the element, and when two such symbols stand side by side, the elements are represented in combination, thus H_2O means that two atoms of the element hydrogen are combined with one atom of the element oxygen, forming one *molecule* of the compound, water. The symbols are also used for the purpose of displaying what is believed to be the order in which the elements in a compound are joined together. Thus the formula for water, H_2O, may be written H.O.H or H-O-H, by which expressions the idea is conveyed that the atoms united together in a molecule of water are not jumbled together irregularly, but that each atom of hydrogen is attached to the oxygen and is unconnected, at least directly, with the other atom of hydrogen. Similarly any number of atoms of carbon may be joined together, and at the same time combined with hydrogen or other elements as in the formulæ for

Alcohol $CH_3.CH_2.OH$ and
Acetic acid $CH_3.CO.OH$.

Such expressions may be expanded if necessary so as to show the manner in which each atom is joined to the others as in the case of acetic acid :—

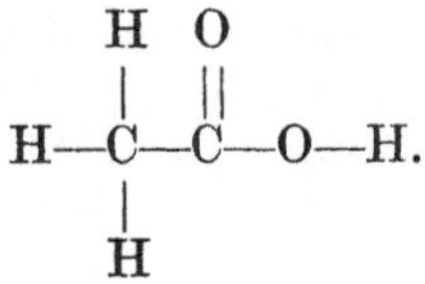

Further there are reasons for believing that a number of carbon atoms joined together do not really range themselves in straight rows as it is convenient to write them on paper, for instance,

$$CH_3.CH_2.CH_2.CH_2.CH_2.CH_3,$$

but that in such a case the chain occupies in space the form of a spiral. Hence there is a tendency to the formation of closed chains or rings as in the case of benzene, C_6H_6, which is represented by the formula below :—

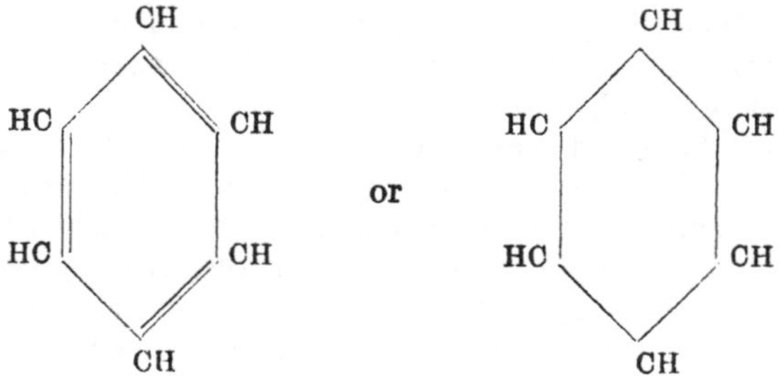

In this case the expression is required so frequently in writing the formula for organic compounds that it is customary to reduce it to the skeleton without symbols thus :—

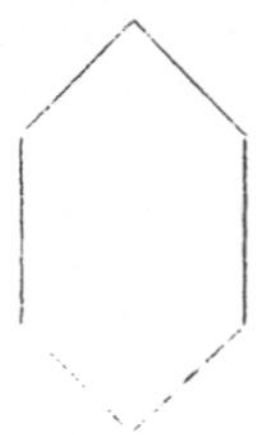

which is generally understood without writing the symbols for the several atoms.

The first synthetic dye was produced accidentally by W. H. Perkin in 1856 when engaged in a research having a different object in view. The product was called *mauve* from the colour which resembles that of the flower of the mallow (French = mauve), and although it has long ceased to be manufactured as a dye for silk and wool, owing to its tendency to fade in sunlight, it has been used in more recent times for colouring postage stamps.

To face page 312.

Mauve was produced by adding potassium dichromate to a solution of aniline in dilute sulphuric acid, whereby a black precipitate was formed, which after removal of impurities by boiling it in coal-tar naphtha, was dissolved in alcohol in which it forms a rich purple-coloured solution.

Dichromate of potassium is an oxidising agent, and naturally other oxidising agents were tried as soon as the process became known. This led to the discovery of the red dye magenta. Strangely enough pure aniline acted upon by these oxidising agents does not yield a dye, but fortunately the benzene of former days was very far from pure, and contained a varying quantity of toluene, the hydrocarbon which stands next to benzene in the series and possesses very similar properties. These hydrocarbons, which are extracted by distillation from coal-tar, form the starting point for the production of this class of colours. The successive steps in the production of magenta are the following.

Benzene acted upon by strong nitric acid is converted into nitrobenzene, $C_6H_5.NO_2$. Toluene, C_7H_8, is similarly converted into nitrotoluene, $C_7H_7.NO_2$, or rather into a mixture of two compounds having the same ultimate composition, though differing in the arrangement of the constituent atoms.

When nitrobenzene is brought into contact with a substance or mixture of substances, such as iron and acetic acid, which is capable of supplying hydrogen, the two atoms of oxygen are removed and two atoms of hydrogen are introduced in place of them, and aniline, $C_6H_5NH_2$, is the result. The nitrotoluenes, under the same circumstances, are affected in a similar manner, and give rise to bases called toluidines, $C_7H_7NH_2$.

Now as the commercial benzene fifty years ago consisted of a mixture of benzene and toluene, the aniline which resulted from using it as the parent material always consisted of a mixture of aniline and ortho- and para- toluidines. Such a mixture acted on by an oxidising agent, yields a deep red mass which contains magenta dye.

The apparatus employed in these operations is simple in principle. The nitration of benzene or toluene is effected by bringing the hydrocarbon into contact with the requisite quantity of a mixture of strong sulphuric and nitric acids, the former being employed to combine with the water generated in the reaction.

The mixture is made in a cast-iron pot, cooled externally by water, and provided with a paddle. A number of these vessels stand side by side, and the contents are stirred by power. After the action of the acid is completed the nitrobenzene or nitrotoluene floating on the acid is separated, and is washed free from acid by mixing it with water in a similar iron or glazed stoneware vessel. It is then separated from the water in which it sinks.

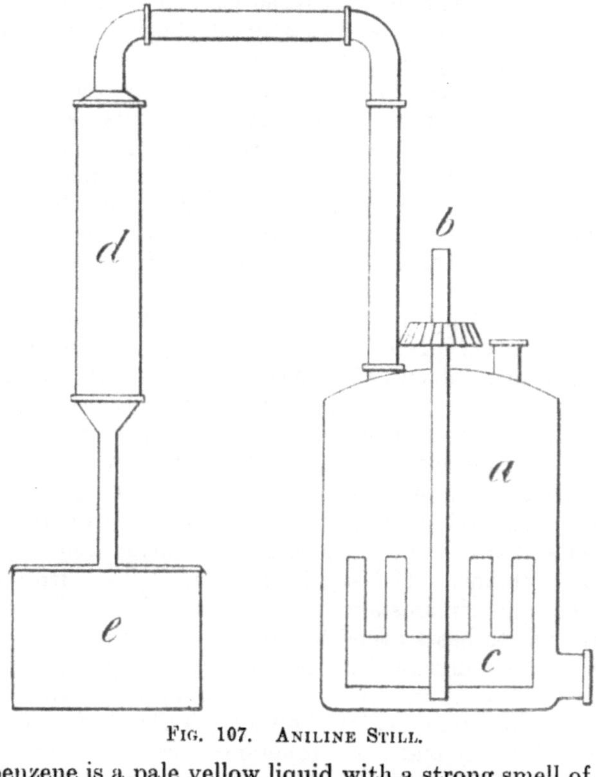

FIG. 107. ANILINE STILL.

Nitrobenzene is a pale yellow liquid with a strong smell of bitter almond oil, and is to some extent used in perfumery under the name of "Oil of Mirbane."

The conversion of nitrobenzene into aniline is effected in a series of vessels, the arrangement of which is shown diagrammatically in the figure.

a is a cast-iron pot provided with a stirrer *c*, the axis of which *b* is a pipe by which steam may be introduced when required.

Nitrobenzene and a little hydrochloric acid are placed in the pot, and iron borings are added to the mixture, which is kept warm by steam and stirred by the paddle till the reaction commences. Once started the chemical action, which is attended by the evolution of heat, proceeds till the reduction of the nitrobenzene is complete. At this stage any aniline evaporating from the mass condenses in the upright pipe and returns to the reduction vessel. Lime is then added to neutralise the acid, and the aniline distilled off through the condenser *d* into the receptacle *e*.

Aniline is a colourless liquid which boils at 182–3° C., and becomes brown by exposure to air. It sinks in water but slowly, as its specific gravity is only 1·024 at 16°. It is slightly soluble in cold water.

The toluidines are produced in a similar way from toluene, and, as already explained, the red magenta is formed by oxidising a mixture of these bases. Several oxidising agents have been and are still used. Of these arsenic acid is one of the most convenient, but the use of this reagent involves the careful removal of arsenic from the finished colouring matter. The presence of even a small quantity of arsenate or arsenite in the dye applied to fabrics which are used for any kind of clothing which comes into direct contact with the skin, leads to irritation of the surface, and even to symptoms of arsenical poisoning.

To make magenta with the aid of arsenic acid a quantity of this acid, in the form of a strong syrupy solution, is placed in a cast-iron pot provided with a stirrer, and a mixture in due proportions of aniline with the two isomeric toluidines is added. The mixture is heated first to a temperature just above the boiling-point of water, and subsequently to 180°–190° C. for several hours. Water and a considerable quantity of oil, consisting of the three unaltered bases, evaporate off and are collected by passing through a proper condenser. The residue is a dark pitch-like mass which is fluid while hot, but after being run off from the pot solidifies into a brittle solid. This contains several colouring matters beside the red, and to obtain the latter free from arsenic the mass is ground to powder, extracted with water heated under pressure, and the solution, after being filtered, is mixed with hydrochloric acid and sufficient common salt to replace the arsenic. On cooling the solution the magenta crystallises out leaving the other colouring matters in the liquid together with the arsenic, partly in the form of arsenic acid, but chiefly as arsenious acid.

Magenta is a compound in which a base, called *rosaniline*, is united with an acid. The product of the operations just described is rosaniline hydrochloride, but in fact the rosaniline produced consists of a mixture of two compounds, one of which, $C_{19}H_{17}N_3HCl$, is called *para* rosaniline, while the rosaniline is a homologous compound containing $C_{20}H_{19}N_3HCl$. These are both red colouring matters which dye silk and wool direct, and both are present together in the common dye.

Another process for making aniline red or magenta avoids the use of arsenic and yields a somewhat larger quantity of the colouring matter. In this method the aniline and toluidine are first converted into their solid salts by dissolving in hydrochloric acid and drying. To this mixture is then added a further quantity of the aniline oil together with nitrobenzene which is the source of the oxygen required, and the whole is heated till the temperature reaches about 190° C., when a small amount of iron borings are added. The action is complicated. Other processes have been introduced which, however, it is not necessary to pursue further.

Strangely enough the rosanilines alone are quite colourless compounds and only form dyes when united with an acid, in the proportions indicated in the formulæ already given. Both these bases are capable of combining with three molecules of hydrochloric acid, altogether forming yellowish brown crystallisable compounds, but these again are not colouring matters.

Magenta is easily soluble in water, forming a magnificent red liquid from which large crystals, having the appearance of green beetle wings, may be obtained by slow deposition.

The use and appearance of substances of this kind have become so familiar that few people of the present generation can imagine the excitement and interest aroused when large frames covered with jewel-like crystals of these dyes were shown in the International Exhibition of 1862 in London. The crowds attracted to the showcases containing these objects were as great as those which gathered round the Koh-i-noor in 1851.

If we refer to Hofmann's interesting Report on the Chemical Products and Processes in the Exhibition of 1862 a few facts may be extracted which serve to show how strangely the condition and distribution of the industry relating to colours from coal-tar have changed in the half-century which has elapsed since that time. In the first place the reporter enumerates the chief colours then recently discovered. These are practically all included under the several headings of paragraphs in the Report, viz. : Aniline Violets (mauve), Aniline Red (magenta), Aniline Yellow (chrysaniline), Aniline Blue, Quinoline Blue, Colouring Matters derived from Phenol (rosolic acid and picric acid), with a short paragraph relating to early results obtained with derivatives of naphthalene. Where one artificial substance capable of employment as a dye was known in 1862, there are probably fifty now available, without reference in either case to the numerous vegetable extracts obtained from wood, leaves, flowers, berries, etc., which produced all the effects known to our forefathers. Nor is it unimportant to notice who were the manufacturers of colours from coal-tar at the time of the exhibition. This can be gathered from the number of awards and medals and honourable mentions to the chief European countries represented. The United Kingdom received twelve medals and four honourable mentions, France received twenty-one medals and five honourable mentions, Germany and Austria together obtained twelve medals and seven honourable mentions. Speaking of aniline red the reporter, a German, makes the following statement :—

"Amongst those who have succeeded best are, in France Messrs. Renard Brothers and Franc and Messrs. Fayolle and Co., licencees of Messrs. Renard of Lyons, who have exhibited aniline reds, violets, and blues of great beauty, and to whom a just tribute of eulogium has been given.

"In Germany M. R. Knosp of Stuttgardt, and in Switzerland Messrs. J. J. Müller and Co. of Basle have also acquired a well-merited reputation.

"But it is in England that the most beautiful products have been obtained ; in proof of which assertion the reporter confidently points to the splendid exhibition of Messrs. Simpson, Maule, and Nicholson which has attracted such general attention. It is only justice to state that while France has had the merit of inaugurating the industrial production of aniline red, England may, thanks to the activity, science, and untiring efforts of Mr. Nicholson, claim the honour of having brought this manufacture to its present high degree of perfection."

It may well be asked why and when did England decline from

this high position ? The answer has been given repeatedly since 1880 when Germany had already succeeded in carrying off a large part of the business of manufacturing colours from coal-tar from England to the Continent. In the repeated warnings which have been issued since that time in no uncertain voice by Meldola, Green, the Perkins (father and son), and many other English chemists, two causes for the change have invariably been indicated, first the neglect of organic chemistry in the universities and colleges of this country, and then the disregard by manufacturers of scientific methods and assistance and total indifference to the practice of research in connection with their processes and products. As to the former no such charge can now be brought against the chemical schools of this country. Manufacturers have been aroused and many have taken steps in the direction of reform, but whether any considerable proportion of the colour industry, so long departed, can be brought back again, time alone will show.

It will now be evident that not only are the dyes not present in coal-tar ready formed, but that the hydrocarbons which are present in coal-tar and are got out of it, as already explained, by distillation, are converted into the colouring matter by undergoing successive chemical changes whereby several intermediate products are formed. The intermediate compounds if merely mixed together would not produce a dye, but when acted upon by chemical agents each one loses a portion of one of its constituents, the hydrogen for instance, and receives something in place of it, and a new compound or association of atoms is formed which has properties different from the properties of the separate materials concerned. Thus a molecule of aniline, $C_6H_5NH_2$, one of orthotoluidine, $C_7H_7NH_2$, and one of paratoluidine, $C_7H_7NH_2$, when attacked by three atoms of oxygen jointly lose six atoms of hydrogen, while the three altered residues combine to form rosaniline. Pararosaniline is produced in like manner when a mixture of aniline and paratoluidine is attacked by oxygen. Equations representing these changes may be written as follows :

$$\underset{\text{Aniline.}}{C_6H_5NH_2} + \underset{\text{Paratoluidine.}}{C_6H_4{<}\begin{matrix}CH_3 & 1\\ NH_2 & 4\end{matrix}} + C_6H_4{<}\begin{matrix}CH_3 & 1\\ NH_2 & 2\end{matrix} + 3O =$$

$$3H_2O + \underset{\text{Rosaniline.}}{C_{20}H_{19}N_3}$$

$$2C_6H_5NH_2 + C_6H_4 < \begin{matrix} CH_3 \; 1 \\ NH_2 \; 4 \end{matrix} + 3O =$$

Aniline. Paratoluidine.

$$3H_2O + C_{19}H_{17}N_3$$

Pararosaniline.

The prefixes *ortho* and *para* and the numerals 1 : 2 and 1 : 4 'ntroduced into the formulæ refer to the relative positions in the molecule of the constituents methyl, CH_3, and amidogen, NH_2. The constitutional formulæ of ortho- and para-toluidines would be represented by the following diagrams, the meaning of which has been already explained :—

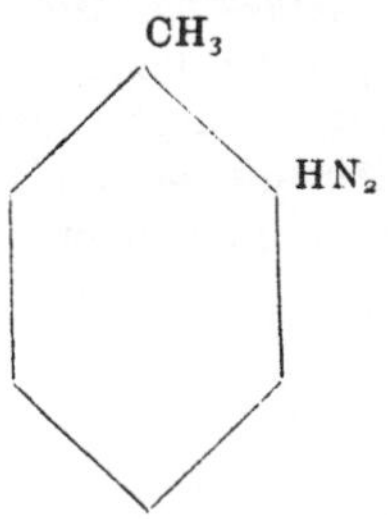

Orthotoluidine.

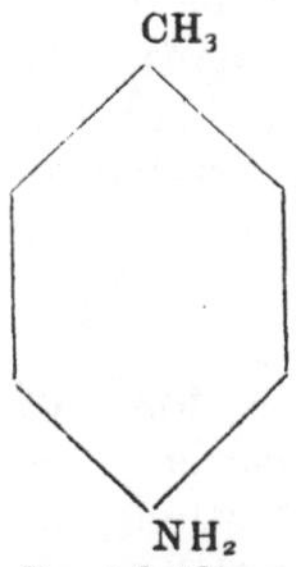

Paratoluidine.

The colouring matters are each built up of three groups of this kind held together by an atom of carbon. They are represented by the following expressions in which every group of C_6 must be regarded as forming a ring or closed chain of carbon atoms.

$$C \begin{cases} C_6H_4 \cdot NH_2 \\ C_6H_4 \cdot NH_2 \\ C_6H_4 \cdot NH_2Cl \end{cases}$$

Pararosaniline hydrochloride.

$$C \begin{cases} C_6H_3(CH_3) \cdot NH_2 \\ C_6H_4 \cdot NH_2 \\ C_6H_4 \cdot NH_2Cl \end{cases}$$

Rosaniline hydrochloride.

There are a number of other colouring matters constructed on the same type ; they are all derivable from one parent hydrocarbon called triphenylmethane, the composition of which is represented by the formula :—

$$HC \begin{cases} C_6H_5 \\ C_6H_5 \\ C_6H_5 \end{cases}$$

Such a substance has no colour and cannot act as a dye. Nevertheless the introduction of certain other constituents, such as NH_2, etc., results in many cases in the formation of the highly coloured compounds, such as magenta. It is obvious, therefore, that the production of a dye stuff depends not so much on the elements which are present, as the order in which they are united together, in other words on the *constitution* of the molecule. An immense amount of research and of speculation has been expended on the endeavour to find general rules which explain the tinctorial property. Here we must learn to distinguish from true dyes substances which exhibit colour, but which are incapable of attaching themselves to fibre, and therefore are not dyes. Thus, for example, azobenzene, $C_6H_5N=NC_6H_5$, is a red substance but it has no dyeing properties. But a compound having a similar constitution, so far as the two nitrogen atoms are concerned, is diamido azobenzene,

$$C_6H_5N=NC_6H_3(NH_2)_2,$$

which, in the form of hydrochloride, forms the orange dye called *chrysoidine.*

Groups of atoms, such as $-N=N-$, are called *chromophors*, as they have the property of producing a dye when introduced into certain compounds called *chromogens*. The chromogens are all compounds which contain the groups C_6 arranged as in benzene and are unsaturated, that is, they combine directly with hydrogen to form saturated compounds, and the latter are colourless. The theory is still a subject of debate, as from time to time examples occur which seem not to comply with the rule. The whole subject is, however, too complex and technical for treatment in these pages.

It would not be possible in the space at our disposal to describe the production of more than a few representative colours. Among these indigo stands in a remarkable position owing to the efforts which have been made during many years to replace the natural by an artificial synthetic product. These efforts have within the last four or five years been crowned with complete commercial success, which will doubtless have important economic results in the near future.

Indigo is a blue substance, insoluble in water, which has long been obtained from the juice of various species of *Indigofera* (Nat. Ord. *Leguminosæ*), cultivated for this purpose in the East.

Fig. 108.—INDIGO. SOWING THE SEED.

Fig. 109.—INDIGO PLANT. STAGES OF GROWTH.

To face page 320.

FIG. 110.—INDIGO. CUTTING THE PLANT.

FIG. 111.—INDIGO. OX-CARTS BRINGING PLANT TO VATS.

To face page 321.

That its use as a dye has been familiar for ages is indicated by the fact that the blue cloths found on Egyptian mummies owe their colour to this substance. Its introduction into Europe from India appears to have occurred in the seventeenth century. The only European plant which yields indigo blue is the woad (*Isatis tinctoria*, Nat. Ord. *Cruciferæ*), which was formerly cultivated to a considerable extent in the eastern counties of England. Its use has much declined of late years, though a small quantity is still grown for use in certain dye processes in Yorkshire.

The indigo plant is herbaceous and grows to a height of about 3 feet. The seed is sown in spring or autumn according to variety and the nature of the soil. The plant is cut just before flowering and is made into bundles which are placed in tanks or steeping vats. The herb is then covered with boards, weighted with stones, and water is added sufficient to cover it. A peculiar fermentation ensues, which lasts from twelve to fifteen hours according to the temperature of the air. From time to time a small quantity of the liquid is taken from the bottom of the vat, and when it exhibits the desired yellow colour the liquor is run off into a separate tank, where it is agitated either by a paddle-wheel or by workmen who stand in the liquid and beat it with paddles. These successive stages in the process are shown in the six illustrations (Figs. 108–113).

In this process oxygen is absorbed from the air and indigo appears as a greenish blue precipitate. This is allowed to settle and is then boiled with water to prevent a second fermentation which would spoil the product. The precipitate is strained off on large canvas filters supported by bamboo canes, and the nearly black mass is pressed, dried, and cut up into cubic cakes (see Figs. 114 and 115). The indigo of commerce is a dark blue solid which on being rubbed or pressed by any hard body presents a bright copper coloured shining surface. It is insoluble in water or alcohol, but dissolves in hot strong sulphuric acid, forming a permanent blue liquid containing indigo-sulphonic acids, which remain in solution when mixed with water and are used in dyeing.

The cultivation of the indigo plant in India has, down to recent years, occupied a very large acreage of ground. It appears from the Agricultural Statistics of India, published by the Department of Revenue and Agriculture (Vol. I, 1904, pp. 2, 3, and 350 ; Vol. I, 1913, pp. 8–9), that in 1884–5 the land devoted to this crop amounted to 897,917 acres. This gradually increased down to 1896–7 when the area under cultivation was 1,583,808 acres, and the weight of indigo produced was 168,673 cwts. or 8433 tons. The value for "fine Bengal" indigo between the years 1812 and 1833 varied from 5s. 6d. to 15s. per pound. The value in 1888 averaged 3s. 9d. per pound, and in 1908 it had come down to 2s. 6d. to 2s. 9d. per pound. But synthetic indigo costing at the same time even less the price of the natural product has been further reduced. The rapidity with which the cultivation of indigo in India has fallen off in recent years is shown by the following table of figures representing the acreage from 1900 down to the most recent estimate available.

ACREAGE OF INDIGO IN BRITISH INDIA

Year.	Acreage.
1900–01	977,349
1901–02	792,179
1902–03	653,801
1903–04	712,049
1904–05	510,289
1905–06	401,138
1906–07	448,594
1907–08	405,905
1908–09	286,354
1909–10	295,706
1910–11	282,757
1911–12	274,925
1912–13	214,500

The yield in 1909–10 was estimated to be 40,040 cwts. or 2002 tons.

Whether the cultivation of indigo has yet reached its nadir remains to be seen. There is naturally some conflict of statement as to the relative merits of the natural and synthetic dyestuffs from the point of view of the dyer. On the one hand, such statements as the following extracted from a paper in the *Kew Bulletin* (No. 8, 1910, p. 285) have been made :—

"There appears to be no doubt as to the superiority of the natural over the artificial product for dyeing purposes, and this is not where the fault lies ; but it does seem very problematical

FIG. 112.—INDIGO. BEATING THE VAT.

FIG. 113.—INDIGO. BEATING VAT WITH WOODEN WHEEL.

To face page 322.

FIG. 114.—INDIGO. BOILERS AND FILTERING TABLE.

FIG. 115.—INDIGO. DRYING HOUSE.

To face page 323.

as to whether good quality indigo can ever be produced under cultivation at so cheap a rate as that at which the synthetic substance is now manufactured. It has been stated that the two products are more effective when mixed in equal proportions, and if this be always true it is possible that it may contribute more than anything else to the support, and perhaps to the expansion of the cultural industry."

On the other hand it is asserted that synthetic indigo is capable of producing all the varieties of shade previously obtained by the use of the natural dye and equally permanent, with the advantage of uniformity of composition and freedom from impurities which saves much trouble to the competent dyer. It has also been urged that indigo as a crop has always been more or less uncertain, and provided that the change does not come about too suddenly, the substitution of crops which supply foodstuffs would be on the whole an advantage to the country.

What has happened in British India represents what has also occurred in other countries in which indigo has been cultivated. It is probable that the position of the planter may be somewhat ameliorated and the growth of indigo continued on the scale to which it has now been reduced. It appears to have been decided that the best chance of improving the yield of the dye is a botanical study of the crop and selection of the best type of plant with improved cultivation. Experiments in this direction are going on at Pusa. In the meantime it is probable that the custom or prejudice existing in the dye-houses will lead to the continued employment of natural indigo in association with the chemical product for securing certain effects for a long time to come.

The artificial production of indigo from constituents of coal-tar has a long scientific history, but in this case, as in so many other cases, success in the laboratory does not necessarily imply success in the factory. For while in the former case the purposes of science and additions to knowledge are the chief objects in view, in the latter the process ultimately to be adopted is determined not alone by practicability, but ultimately by the *cost*. The various synthetic methods which have been proposed since 1875 are to be found in the chief textbooks of organic chemistry. Here we can only review those which have had a practical success and form the present source of the synthetic dye.

One other point for consideration has also had a large influence in the choice of the process adopted. The question is which of the hydrocarbons present in coal-tar is to be chosen as the starting point, supposing the operations leading to the dye being about equally suitable? This can only be answered by calculating the probable demand for the synthetic product, and ascertaining the probable amount of the requisite coal-tar hydrocarbon available. These considerations it was, which, according to Dr. H. Brunck, director of the great colour works at Ludwigshafen, led to the abandonment of the otherwise successful process which depended on toluene, in favour of a process which starts from the far more abundant naphthalene obtained from coal-tar.

The raw material of one of the modern processes is then naphthalene, and the successive steps can be most concisely indicated by the following formulæ.

Naphthalene heated with fuming sulphuric acid and a small quantity of mercury is oxidised into phthalic acid. The sulphuric acid is reduced to sulphur dioxide which, by the modern contact process, is converted back again into sulphuric acid. It is therefore the oxygen of the air which thus indirectly acts on the naphthalene.

Naphthalene		Phthalic acid
$C_{10}H_8$	gives	$C_6H_4(CO_2H)_2$

Phthalic acid when heated (sublimed) loses water and is converted into

Phthalic anhydride

$$C_6H_4{<}^{CO}_{CO}{>}O$$

By heating this substance in presence of ammonia it is converted into

Phthalimide

$$C_6H_4{<}^{CO}_{CO}{>}NH$$

This compound under the simultaneous action of alkalis and chlorine in the form of alkaline hypochlorite yields

Anthranilic acid

$$C_6H_4{<}^{NH_2}_{CO_2H}$$

The next step is the conversion of anthranilic acid into phenyl-glycine-ortho-carboxylic acid by interaction with monochloracetic acid. This reagent has to be manufactured separately, but the electrolytic soda industry yields the necessary chlorine, together with caustic soda which is also required in various stages of the process.

Lastly, the phenyl-glycine acid is melted with alkali or a substitute for it, and the solution of the mass in water is oxidised by exposure to a stream of air whereby indigo blue is precipitated.

Phenyl-glycine-ortho-carboxylic acid

$$C_6H_4<\begin{matrix}NH{-}CH_2{-}CO_2H\\CO_2H\end{matrix}$$

produces first indoxyl

$$C_6H_4<\begin{matrix}NH\\CO\end{matrix}>CH_2$$

and by oxidation indigo blue

$$C_6H_4<\begin{matrix}NH\\CO\end{matrix}>C=C<\begin{matrix}NH\\CO\end{matrix}>C_6H_4$$

But the introduction of sodamide, NH_2Na, an alkaline substance which also acts powerfully as a dehydrating agent, has led to the development of another process which runs on simpler lines starting from benzene as the primary material.

When aniline, $C_6H_5NH_2$, is treated with chloracetic acid phenyl-glycine, $C_6H_5.NH.CH_2.COOH$, results. And this compound, heated with sodamide or with sodium in the presence of ammonia according to a process adopted by Meister, Lucius and Brüning, and now worked in England, gives a good yield of indigo. The intermediate compound is indoxyl or its sodium derivative as already explained.

Indigo being insoluble in ordinary aqueous solvents two special methods have to be used in applying it as a dye. One which makes use of the solution in sulphuric acid has already been referred to, but the other, of very ancient origin, is dependent on the reduction of the blue to *indigo white*, a colourless substance, which forms soluble salts with alkaline solutions.

The vat is prepared by mixing the blue with a solution of lime or caustic potash or soda and green vitriol (sulphate of iron), or with vegetable matter such as bran. In some of the modern processes the iron salt is replaced by metallic zinc or by a peculiar hyposulphite of sodium specially manufactured for the purpose. The result in any case is a yellow liquid, which on exposure to the air absorbs oxygen and reproduces the blue. Hence for dyeing cotton the cloth is immersed in the alkaline liquid, and when saturated it is exposed to the air and the blue is deposited in the insoluble form within the fibre.

Some of the derivatives of indigo produced by the introduction of bromine into the molecule are important dye-stuffs which are known in the trade as Ciba dyes.

In this connection an interesting discovery has resulted from the modern investigation of the purple dye extracted as in ancient times, from a species of *murex*, a mollusc found in the Mediterranean, and generally referred to as Tyrian purple. It is now known to be dibromindigo with the following formula :—

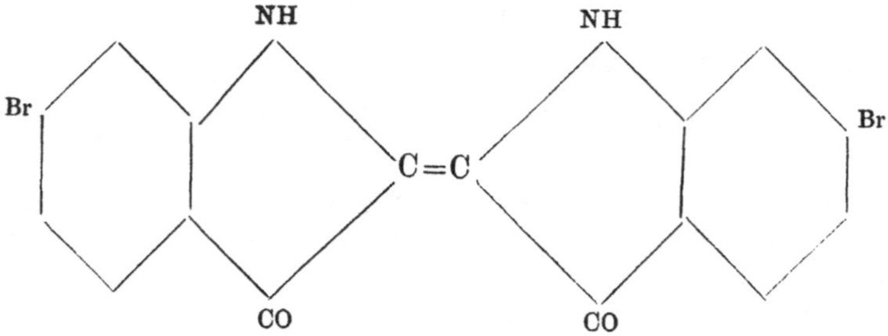

The next dyestuff which may be described is very different in constitution from indigo, but is interesting not only on account of its great practical importance, but because it has a very similar history. The substance referred to is the red colouring matter of the madder root.

The madder plant is a herbaceous perennial, *Rubia tinctorum* (Nat. Ord. *Stellatæ*), very nearly allied to the common goose grass or cleavers of our hedges. The *R. peregrina*, which is regarded by Hooker (Bentham's *British Flora*) as probably a mere variety of *R. tinctorum*, is found in the southern and western parts of Britain and of Europe. The use of the root as a dye can be traced to very ancient times, as it is not only mentioned by Pliny, but the red dye-stuff has been recognised in the mummy cloths of Egypt. It has been cultivated for centuries in the Levant, and in 1766 was introduced into the south of France by Jean Althen, to whom a statue was erected at Avignon.

The chief colouring matter in madder is called *alizarin*, it is accompanied by several others of which the most important is named purpurine. Alizarin was isolated for the first time by Robiquet and Colin in 1828. The composition of this substance remained for many years a mystery. It was at one time supposed to be a derivative of naphthalene, and many attempts were made to produce it synthetically from that substance, but it was not till about 1868 that its true nature was discovered and its connection with anthracene was established. Anthracene is a hydrocarbon, $C_{14}H_{10}$, which occurs among the least volatile portions of coal-tar, and up to this time it had been totally neglected and left in the pitch. Its relation to anthracene being once known a key was obtained to its constitution, and in 1869 patents were taken out for its production by Caro, Graebe, and Liebermann, followed only one day later by W. H. Perkin.

This was the first natural colouring matter to be produced synthetically and one of the most important, inasmuch as alizarin and its derivatives and associates are employed in the production of cotton prints all over the world, and are capable of giving a great variety of colours. Here it must be explained that alizarin and the allied colouring matter purpurine are often referred to as *adjective* dyes, because they do not dye either animal or vegetable fibre without previously impregnating the fibre with some basic substance, such as alumina or oxide of iron, chromium, or some other metal. The colouring matter unites with such substances forming chemical compounds called *lakes* which are insoluble in water. The metallic base introduced is called a *mordant*, and each colouring matter produces a different colour on the cloth by varying the mordant. Thus alizarin with

Iron oxide	gives	a	violet colour
Chromic oxide	,,		brown ,,
Aluminium ,,	,,		bright red ,,

Dyes which attach themselves directly to the fibre are spoken of as *substantive* dyes.

A large amount of knowledge has been gained in recent times as to the chemical constitution of the substances of which wool and silk fibres are composed, and the facts have given much assistance toward the conception of a comprehensive theory of dyeing. For reasons to be explained later no one theory has yet been completely established, and much remains in obscurity.

It would perhaps be advisable at this point to remind the non-chemical reader of the meaning attached to the words acid and base.

Acids are substances such as acetic acid in vinegar, tartaric, citric, and malic acids found in fruit, and the mineral acids, sulphuric, nitric, and hydrochloric acids. These are all soluble in water, and when sufficiently diluted have a sour taste and cause chalk and other carbonates to effervesce when mixed with them.

Basic substances are those which when mixed with an acid in due proportion destroy the acid taste and give rise to a new compound called a salt. Basic substances are divisible into two classes, of which one is represented by ammonia, aniline, and other organic bases. These combine directly with acids. The other class of basic compounds includes hydrated oxides of metals, such as caustic soda, lime, alumina, etc. When these are mixed with an acid they produce a salt, and the oxygen of the base unites with the hydrogen of the acid forming water.

The word salt is used in a very wide sense by the chemist, and though common salt is the most familiar of all salts, they are not all, like it, soluble in water and neutral, that is neither sour nor soapy to the taste.

Wool and silk are known to be composed chiefly of substances which are very complex in constitution, but when decomposed they yield peculiar compounds which behave under one set of conditions as acids, and under another as bases. The simplest example of such compounds is *glycocoll* (now commonly called glycine) or amino-acetic acid, $NH_2{\cdot}CH_2{\cdot}CO_2H$, which will produce a salt either by combining with a base or, in virtue of the ammonia residue, NH_2, which it contains, by combining with an acid such as hydrochloric acid.

This peculiar property is shared by the substances of which silk and wool consist, and they are therefore able to enter into chemical combination with dyeing matters of both an acid and basic character.

When, for example, silk or wool is dyed with magenta, which is a salt of a basic dye, the colour base leaves the mineral acid with which it is combined and unites with the acidic constituent of the fibre, while the mineral acid is left in the bath. This may be expressed diagrammatically as follows :—

$$\bar{W}<^{CO_2H}_{NH_2} + \bar{B}HCl \text{ give } W<^{CO_2H\bar{B}}_{NH_2} + HCl$$

Wool fibre	Dye hydrochloride	Dyed wool	Mineral acid

In a similar manner the process of dyeing by acidic colours, such as picric acid, may be represented. In this case a combination occurs in which the colour acid, $H\bar{A}$, unites with the ammonia residue of the fibre, thus :—

$$W<^{CO_2H}_{NH_2 \cdot H\bar{A}}$$

Dyed wool.

Such has been the hitherto-accepted theory of the dye-bath. During recent years, however, the study of the phenomena of surface attractions, not usually regarded as chemical in nature, has led to discoveries which require a modification in such theories.

It has long been known that many substances such as charcoal have the property of withdrawing colouring matters from solution, and the fact has long been turned to practical account in the refining of sugar. The syrups from which white sugar is to be obtained are filtered through thick beds of bone charcoal, where the brown uncrystallisable substances present are *adsorbed* into the substance of the charcoal, while the sugar is retained in solution. This power possessed by charcoal seems to be connected with the extent of surface of the particles of which it is composed, and it is shared to a greater or less extent by other substances which can be got into a fine state of division, notably by metals such as platinum and palladium, metallic oxides and hydroxides such as alumina, and even by fine sand and powdered glass though in an inferior degree.

The fibres of silk, wool, and cotton consist of substances possessing more or less the character of colloids (Chap. XIII). They are capable of taking up considerable quantities of water, not, however, in any definite proportion corresponding to the formation of chemical compounds. In the imbibition and separation of the water in such cases very great pressures are produced. When immersed in the dye-bath the deposition of the colouring matter takes place in a somewhat irregular manner, some of it forming a layer external to the fibre and some penetrating into the interior. This can be seen under the microscope. The probability therefore is that the attachment of the dye-stuff to the fibre is due in the first instance to mere surface attraction or adsorption, and that the colouring matter is then retained, partly, at least, by chemical combination. Some dyes can be removed from dyed fabrics by ordinary neutral solvents such as alcohol. This does not prove that chemical combination has not taken place between the dye-stuff and the fibre, and that the union is comparable with mere surface adhesion; it only seems to indicate that the saline combination formed is weak in character. The picric acid combination with the basic elements in wool fibre, for example, may in fact be similar to the combinations formed by weak bases or acids which are more or less completely decomposed by water. Thus urea combines with nearly all acids forming crystalline compounds which, however, possess a definite composition only when deposited in the presence of excess of acid. Any attempt to recrystallise from pure water leads to the reproduction of the base, while the acid passes into the liquid.

From these facts it appears that the process by which colouring matters become attached to vegetable or animal fibres and to mordants is more complicated than was supposed, and the phenomena of the dye-bath are for the present not fully understood.

Linen and cotton consist of *cellulose*, a compound or mixture of compounds which possesses neither acid nor basic character, and accordingly dyes do not attach themselves to the fibre in consequence of chemical combination of the kind just referred to. Cotton, however, is dyed by utilising the action of mordants as already explained in connection with alizarin, and there are substantive cotton dyes which work without the application of a mordant.

Safflower (*Carthamus tinctorius*) yields a natural red colour which is one of these, but many artificial dyes made from coal-tar are now used as direct cotton dyes. The first to be discovered was Congo red.

In many cases of this kind the retention of the colour by the vegetable fibre is probably due to the fact that the fibres of cotton and flax are hollow, and a portion of the dye becomes imprisoned within these microscopic tubes, and by a peculiar surface action of the cellulose it is retained there. There is no formation of a lake, and the cotton dyes are less resistant to soap and other agents than those which have been fixed by a mordant.

A large and important class of dye-stuffs, known as the azo-dyes, must not be overlooked, though it is impossible to explain fully to the non-chemical reader the nature of their constitution. They may be regarded as made up of two proximate constituents, one of which is a compound formed by the action of nitrous acid on a base such as aniline, naphthylamine, benzidine, while the other is a phenol or its sulphonic acid, or a base (amine). These two being brought together in solution " coupling " occurs, and the new colouring matter is produced. This always contains an association of nitrogen atoms

$$-N=N-$$

which may exist in the molecule several times over, and serves as the link to which are attached the two proximate constituents of the colour molecule. These colouring matters have consequently a rather complex constitution, which may be illustrated by the formula of Congo Red mentioned above.

$$\begin{array}{l} C_6H_4-N=N-C_{10}H_5<^{NH_2}_{SO_3Na} \\ | \\ C_6H_4-N=N-C_{10}H_5<^{NH_2}_{SO_3Na} \end{array}$$

In some cases the coupling process may be effected in the fibre itself, the diazo-compound being formed by immersing the fabric to be dyed, first in a solution of the base, then in a solution of nitrous acid, and lastly in the bath in which coupling takes place with a phenolic or basic substance.

The following table contains a synopsis of the more important modern synthetic dye-stuffs classified according to their character. It must be understood that the substances mentioned in the table are merely representative. Many thousands of coloured substances capable of acting as dyes are known or foreseen by theory, and several hundreds are already or have been practically manufactured on a large scale.

MODERN OR SYNTHETIC DYES

Class I. Dyes with a basic character. These colours are applied directly to wool and silk. They are fixed on cotton by means of an acidic mordant.

Mauve	discovered 1856
Magenta	,, 1860
Bismarck Brown	,, 1865
Chrysoidine	,, 1875
Methylene Blue	,, 1876
Malachite Green	,, 1878
Victoria Blue	,, 1883
Auramine Yellow	,, ,,

Class II. Dyes with an acidic character. These colours are applied directly to wool and silk from an acid bath.

Picric acid	discovered 1771
	applied 1855
Aniline Blue	discovered 1862
Azo-scarlets	,, 1876
Naphthol Yellow	,, 1879
Acid Green	,, ,,
Tartrazine Yellow	,, 1884
Wool Blacks	,, 1885
Acid Magenta	,, 1887

Class III. Acidic dyes requiring a metallic mordant. These dyes are fixed to mordanted fabrics, the colour produced being due to an insoluble salt. (Dye+mordant=lake.)

Alizarin Red discovered 1868

and other dyes of the alizarin group, including Alizarin Blue and Alizarin Green, Gambine Yellow, and certain other Azo-dyes.

Quinone-oxime Dyes discovered 1875

Class IV. Dyes with a saline character. Colours dyeing unmordanted cotton or linen directly in neutral or alkaline baths.

Congo Red Series	discovered 1884
Primulin	,, 1887
Cotton Black (Diamine Black)	,, 1889
Chlorazol Blue	,, 1898

Class V. Pigment dyes. Colour developed on the fibre.

Mineral colours, e.g. Chrome Yellow.		
Aniline Black	discovered	1862
Ingrain Azo-dyes (Ice colours)	,,	1880
Sulphide Dyes	,,	1893
Synthetic Indigo	,,	1897
Indanthrene Dyes	,,	1901
Thio-indigo	,,	1906

(Adapted from "Modern Dyes and Dyeing," a paper read to the Royal Dublin Society by Professor G. T. Morgan, F.R.S., 1914.)

CHAPTER XXII

DRUGS

It is perhaps worth while, before proceeding further, to review very briefly the history of the processes by which the chemist has gradually learnt how to build up very complicated substances by bringing together the elements of which they are composed, in other words to produce such substances by what is called synthesis. The first case of the production of a compound previously known only as a result of processes going on in organic living matter is that of urea.

This is the chief nitrogenous excretory product in mammalia, a smaller amount being thrown off by other animals such as birds. The source of the nitrogen is the protein constituents of the food, and a man on ordinary diet excretes on the average 30 to 35 grams (1 oz. to $1\frac{1}{4}$ oz.) per diem, almost entirely in the urine.

By the transformation of ammonium cyanate, $HN_4.CNO$, a salt which can be made from wholly *inorganic* materials, urea $CO(NH_2)_2$, was obtained by Wöhler in 1828. As the formula indicates, this change is due to a mere rearrangement of the elements present in the molecule, without addition or subtraction of anything. Such a transformation is called in chemical language an isomeric change. Comparatively little notice was taken of this remarkable discovery for many years, and it was reserved for later times to recognise its significance. Forty years later a Russian chemist, Basaroff, discovered that urea might be formed by the simple action of heat on ammonium carbonate, whereby it loses the elements of water in two stages and leaves a residue which is urea. Here again a rearrangement takes place, but it is accompanied by elimination of hydrogen and oxygen in the form of water.

$O:C<^{ONH_4}_{ONH_4}$	$O:C<^{ONH_4}_{NH_2}$	$O:C<^{NH_2}_{NH_2}$
ammonium carbonate.	ammonium carbamate.	carbamide or urea.

The production of urea in the animal body is, apparently, closely related to this change, the primary materials being the ammonia and carbonic acid of the blood which unite to form ammonium carbonate or carbamate.

Another organic compound produced from inorganic substances in the chemical laboratory was acetic acid, which was synthesised by Kolbe in 1845. Again very little attention was given to the discovery owing to the state of ignorance then prevalent in the domain of organic chemistry.

It was only some years later that a systematic study of synthetical processes was undertaken by Berthelot the famous French chemist. He showed how, by starting from the elements and from mineral substances, carbon can be combined step by step with hydrogen, then with oxygen, and again with nitrogen, producing thereby organic compounds, some identical with certain products of nature, others only analogous thereto, but at the same time serving as starting points for the formation of natural organic compounds. A single example taken from Berthelot's work will suffice by way of illustration. By heating carbon (coke or charcoal) in the electric arc surrounded by an atmosphere of hydrogen acetylene C_2H_2 is formed. By an easy process acetylene can be made to combine with more hydrogen so as to produce ethylene, C_2H_4. Ethylene dissolves in concentrated sulphuric acid, and the compound thus formed when mixed with water unites with the elements of water and, distilled, yields *alcohol*, C_2H_6O. The alcohol thus formed is identical in every respect with alcohol produced by fermentation of sugar. The synthetic process is so practicable that a company was at one time actually formed with the object of manufacturing alcohol from common coal-gas, of which ethylene is a constituent. This was fifty years ago, but the development of the same idea as that which was the basis of Berthelot's experiments has led

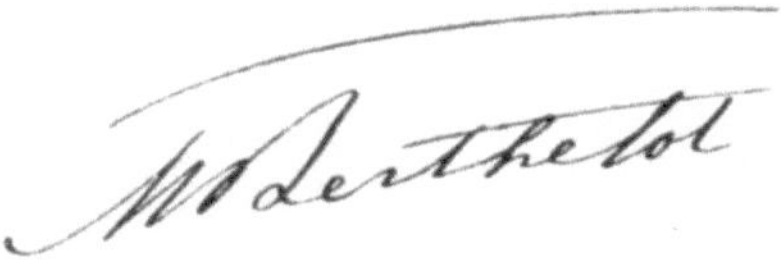

To face page 334.

in later times to the building up of numberless chemical compounds previously known only as limited products of animal or vegetable life. The first example of the application of this principle on the large scale was the manufacture of salicylic acid, by a method discovered by Kolbe in 1874 in which phenol (carbolic acid) present in coal-tar is the starting point.

The phenol is dissolved in caustic soda producing sodium phenate, C_6H_5ONa, and this compound, saturated with carbon dioxide gas under pressure at a slightly elevated temperature, is converted into sodium salicylate, $C_6H_4(OH)$, COONa, from which, of course, the acid is easily made.

Since that time the number of syntheses turned to practical account is very large. Two examples have already been mentioned in the two dye-stuffs, alizarin and indigo, of which the history has already been given. Many other cases will be referred to in the following chapters, where it will be noticed that though the hydrocarbons extracted from coal-tar are the fertile parents of a whole host of new substances, others are actually derived from the more simple combinations of the elements themselves, starting from carbon itself and bringing it into a state of union with hydrogen or with the gases of the atmosphere, oxygen and nitrogen.

With this by way of preliminary we may now proceed to enumerate, rather than describe, a few of the more prominent among medicinal agents which are the products of the chemical laboratory derived from materials of inorganic origin. These are commonly referred to as synthetic drugs, to distinguish them from those which, like quinine, morphine, strychnine, aloin and others, are provided by nature, and which constitute the active principles of plants which have long supplied active remedies in the treatment of disease. These principles exist ready formed in the plant, and are not in any way transformed in the chemical laboratory, but are merely separated in a pure state by suitable solvents or otherwise from the vegetable tissues which contain them.

Before proceeding to describe the origin of some of the most modern of chemical drugs the reader may be reminded that previous generations have already enjoyed the use of some of the agents originating in the chemical laboratory. The most familiar of anæsthetics, "the gas" used by every modern dentist, was breathed for the first time by Sir Humphry Davy so long ago as

1798, and in the earliest of his works he describes his "Researches Chemical and Philosophical chiefly concerning Nitrous Oxide and its Respiration" (1800).

Ether has been known from the times of the alchemists and from its being produced by the action of strong sulphuric acid on alcohol, it was called in those days *oleum vitrioli dulce.* Its use as an anæsthetic belongs exclusively to quite modern practice; it is generally associated with chloroform.

The other famous anæsthetic, chloroform, was discovered by Liebig in 1832, but its remarkable physiological action was recognised in 1847 by Sir James Young Simpson, who thus conferred on suffering mankind for all future time a benefit of incalculable value.

The alkaloids of opium and those of cinchona bark and many other vegetable principles had also been introduced into regular medical practice long ago. The characteristic of our own time in respect to medicine arises from the great advances which have been made in the knowledge of physiology and theoretical chemistry, and recognition of the interdependence the one on the other.

A brief account of some of the more prominent among chemical medicines now follows.

Acetanilide, known as *antifebrine,* is produced by the action of acetic acid on aniline, and is represented by the formula $C_6H_5NH.C_2H_3O$.

Acetyl Salicylic Acid, still known as *aspirin,* is the acetic derivative of *Salicylic Acid,* which is itself orthohydroxy benzoic acid, $C_6H_4<^{CO}_{OH_2H.}$ This important compound exists in the meadowsweet and other plants, but is manufactured by heating sodium phenate, C_6H_5ONa, in the presence of carbon dioxide, as already explained. From the product dissolved in water salicylic acid is precipitated by the addition of hydrochloric acid.

Another acetyl compound is *phenacetin,* which is the para acetamino—derivative of phenetol $C_6H_5.OC_2H_5$. The formula is therefore:

$$C_6H_4<^{OC_2.H_5}_{NH.C_2H_3O}$$

Picric Acid is a pale yellow crystalline substance produced by the action of nitric acid on phenol. It is trinitrophenol, C_6H_2

$(NO_2)_3.OH$, and is extensively used as a yellow dye for silk and wool (p. 330) and as an explosive (p. 386). In medicine it is chiefly employed as a lotion for burns.

A somewhat more complicated compound is *antipyrine*, called in the *British Pharmacopœia phenazone.* This is obtained by a succession of steps which begin with aniline and ultimately give the compound which, in chemical language, is dimethylphenyl pyrazolone, and the formula is

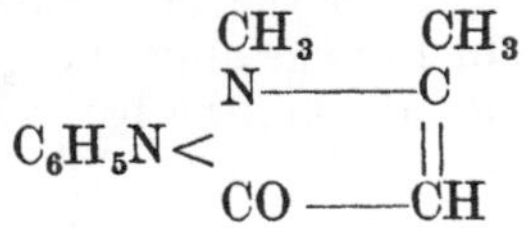

Novocaine is one of the modern local anæsthetics. This substance is the hydrochloride of diethylamino-ethyl-*p*-aminobenzoate

$$NH_2.C_6H_4.CO.O.C_2H_4.N(C_2H_5)_2HCl.$$

One of the most remarkable compounds which has come into use during recent years is *formaldehyde.* This is obtained on a large scale by bringing the vapour of methyl alcohol (wood spirit) mixed with air into contact with heated platinum or copper. The formaldehyde produced is a gas, but dissolves readily in water, and a solution containing nearly 40 per cent is sold under the name of "formalin." This is used as a disinfectant and antiseptic. A very minute quantity of it added to milk, for example, will prevent change for many days. It has also the remarkable property of rendering gelatine in any form, such as glue, insoluble in water, whence many applications of this property to technical purposes. When brought into contact with ammonia it is converted into a solid crystalline substance, hexmethylene tetramine $(CH_2)_6N_4$, used in medicine under the name *hexamine,* urotropin or formin.

A well-known soporific goes under the name *sulphonal.* It is dimethylmethane diethylsulphone

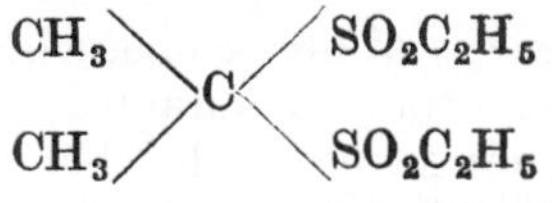

Veronal is the name of a compound which, under the new designation *barbitone*, has found its way into the *British Pharmacopœia.* Its chemical nature is indicated by the name diethylbarbituric acid. Barbituric acid, otherwise known to the chemist as malonyl-urea, has the formula

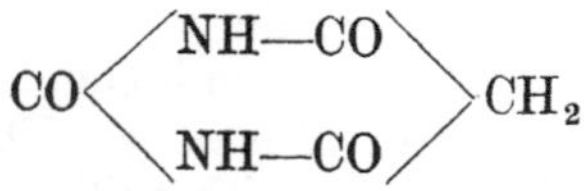

Its sodium salt is used medicinally under the name *medinal.*

Malonyl-urea has an interest apart from its use in medicine, arising from its relationship on the one hand to uric acid, an important excretory product of the animal organism, and on the other to theobromine and caffeine, the bases found respectively in cocoa and tea or coffee.

Formula of

Uric Acid.

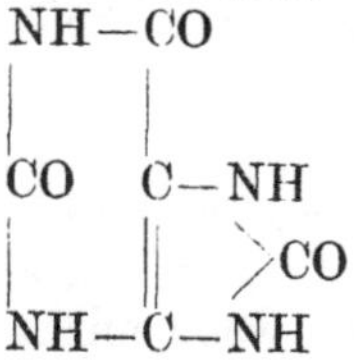

Theobromine.

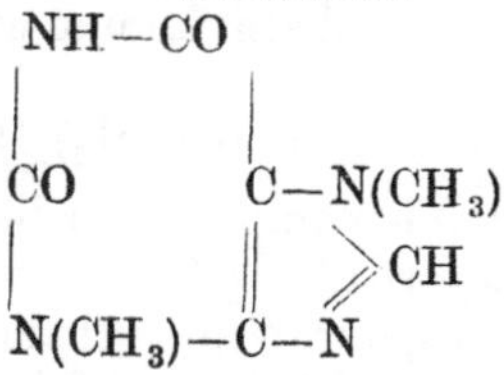

Caffeine.

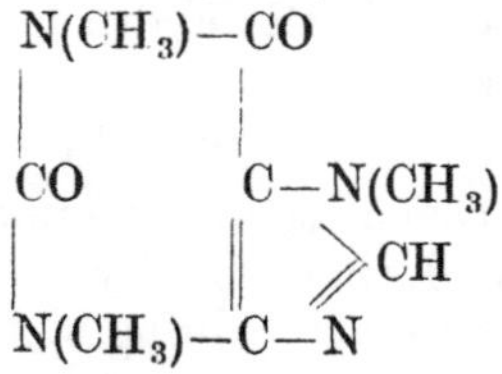

All these compounds have been produced synthetically from purely chemical materials and independently of animal or vegetable agency.

Among the many synthetical products which have become familiar in our own time is *saccharin,* a compound which is reputed to have a sweetening power four to five hundred times the sweetness of cane sugar. Saccharin, or *gluside* as it is called in the *British Pharmacopœia,* is orthosulphamido benzoic anhydride

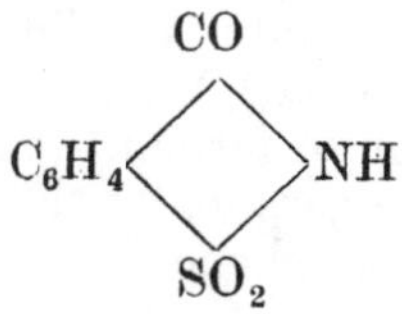

It is produced by a series of operations which have for starting point the hydrocarbon toluene obtained from coal-tar. It is used as a substitute for sugar in the diet of patients suffering from diabetes and other disorders.

All the preceding compounds and many others are officially recognised as medicinal agents by the General Medical Council by whom the *British Pharmacopœia* (1914) is issued. But many other chemical compounds used for medicinal or dietary purposes have been the subject of experiment. Some have had their utility established and have been adopted with practice, while many others, after a brief notoriety, have returned to oblivion.

The discovery of new remedies depends more and more on a combination of chemical and physiological knowledge. No better illustration of this principle could be adduced than the case of the remarkable compound "salvarsan," or 606, the use of which was introduced into medicine by the late Professor Ehrlich.[1]

Salvarsan is an artificial chemical compound containing the element arsenic in such a condition that it does not produce the ordinary effects of arsenical poisoning. It possesses the property of seeking out and destroying the specific organism of syphilis, the *spirochœta pallida*.

Salvarsan does not represent the first attempt to use arsenical compounds for medical purposes. Common white arsenic, the arsenious oxide As_4O_6, has long been recognised as a valuable alterative and tonic medicine when given in minute doses. It is also known to act as a dangerous poison in quantities exceeding a small fraction of a grain. Some fifty years ago cacodylic acid (dimethylarsinic acid)

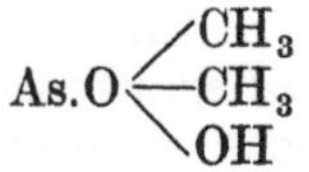

was tried in cases of tuberculosis and arsenic acid itself was reported to have some value. Later a number of arsenical organic compounds were prepared by the French chemist Béchamp and others. Among the rest a substance named "atoxyl" was introduced into medicine. Its constitution was, however, unknown and misrepresented till, in 1907, Ehrlich, in

[1] Paul Ehrlich was born of Jewish parentage at Strehlen, in Silesia, in 1854. He studied medicine in the Universities of Breslau and Strasburg, and graduated in the latter. He devoted himself to researches connected with the nature, origin and treatment of diseases which are attributable to the presence in the body of specific organisms such as tubercle, diphtheria and syphilis. His idea appears to have been that each cell in the body, including bacterial parasites, has a specific affinity for some particular substance. The blood stream may be compared to a river containing a variety of fish, and the problem is to find a drug which when introduced into this stream will kill the noxious while not injuring the normal inhabitants.

Ehrlich died on August 20, 1915.

conjunction with A. Bertheim, proved that atoxyl is the sodium salt of para amino phenyl-arsinic acid, the formula of which is

$$AsO \begin{cases} ONa \\ OH \\ C_6H_4.NH_2 \end{cases}$$

The use of atoxyl has been practically abandoned in favour of the more complicated dioxydiaminoarsenobenzene,

$$\begin{array}{l} As—C_6H_3(NH_2)OH \\ \| \\ As—C_6H_3(NH_2)OH \end{array}$$

the hydrochloride of which is *salvarsan*. The rapid rise into notoriety of this remarkable substance is known to all the world, but it appears to be still doubtful whether it is effectual in all cases, and its action occasionally becomes poisonous. This is probably partly due to the fact that on exposure to the air it readily undergoes oxidation yielding a more poisonous compound.

A large number of researches have been carried out on aromatic compounds containing the elements phosphorus, arsenic, and antimony, which in the periodic scheme are members of the same family as nitrogen. Some of these may hereafter be found to possess medicinal properties similar to those of salvarsan and, it is to be hoped, less dangerous. Announcements from time to time appear in the newspapers, of which the following is an example, 22nd March, 1916:—

"At the last meeting of the Academy of Sciences, a Paris telegram states, Professor A. Laveran described a new specific for syphilis, the discovery of Dr. Danysz, of the Pasteur Institute, who claims that it is by far the most effective yet found. The remedy is a preparation based upon a mixture of arsenic, antimony, and silver, which the discoverer has named '102' or 'Margol.'"

Time alone can show whether these expectations are justified.

On a previous page several of the natural drugs provided by the vegetable kingdom were mentioned. Of these the most important by far are those which are familiarly known as "alkaloids" inasmuch as the majority of them possess very powerful physiological action, and in many cases act as violent poisons when introduced into the animal economy either by the mouth or by hypodermic injection into the circulatory system.

It is only necessary to mention strychnine, morphine, and atropine, all of which are used in medicine. These and many other substances of the same class have been known for a long time, approaching a century. But beyond the fact that they contain beside carbon, hydrogen, and commonly also oxygen, together with nitrogen, little was known until recent times as to their chemical constitution. They agree in possessing the power of uniting with acids forming definite and usually crystalline salts. This property is connected with the nitrogen they contain, and down to about forty years ago they were assumed to be derivatives of ammonia, and the name *alkaloid* applied to them all had reference to the basic or alkaline character exhibited more or less strongly by every one. A considerable number of basic substances have been discovered in animal tissues or in products of decomposition and in a few cases these are identical with alkaloids derived from vegetable substances. Adenine, for example, is a base of comparatively simple composition with the formula $C_5H_5N_5$, which occurs not only in the pancreas but in small quantity in tea, beetroot, and shoots of bamboo. Its constitution is perfectly well known, not only from a study of its products of decomposition, but from the fact that it has been produced synthetically in the laboratory.

Another case of a natural alkaloid which has been produced by artificial processes is coniine, the poisonous principle of hemlock (*Conium maculatum*). This is a colourless, oily substance, having the formula $C_8H_{17}N$, which is obtainable from the hemlock plant or fruits by distillation with a solution of sodium carbonate. It rotates the plane of polarisation to the right. The artificial product is optically inactive as it consists of equal quantities of two stereo-isomeric bases, the one rotating to the right, the other to an equal extent to the left. These were separated from each other by Ladenburg by fractionally crystallising the tartrates, and the artificial right-handed base was found to be identical with the natural.

It will be noticed that the two examples of alkaloids mentioned are devoid of oxygen. When this element is present the problem presented is far more difficult, and, notwithstanding the progress which has actually been accomplished within recent years, as the result of researches by a number of distinguished chemists, there are no cases in which complete knowledge has been yet achieved except in regard to the bases found in tea and cocoa, to which reference has already been made.

The complete synthesis of such an alkaloid as quinine, strychnine, or morphine will doubtless be accomplished in the not far distant future, but at present all that can be put forward in such a case is a tentative expression which embodies more or less completely the results of operations in which the molecule is broken up into recognisable parts. The solution of problems of this kind is, however, less important from one point of view than was formerly the case. For physiologists have learnt to make use of chemistry more freely than in earlier times, and, as already mentioned, a large number of laboratory products are now at the disposal of the physician, which make the ordinary practice of medicine to some extent independent of the supply of natural drugs. Probably this tendency will continue to be developed until in time chemical constitution and physiological action will be so completely correlated that the requirements of medicine will be immediately supplied by the chemical laboratory.

CHAPTER XXIII

PERFUMES AND ESSENTIAL OILS

FROM drugs we may pass by an easy transition to perfumes and flavouring materials. It was among these things that one of the earliest triumphs of synthetical chemistry was celebrated when Perkin, the discoverer of the first coal-tar dye, contrived a process by which salicylic aldehyde could be transformed into *coumarin*. This was in 1868, and since a method was found in 1876 by which salicylic aldehyde could be produced from phenol, the synthesis may be regarded as complete, for, if necessary, phenol can be made from benzene, and benzene from acetylene, and the last can be formed by uniting carbon and hydrogen.

Coumarin is the fragrant substance to which the perfume of the Tonquin bean, of woodruff, and some other plants is due, and artificial coumarin is now an article of manufacture without the aid of the plant.

It was not long before a second step of the same kind was taken, for in 1876 a method was discovered for the synthesis of *vanillin*, the sweet-smelling constituent of the vanilla pod, so long used in confectionery. Here again it would be possible to proceed from the elements carbon, hydrogen, and oxygen. This, however, would necessitate several roundabout processes, and fortunately nature provides, in the substance called eugenol, a convenient and not too expensive material. Eugenol is the chief constituent of oil of cloves, and by acting on it with oxidising agents vanillin is produced.

The reader may be reminded that many years before such achievements could be placed on record the chemist had already learned that some of the fragrant essences so lavishly provided in fruit and flower and leaf could be reproduced by purely laboratory operations. As soon as organic chemistry began to be seriously studied nearly a century ago, among the earliest results was the production of what used to be called compound ethers, by the action of various acids on common alcohol, on the alcohol from wood spirit, and on the alcohol from fusel oil separated in the rectification of whiskey. Among these products were speedily recognised such odours and flavours as those of the pineapple, the jargonelle pear, and others. Pineapple owes its fragrance to ethyl butyrate, the pear to amyl acetate, winter-green (largely used in the United States) to methyl salicylate, while the strawberry and raspberry contain mixtures of several such ethereal compounds. These are now common articles of commerce.

These, however, were not alone, for already in those early days the odour developed when bitter almonds are crushed with water was found to be due to the formation of another kind of substance already mentioned in previous pages, namely, benz-aldehyde. Similarly the flowers of the meadowsweet contain salicylic aldehyde, the barks of cinnamon and cassia yield cinnamic aldehyde, the hawthorn and many garden flowers secrete other characteristic aldehydes. Nor were the older chemists altogether ignorant of the constitution of the essences to which the pungency of mustard, garlic, onions, and horse-radish are due. These are also ethereal salts or compound ethers which are characterised by the presence in them of sulphur associated with a radicle called *allyl* in reference to their frequent presence in plants of the genus *Allium*, belonging to the onion tribe.

There is perhaps no department of applied organic chemistry which has attracted during the last thirty years a larger number of workers, nor one in which a larger amount of definite progress has been achieved, for although the preparation of perfumes from plants is an industry which dates back many centuries, any knowledge of the composition of the "essential oils" has been derived almost wholly from chemical researches conducted within living memory. Before proceeding to the most recent advances it will be in the interest of those who are quite unacquainted with the technology of the subject to explain briefly what is understood by an essential oil. An oil is usually understood to be a liquid fat which is practically insoluble in water and which floats on that liquid. When boiled with an alkaline liquid, such as solution of caustic soda, it slowly dissolves forming a solution of soap. And if a drop of oil is placed on paper it forms a translucent spot which is permanent, for common oil does not evaporate away when exposed to the air.

An essential oil is distinguished from the fatty oils first by a strong and characteristic odour ; it usually floats on water, but it is slightly soluble, for the odour is commonly communicated to the water, as in such instances as the familiar rose-water or orange-flower water. An essential oil is usually changed by contact with caustic alkali, but it does not produce a soap. And lastly if a drop of essential oil is placed on paper the translucent stain disappears gradually as the oil evaporates away.

Most commonly, though not invariably, an essential oil is a mixture of two chief ingredients. One of these is a terpene—a compound of carbon and hydrogen only—the other is usually a compound of carbon and hydrogen with oxygen, and consists of an aldehyde, a ketone, a compound ether or "ester" or something else. To the latter ingredient the characteristic odour of the oil is mainly due.

Some essential oils consist of one constituent only with only slight impurities. Such, for instance, are the essential oils following :—

Name of oil.	Composed almost entirely of
Turpentine, American . .	Dextro-pinene $C_{10}H_{16}$.
,, French . .	Lævo-pinene $C_{10}H_{16}$.
Bitter almond . . .	Benzaldehyde $C_6H_5.CHO$.
Apricot and peach kernel .	Benzaldehyde $C_6H_5.CHO$.
Cherry laurel leaf . . .	Benzaldehyde $C_6H_5.CHO$.
Cinnamon bark . . .	Cinnamic aldehyde $C_6H_5.CH : CH.CHO$.
Cassia ,, . . .	Cinnamic aldehyde $C_6H_5.CH : CH.CHO$.
Mustard seed	Allyl isothiocyanate $SC : NC_3H_5$.

The extraction of essential oils from the plants which contain them is accomplished in most cases by a process of distillation with water. The essential oil usually boils at a much higher temperature than water, but the vapour rises with the steam and both are condensed together, the oil then floating to the surface of the water from which it may be separated. The principle of the process may be easily understood by reference to the diagram.

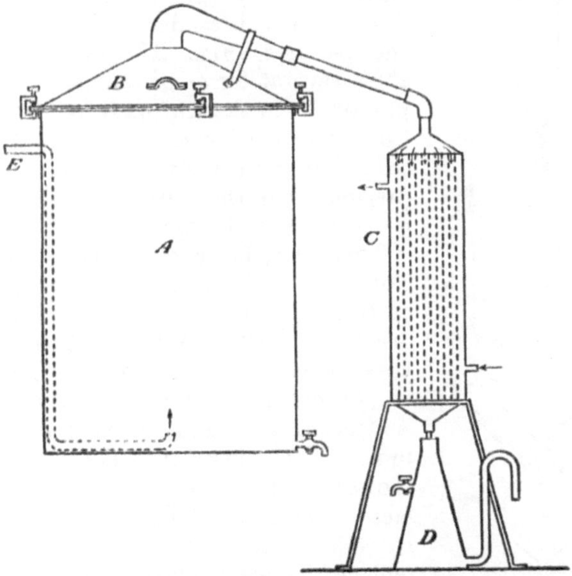

FIG. 116. DISTILLING AND CONDENSING APPARATUS.

The body of the still A is generally cylindrical and is of rather large dimensions on account of the usually bulky character of the plant material to be operated on. B is the still head or cover which is usually removable. C is the condenser supplied with a stream of cold water which enters at the bottom by the pipe indicated by the arrow, and being warmed by the steam pipes within escapes at the upper pipe to the drain. D is the receiver in which both essential oil and condensed water are collected, the former remaining above in a separate layer, and the water retaining a small quantity of dissolved essence running off continuously by means of the siphon pipe into another receptacle. E shows where a pipe conveying steam may be driven into the still so that the steam may pass through the leaves, flowers, or other matters which are supported just above by means of a wire screen resting on a frame, not shown. The condenser may consist of a number of tubes passing into a conical chamber at top and bottom as shown in the diagram, or it may take the form of a spiral pipe coiled up in the vertical cylinder so as to be surrounded by the cooling water. This description will render intelligible some of the pictures (Figs. 118 and 119) which show several forms of still used in actual practice. In the still used for the production of otto of rose (Fig. 119) it will be seen that with a view to economy of the precious otto, the water from the receivers is returned to the stills as it comes over.

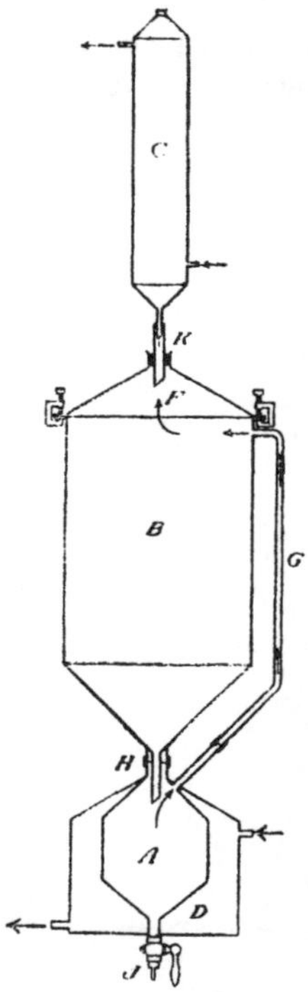

FIG. 117. CONTINUOUS EXTRACTION APPARATUS.

There are, however, other essential oils which cannot be extracted from the flowers or fruit containing them by a process of distillation without injury to their delicacy. In such cases as the violet, for example, the flowers are macerated in hot lard, which is afterwards pressed out retaining the perfume. The illustration (Fig. 122) which shows this process in operation at one of the factories at Grasse sufficiently explains itself. In some factories the stirring by hand is replaced by mechanical arrangements.

According to another plan the scent may be extracted by immersing the flowers in light petroleum spirit, which can afterwards be separated and distilled off leaving the essence behind. The accompanying diagrammatic representation of the apparatus employed in this process will render it easily understood. The percolator B is a vessel in which the flowers can be placed so that they are continuously exposed to a stream of the volatile liquid, which is driven in vapour from the receiver A up the side pipe G, and condensed again in the condenser C placed above. The arrows show the direction which the vapour takes, after it has been produced in the vessel A by the heat applied by means of steam to the surrounding jacket or steam bath D.

FIG. 118.—MODERN STILLS AT SCHIMMELS WORKS, MILITZ, NEAR LEIPZIG.

To face page 346.

FIG. 116.—DISTILLATION OF OTTO

To face page 347.

The solution of the perfume is finally drawn off by the tap at the bottom of A and submitted to distillation in a separate still.

In the illustration which follows (Fig. 123 facing page 349) showing the extraction plant in practice, the percolators, etc., are shown on the right, while the recovery of the solvent is effected by means of the stills on the left.

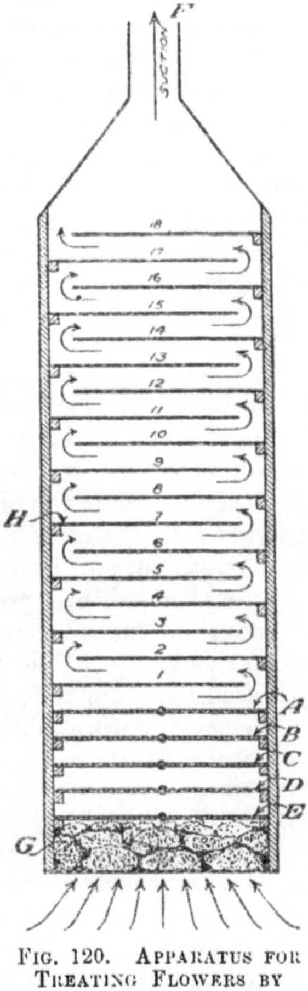

Fig. 120. Apparatus for Treating Flowers by Enfleurage Process.

In other cases, such as the jasmine, there is reason for believing that the flowers continue to generate and emit the perfume for some time after they have been gathered. Hence it is desirable to leave them for some time in contact with the agent, generally a solid fat, which absorbs the perfume during many days. This process of *enfleurage*, as it is called, was originally conducted by laying the flowers on the surface of a thin layer of lard spread on glass plates, renewing the flowers at intervals until the fat was duly charged. According to this plan the perfumed fat had to be melted and strained to free it from remains of petals and other impurities. Contact of the flowers with the fat has been avoided by the more modern apparatus shown in the diagram. The box about 2 feet square and 6 feet high is constructed so as to be practically air tight. It is fitted with a number of glass plates, H, which are arranged so that they can be easily withdrawn and replaced. The flowers are placed on five or six trays, A, B, C, D, E, at the bottom, and beneath them are sponges or cloths, G, wetted with water. Air can be drawn in as shown by the arrows through the perforated bottom. It carries with it sufficient moisture to prevent the flowers from drying too much, while the vapour of the perfume is carried successively over the fatty surfaces above.

By whichever method the fat is charged the perfume is extracted from it by shaking it with strong alcohol. The extract thus obtained is superior to that which is prepared by the hot maceration method previously described.

These processes are carried on in the south of France, especially at Grasse, where large quantities of flowers are grown for treatment in the factories of the neighbourhood.

In the case of the orange, lemon, bergamot, and other fruits of the orange tribe the essential oil resides in the rather large visible receptacles on the surface of the fruit. These are easily burst by pressure. If a bit of fresh orange peel is squeezed close to a flame it will be noticed that the expelled juice takes fire. The process employed in such case consists in pricking or squeezing the rind of the fruit and collecting the oil which runs out.

The hand processes are still preferred, probably in part owing to the difficulty of inducing the Italian peasantry to change their customs. The pricking process employs a copper saucer-shaped vessel, called an *écuelle*, the inside of which is covered with short spikes, while the oil as it exudes runs into a hollow handle inserted into the middle of the cup. In the antiquated sponge process, still largely used, the peel of the fruit is removed in thick slices, which are then pressed flat by the fingers against a piece of sponge. The oil glands are burst by the pressure and the sponge soaks up the oil, together with some juice, which is then squeezed out from time to time into a bowl, and finally filtered.

The use of perfumes is a form of luxury which in ancient times was probably limited to the rich, but in our own day they are used more or less unconsciously by everybody. For while nearly all women delight in perfumes, few men deliberately scent their persons or their clothes, but they cannot escape the use of soap, which in the form of toilet soap invariably contains some kind of essential oil. The extent to which the use of soap is increasing in all civilised countries may be judged from the case of the United States. The census of production in that country for 1904 showed a production of 605,000 tons, while in 1909 the output was 775,000 tons, an increase in five years of 27·4 per cent. Something of the same order has taken place in the countries of Europe.

The consumption of essential oils is, however, not the greatest in the form of perfume. Immense quantities are used in making

FIG. 121.—GATHERING VIOLETS UNDER SHADE OF OLIVE TREES, GRASSE.

FIG. 122.—PREPARING FLOWER POMADE BY HOT MACERATION. BRUNO COURT, GRASSE.

To face page 348.

FIG. 123.—MODERN FACTORY OF JEANCARD FILS & CO., LA BOCCA, NEAR CANNES. THIS SHOWS THE PLANT USED FOR THE MANUFACTURE OF CONCRETE FLORAL ESSENCE BY EXTRACTION WITH PETROLEUM ETHER.

To face page 349.

drinks like lemonade and alcoholic liqueurs of all kinds, also in confectionery and cookery. There is also an enormous demand for certain oils as medicinal agents, some for external use, others to be administered by the mouth, and many of them are included in all the pharmacopœias.

The following figures will give some idea of the magnitude of the industry connected with the production, the buying and selling of these materials.

German Foreign Trade in Essential Oils

Year.		Imports.		Exports.
1911	..	1,251,500	..	599,300 kilos.[1]
1910	..	1,524,300	..	547,600 ,,
1909	..	755,800	..	512,600 ,,
1908	..	911,300	..	390,800 ,,
1907	..	1,498,600	..	491,700 ,,

German Foreign Trade in Synthetic Odoriferous Substances

Year.		Imports.		Exports.
1911	..	17,300	..	492,800 kilos.
1910	..	17,900	..	428,800 ,,
1909	..	16,100	..	417,100 ,,
1908	..	11,300	..	280,000 ,,

(Statistisches Warenverzeichnis, Nos. 353–4.)

The following figures from the United States convey a further idea of the value of the essential oil business. In these figures the most prominent export is oil of peppermint, obtained from the *Mentha piperita*, which is cultivated very extensively in America, as well as in the chief countries of Europe, including Great Britain.

Total Value of Essential Oils

Year.		Imports.		Exports.
1913	..	5,619,921	..	748,105 dollars.
1912	..	4,116,641	..	850,923 ,,
1911	..	2,864,251	..	798,484 ,,
1910	..	2,446,716	..	675,030 ,,

France has long been a home of this industry, especially in the south, where large areas of land are devoted to the cultivation of

[1] A kilogram is $2\frac{1}{5}$ pounds.

the violet, tuberose, jasmine, as well as orange for the sake of the flowers beside the fruit. The slopes of the Alpes Maritimes and the Basses Alpes are in many places covered with wild lavender, both *Lavandula vera* and *L. spica*, from both of which the essential oil is distilled by means of portable stills which are carried from place to place. There are also factories where the plant is dealt with on a larger scale.

The following figures show the value of imports into and exports from France both of essential oils and synthetic perfumes. The imports of the latter have been chiefly derived from Germany.

Essential Oils

Year.	Imports.	Exports.
1913	2,827,100	33,812,600 francs.
1912	2,880,500	38,740,000 ,,
1911	2,687,600	32,802,000 ,,

Synthetic Perfumes

Year.	Imports.	Exports.
1913	1,378,000	164,000 francs.
1912	1,424,000	192,000 ,,
1911	1,372,000	168,000 ,,

As may be imagined the prices of the individual oils differ greatly. Among the most costly are natural otto of rose of which the price *wholesale* is according to an American price list from 5 to 5½ dollars per *ounce*, while bergamot is only 3¾ to 4 dollars per *pound*.

The best neroli from orange flowers is from 50 to 75 dollars per pound, while peppermint is less than 2½ dollars per pound.

It is interesting to glance at the pages of one of the trade journals in which essential oils and perfumes are advertised, for there one may trace evidence of the progress made in our own time in the application of chemical knowledge, and the extent to which the artificial are now competing with the natural essences. Other evidence has already been given in the value of the imports and exports of such materials into France. The table last quoted shows that though that country retains its position as the greatest producer of natural perfumes, the amount of imported synthetics already reached in 1913 a noteworthy figure. Whether this is likely to continue in the future

Fig. 124.—HAULING PEPPERMINT TO DISTILLERY IN MICHIGAN, U.S.A.

To face page 350.

Fig. 125.—INTERIOR OF JONONE DISTILLATION FACTORY AT ARGENTUEIL, NEAR PARIS.

Fig. 126.—LAVENDER FIELD IN VICTORIA, AUSTRALIA.

To face page 351.

is at least doubtful considering the present relations of France and Germany. Synthetic products, however, will certainly continue to be made, and the question where (?) can only be answered by time.

This synthetic industry is a development which follows naturally on the pursuit of knowledge in pure scientific chemistry without regard to possible applications. And much of the knowledge thus accumulated during the last forty years or more was no doubt regarded by the " practical " man in years gone by as useless. The story is a long one, and it would be unsuitable to a book designed for general reading to attempt to set forth the successive steps which have led up to the position which enables the manufacturer to place on the market substances which can successfully take the place of the perfumes derived from the rose, violet, lilac, lily of the valley, heliotrope, and many other flowers.

Though many of the discoveries were originally made in German laboratories the later developments have gone forward elsewhere, and many of the leading French perfumery houses are devoting attention and capital to the subject.

The methods by which coumarin and vanillin have been produced were described at the beginning of the chapter. To these we may now add two other examples which on account of their scientific interest as well as their commercial importance cannot be overlooked The first of these is the substance known as *terpineol*, a crystalline solid of which there are two varieties having a pleasant odour. It is the basis of the lilac and lily of the valley artificial perfumes, and is now manufactured by the ton. The original process started with turpentine oil, $C_{10}H_{16}$, which mixed with an alcoholic solution of nitric acid is converted into a beautiful crystalline compound called terpin hydrate, $C_{10}H_{20}O_2.H_2O$. When this is distilled with water and a small quantity of almost any acid it loses the elements of water and terpineol, $C_{10}H_{18}O$, passes over in the form of a syrupy liquid. This can be crystallised by cooling. Other processes are now used.

The chemical structure of terpineol, which is a kind of alcohol, has been the subject of many researches, but it is now fully understood, as it has been produced synthetically from compounds of known constitution by Professor W. H. Perkin in 1904.

The perfume of the violet was the subject of research by Tiemann and Krüger so long ago as 1893. They found it impossible to obtain sufficient material for their work from the flowers, but as this characteristic fragrance is possessed by the dried root of iris (orris), the latter was used by these chemists as the source of the fragrant oil on which their experiments were made. To this substance when purified they gave the name *irone*. It belongs to the class of compounds, called in chemical language ketones, and its composition is expressed by the formula $C_{13}H_{20}O$.

Protracted study of this compound led to the synthetical production of another substance to which the name *ionone* is given. This has the same ultimate composition, namely $C_{13}H_{20}O$, and closely similar properties, including especially the fragrant odour of the violet. No time was lost in applying these facts to the manufacture of artificial essence of violet. Ionone is made from citral, an aldehyde existing in considerable proportion in essences of lemon, and citron, and in lemon grass oil, to which the characteristic lemon odour of these materials is due.

Citral and acetone heated together in presence of an alkali condense together to form a compound called *pseudo-ionone*, which when boiled with dilute sulphuric acid yields a mixture of two isomeric substances, α and β ionone. The relation of these compounds to one another is shown in the following constitutional formulæ, from which it will be seen that the arrangement of the atoms in each is essentially the same.

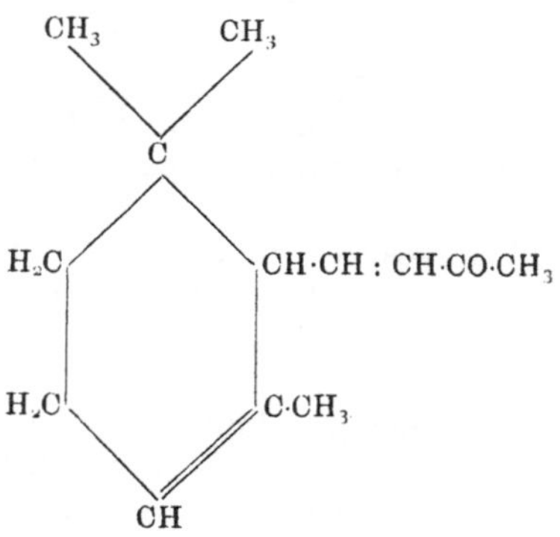

α. Ionone.

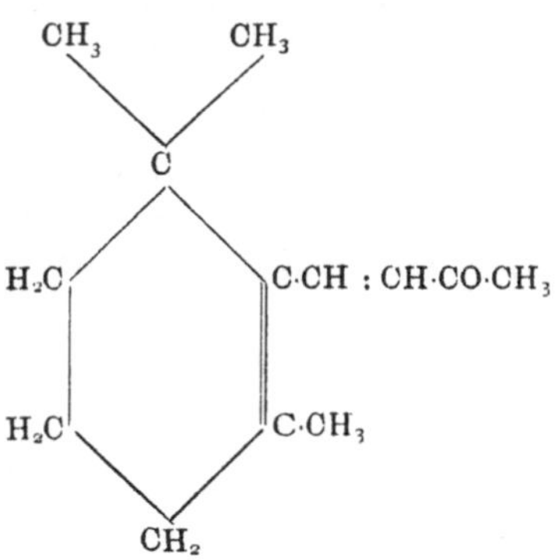

β. Ionone.

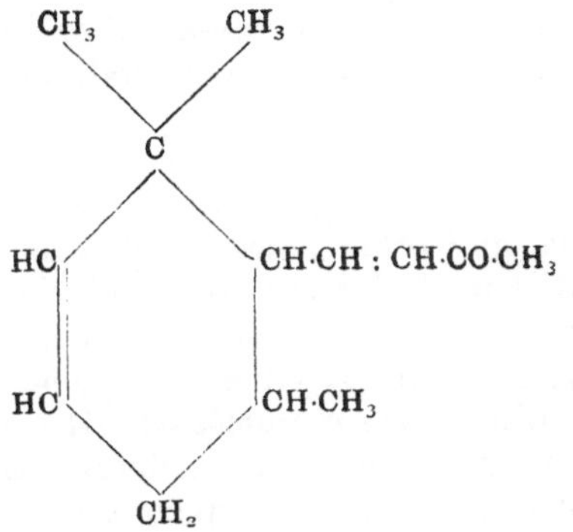

Irone.

The odours of the α and β modifications of ionone are described by an expert as distinguishable by the practised nose. "The alpha-ionone is excessively sweet, having the light fragrance of the violet, whilst the beta is of a heavier type, more suitable for soap manufacture. Many of the artificial violet products sold are blends of these two bodies with natural ingredients, such as extract or oil of orris, essence of cassia, etc.

"Preparations are also made from both violet flowers and violet leaves which are particularly useful as bases. A new series of preparations has recently been brought out termed Raldeines, which are methyl derivatives of ionone" (J. C. Umney, *Perfumery and Essential Oil Record*, July 9, 1912).

Many other perfumes are produced artificially. The following will perhaps be considered sufficient by way of example. The flowers of the May or Hawthorn (*Cratægus oxyacantha*) are believed to owe their fragrance to the presence of anisic aldehyde C_6H_4 $1(CHO){\cdot}4(OCH_3)$, though at present direct evidence is not on record. Otto of rose is a mixture of substances of which geraniol, $C(CH_3)_2 : CH{\cdot}CH_2{\cdot}CH_2{\cdot}C(CH_3) : CH{\cdot}CH_2OH$, is the principal ingredient. It contains also another interesting compound which can be made artificially by a process which has been patented. This is phenyl-ethyl alcohol, $C_6H_5{\cdot}CH_2{\cdot}CH_2OH$, which is obtained from phenyl-acetic acid ester by reduction by sodium. It is soluble in 60 parts of water, hence very little remains in the otto, the greater part remaining in the rose water, which has a peculiar odour of its own.

The perfume of the hyacinth is a very peculiar one, and is supposed to be due to the presence of cinnamic alcohol, $C_6H_5 \cdot CH : CH \cdot CH_2OH$. This is a crystalline, though volatile, solid.

The methyl ester of anthranilic acid $C_6H_4(NH_2)CO \cdot OCH_3$ is found in neroli from orange flowers, tuberose, ylang-ylang, jasmin, gardenia, and other flowers. This also is a crystalline compound, and is made by the action of the acid on methyl alcohol in the presence of hydrogen chloride.

Methyl salicylate has already been mentioned as derived from the wintergreen (*Gaultheria procumbens*). It occurs in a number of plants, but in most of them, including the gaultheria, it appears to exist in the form of a glucoside. The yield of oil is increased by wetting the plant and keeping it for a few hours before distillation, when a fermentation sets in by which the glucose is destroyed.

Ethyl and amyl salicylates are made and used in perfumery, but they are not known to be contained in any natural essential oils.

The odour of the heliotrope is said to be due to piperonal

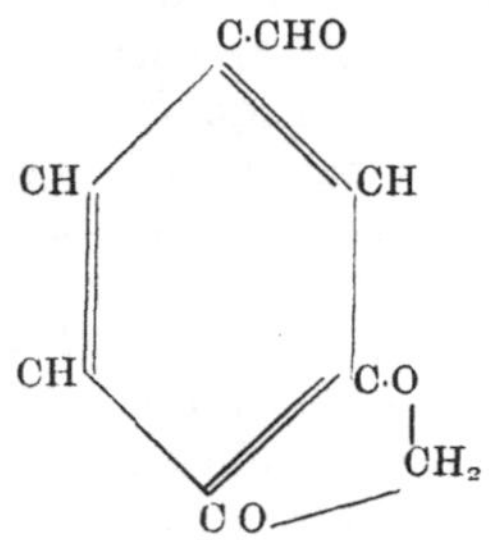

which has long been known.

The only important compound not referred to so far is camphor. This substance has been known from very early times, and before the sixth century was brought to Europe by the Arabians. In China and the East generally it has always been regarded as a valuable medicine, and is familiar enough in modern use both as a medicinal agent and antiseptic, as well as for other purposes, for example in the manufacture of celluloid. The greater part of the camphor of European commerce is obtained from the island of Formosa, and is distinguished as Japanese or

Chinese camphor to distinguish it from Borneo camphor which is obtained from Borneo and Sumatra. The former is obtained from the *Laurus camphora*, the latter from *Dryobalanops camphora*, both large trees. Their chemical relations are indicated by the following formulæ :—

Chinese Camphor.	Borneo Camphor.
$C_{10}H_{16}O$	$C_{10}H_{18}O$
or	or
$C_8H_{14}\left\langle \begin{array}{l} CO \\ CH_2 \end{array} \right.$	$C_8H_{14}\left\langle \begin{array}{l} CH{\cdot}OH \\ CH_2 \end{array} \right.$

Camphor has always been procured by the crude and wasteful method of cutting up the wood, in which the camphor exists in crystals, and distilling it with water in stills of very primitive construction.

It is purified by resublimation and is obtained in large hemispherical masses called bells, or being obtained in crystalline powder is then compressed into cakes. Common camphor is a natural constituent of several essential oils, especially those of lavender, rosemary, and sage. Borneo camphor does not come into European commerce, but it is preferred in Eastern Asia, where it commands a high price, and is used chiefly for making incense and generally for ceremonial purposes.

These two substances camphor and borneol are easily converted the one into the other, having between them a difference of only two atoms of hydrogen and standing toward each other in the relation of ketone (camphor) and secondary alcohol (borneol). Hence camphor is easily made from borneol by the action of oxidising agents, nitric acid, for example, while borneol can be produced from camphor by the action of sodium on an alcoholic solution of camphor.

Camphor has been the subject of very protracted investigations, as its constitution was for many years somewhat mysterious. These difficulties have now been cleared away, and the knowledge now existing of the relation of camphor to the terpenes has enabled chemists to contrive a process by which it can be made from oil of turpentine (pinene $C_{10}H_{16}$).

The artificial camphor is identical in every respect with natural camphor, except that it is optically inactive while all the natural products rotate the polarised ray. The manufacture of camphor has been the subject of several patents, and for a time synthetic camphor was found in the European markets, but has disappeared again in consequence of a reduction of price in the Japanese product.

The history of essential oils would not be complete without at least a passing reference to the extensive class of hydrocarbons called terpenes. The most prominent and abundant of these compounds are the two pinenes which are the chief constituents of the oil or spirit of turpentine. The one obtained from the United States is known as American or as English turpentine, and is obtained by distillation from the resinous exudation from the *Pinus australis* and *Pinus Tœda*. It rotates the polarised ray to the right. French turpentine is a similar product from *Pinus maritima*, but is lævorotatory. Spirit of turpentine is familiar enough as a colourless inflammable liquid with a peculiar smell. It is consumed in large quantity as a solvent and diluent in common paint, and some varnishes. It is also used as an external application in rheumatism and other disorders in which a stimulant is required.

The terpenes have been the subject of investigation in the hands of many chemists, but their constitution is now completely understood, and several of them have been produced synthetically since the beginning of this century.

A complete account of the chemistry of the terpenes will be found in a course of five lectures given in 1912 before the Pharmaceutical Society of Great Britain, by Sir William Tilden, Professor W. H. Perkin, and Mr. J. C. Umney.

CHAPTER XXIV

VEGETABLE FIBRE AND PRODUCTS FROM CELLULOSE

All plants in the earliest stages or most primitive forms are composed of *cells*, that is minute membranous bags spheroidal in form, and having no mouth or opening. As the plant reaches a more advanced stage of development these cells change in form, many assuming elongated shapes, becoming tubes often with tapering extremities. All the vegetation which clothes the surface of the earth consists therefore of a mixture of such minute hollow elements, which closely packed together form the tissues of the plant, the soft parts, as of leaf, flower, and fruit, being composed of more or less rounded cells, while the wood and veins of the leaves and other parts consist of fibres. The cells and fibres contain sap, which is water holding in solution gum, sugar, albuminous and saline matters, together with solid deposits of starch, green colouring matter (chlorophyll), crystalline solids, and resinous incrustations. Now when a mass of vegetable tissue, say sawdust, has been boiled with water, with caustic soda, with alcohol, and other solvents a mass of colourless, odourless, and tasteless material remains, composed of the membrane which forms the wall of the cell and consists of *cellulose*. This is the universal basis of vegetable tissue as already explained. Cellulose is found naturally in a nearly pure form as cotton, the hair from the seed of several species of gossypium, which grows in tropical and sub-tropical climates. Examined under the microscope cotton is seen to consist of long translucent fibres more or less flattened and twisted.

In the form of cotton wool, and woven in the various cotton and linen fabrics, as well as in paper, cellulose is familiar. But regarded from the chemical point of view the question is more difficult. Its composition is expressed by the formula $(C_6H_{10}O_5)n$, but it seems not improbable that the substance called cellulose may consist of a mixture of two or perhaps more substances of the same ultimate composition, with some difference of constitution. Cotton wool duly washed and purified may be regarded as normal cellulose, of which, however, the value of n in the above formula is unknown, that is the molecular weight of the compound is unknown. Though insoluble in all ordinary neutral solvents cellulose behaves towards acids as a kind of alcohol, yielding sulphates, nitrates (gun-cotton), acetates, and benzoates when acted on by the respective acids.

There are, however, several liquids which possess the property of dissolving cellulose without obvious change of composition, so that by appropriate treatment of the solution a substance having the same composition as cellulose may be recovered in the form of a gelatinous mass. Very important practical applications of these facts have been made within comparatively recent times. Thus it has long been known that a solution of copper oxide or hydroxide in solution of ammonia will dissolve cellulose, and that when the liquid is neutralised by an acid or mixed with various other liquids the cellulose is thrown down again as a gelatinous precipitate. The solvent which is known as Schweizer's reagent is made by immersing copper turnings in solution of ammonia and bubbling air through it. The reaction with cellulose has been turned to account in the manufacture of *Willesden Paper*, which consists of a coarse paper the surface of which is gelatinised and rendered waterproof by moistening it with the copper-ammonia solution. Within the last few years this solution of cellulose has been employed in one of the processes for the production of artificial silk, to be referred to a little later.

The action of caustic alkalis on the fibre of cotton was observed and applied about 1850 by John Mercer, a well-known calico-printer, and the product has long been known as "mercerised" cotton. The effect of the alkali is to untwist the naturally twisted flattened tubes of which cotton fibres consist, and thicken their walls. The fabric thus treated presents a somewhat silky gloss and is increased in strength.

Another very remarkable reaction of cellulose was discovered more than twenty years ago by Messrs. Cross and Bevan, well-known authorities on this department of applied chemistry.

If cellulose in any of its forms is treated with a concentrated solution of caustic soda, and the altered (mercerised) cellulose thus obtained is exposed to the action of carbon bisulphide, a yellowish mass is formed in an hour or two which swells up enormously on mixing with water and finally dissolves completely. This soluble compound appears to consist of a peculiar cellulose xanthate, to which the formula $NaS{\cdot}CS{\cdot}O{\cdot}C_{12}H_{19}O_9$ has been attributed. From a solution of this compound cellulose is again precipitated by acids, by heat, or simply by long standing, in the form of a gelatinous mass appropriately termed "viscose." This also is applied to the production of artificial silk, the annual output of which was stated in 1914 to be 14 million pounds, the capital employed being £5,000,000. Viscose is also largely used in the manufacture of photographic films.

Another liquid which is capable of dissolving cellulose without breaking up the molecule is a concentrated solution of zinc chloride. This solution was at one time used in the production of carbon filaments for electric glow lamps.

Strong sulphuric acid dissolves cellulose forming a mixture of sulphates, but if the acid is diluted with about half its volume of water the cellulose does not dissolve, but is superficially hydrated and gelatinised. Paper dipped into acid of this strength and subsequently washed free from the acid and dried produces *parchment* paper, a tough translucent material extensively used for a great variety of purposes.

So far only those agents have been considered which, while affecting the structure of the fibre by removing impurities or by adding to the cellulose the elements of water, do not destroy the integrity of the molecule which may be practically represented by the symbols $C_{12}H_{20}O_{10}$, though the molecule is probably much more complex. But acids when concentrated or allowed to act for a long time are capable of breaking up this association of carbon hydrogen and oxygen, by causing the assumption of the elements of water and a subsequent disruption of the molecule. The product is *glucose*, identical with the substance called grape sugar which is widely distributed in the vegetable kingdom. It is especially found in fruits and other sweet parts, where it is usually accompanied by another compound called fruit sugar having the same composition, and common cane sugar. Glucose is manufactured from starch by boiling it with dilute sulphuric acid, and when the change is complete, neutralising the liquid with chalk, filtering from the gypsum formed, and then evaporating the purified and decolourised syrup in vacuum pans. Woody fibre treated in the same way also yields glucose.

The chemist has been very busy in this kind of work and especially also in the development of processes which lead up to the production of paper. We may consider in outline the preparation of cellulose fibre from the various coniferous woods growing most abundantly in the northern countries of Europe, Sweden, Norway, and the shores of the Baltic. The processes involved are divisible into two classes in which the details vary, but they are both applicable to other raw materials, such as esparto grass and straw. In the one case caustic soda is the agent employed, in the other a bisulphite of lime or magnesia.

According to the former method of procedure, the raw material, pine or fir wood deprived of bark, is heated with a rather strong solution of caustic soda in a boiler which bears steam pressure, and of which the contents can therefore be heated considerably above the ordinary boiling-point of water. The chemical changes which take place are very complex, but the result is that the greater part of the cellulose remains unaltered or only hydrated, while the encrusting ligno-cellulose, etc., is dissolved out. About one-third of the weight of white pine wood is left in the form of a pulp, which after washing free from soda requires to be bleached. The bleaching is effected largely by the use of so-called chloride of lime or bleaching powder, but in some of the modern processes chlorine is produced electrolytically from magnesium or sodium chloride solution in which the pulp is immersed.

In the second class of processes the acid bisulphite solution required is obtained by passing through or over a milk of lime or magnesia, the sulphur dioxide gas formed by burning sulphur or iron pyrites in suitable kilns. The wood is digested with this solution for many hours at temperatures running from 120° to 140° C. The yield of pulp is greater than in the case of the soda process, and amounts to about 40 per cent. The brownish product is then bleached as already described.

The bleached pulp, however produced, is next subjected to a process of "beating" by which the individual fibres are separated and a perfectly smooth pulp is produced. With this is incorporated the size and the various colouring matters or loading material, such as china-clay, which are required according to the quality of the paper to be manufactured. The pulp is then ready for the paper-making machine.

In all these operations there is full opportunity and need for chemical study and supervision in improvement of processes or recovery of waste products, but in a superficial sketch it is impossible to supply the details which, moreover, are to be found in the several technical treatises to which the reader interested in these matters is referred. (See Thorpe's *Dictionary of Applied Chemistry*, Arts. *Cellulose*, *Paper*, *etc.*)

For the production of the cheaper kinds of paper a large quantity of wood pulp is produced without chemicals by mechanical crushing in a stream of water which carries off the pulp as it is produced.

The importance of this branch of manufacture can be roughly estimated from the figures to be found in the official publications of the Board of Trade. We learn from the Report of the First Census of Production of the United Kingdom (1907) that the value of the paper produced in the United Kingdom amounted at that time to about 13½ million pounds sterling.

From the annual statement of the Board of Trade for 1913, the following further figures have been gathered which show, at any rate, that the consumption of paper in Great Britain is immense. If as someone has suggested the use of paper is a measure of the degree of civilisation in a people, we may lay claim to a high place among the nations on this ground alone.

1913 IMPORTS

Paper.	Value.
For Printing and Writing	£2,343,934
For Packing and Wrapping—	
Packing	2,837,238
Strawboard	978,334
Millboard and Wood Pulp Board	665,977
Total	£4,481,549

Paper-making material.	Value.
Linen and Cotton Rags	312,351
Esparto and other Fibre	743,354
Pulp of Wood (Chemical dry)—	
Bleached	221,565
Unbleached	3,031,677
Mechanical and other Pulp of Wood, mostly from Sweden and Norway	1,406,128
Total	£5,715,075

The grand total of imported material, namely £12,540,558, therefore approaches the amount manufactured in Britain.

For all ordinary textile purposes, as we have seen, the natural fibre of vegetable matter, consisting essentially of cellulose, is the basic material. It may be twisted into threads, and the threads woven by the art of the weaver into fabrics of multitudinous designs, or the fibre may be beaten by the paper-maker till it is reduced to very tiny fragments which when stirred up with water form a smooth pulp. But the fibre is still there and its structure remains fundamentally unaltered.

Cellulose may, however, be obtained in the form of continuous threads applicable to all textile purposes in which its natural structure completely disappears. To effect this it must be converted into a soluble substance by the action of some chemical agent, and the product is *artificial silk*.

This interesting result belongs practically to the twentieth century, for although processes were invented and patents taken out so long ago as 1890, the use of artificial silk as a weaving material has not become commercially important till within the last few years.

As a matter of history the first process employed was based on the already long known properties of the nitrocelluloses. (See also Explosives.) When cotton is immersed in a mixture of nitric and sulphuric acids with a little water, a mixture of cellulose nitrates is formed which retains the form of the cotton, but differs from it in being soluble in a mixture of alcohol and ether. The resulting viscous solution when evaporated leaves a colourless film insoluble in water, the *collodion* of the photographer. If such a viscous solution made of suitable strength is forced through minute holes or fine glass jets either into water or into a warm atmosphere, the alcohol and ether are removed and fine threads are obtained which may be wound on a spool much in the same way as in winding silk from the cocoon. The fibre thus produced, however, has the great disadvantage of being dangerously inflammable. It was therefore reduced by passing it through a solution of ammonium sulphide by which the nitrate groups it contains are removed and a substance having the composition of cellulose is reproduced. This process has survived only to a limited extent as a manufacturing process, but by the employment of other and cheaper solvents artificial silk is manufactured and finds application in a variety of ways.

The cuprammonium process based on the employment of the solution of copper oxide in ammonia already described is one of these. The threads of cellulose solution are forced through jets into dilute sulphuric acid which removes the copper and reproduces the solid cellulose. But another and more successful process employs the *viscose* reaction of Messrs. Cross and Bevan which has already been explained.

Yet another process based on the formation of a cellulose acetate is employed for the production of threads, but more particularly, films of cellulose for use in the cinematograph and for other purposes.

It is obvious, from the brief description which has been given, that the threads of cellulose thus produced and which when spun form artificial silk, are entirely devoid of structure. Instead of being hollow, as are natural fibres, they are solid cylindrical threads, and as such present in the woven form an appearance different from that of cotton or linen. The lustre of artificial silk is greater than that of natural silk, and in the dye-bath it takes up colouring matter freely. One defect it has and that is a considerable loss of strength when wetted, which, however, is recovered on drying. The fibres are said to be strengthened by immersion in a solution of formaldehyde, which is supposed to condense the cellulose molecule.

Celluloid, formerly called xylonite, is another useful product of which the basis is cellulose. A nitrocellulose, made usually from tissue paper, is mixed with camphor dissolved in some solvent such as alcohol. After the evaporation of the solvent the mass remains plastic when warm, but solidifies on cooling and can be turned on a lathe. The process was invented by Daniell Spill, of Hackney, some forty years ago.

CHAPTER XXV

RUBBER

The substance long known as india-rubber is familiar enough, but down to a period about forty years ago the demand for it was comparatively moderate. Its use for waterproofing was known long before that time, and the great increase in the commercial application of rubber dates from the introduction of the rubber tyre as applied to bicycles and later to motor vehicles of all kinds. This increase in consumption has naturally led not only to the cultivation of the plants from which rubber is obtained, but to extensive chemical investigations into its properties and constitution which have culminated in the artificial production of what is always referred to as "synthetic rubber." Synthetic rubber has not become as yet a commercial article.

Rubber is produced by the coagulation of the *latex* or milky juice secreted by many plants. Those which yield commercial rubber flourish only in tropical or sub-tropical regions and belong to several natural orders. Of these the most important is *Hevea brasiliensis* (N.O. *Euphorbiaceæ*), which yields Para rubber, of which the amount constitutes about two-thirds of the total rubber of commerce. Other plants of the same order are the *Manihot* and *Sapium*, which furnish a portion of the wild rubbers of Brazil.

Rubber is also obtained from different species of *Funtumia* and *Landolphia* (N.O. *Apocynaceæ*) growing in Africa.

Ficus elastica (N.O. *Urticaceæ*) is a native of India and the

Malay States, which yields rubber, but less abundantly than the *Hevea*.

Down to the year 1875 no attempts had been made to provide for the demands of the rubber market artificially, and all the rubber up to that time and for some years later had been derived from the trees growing wild in the Brazilian forests. The idea of cultivating rubber plantations in the British Indian possessions was then carried into effect, and starting with large scale experiments in Ceylon, the plantations of, chiefly, *Hevea* have extended into the neighbouring countries and some other parts of the world.

The total acreage of rubber plantation was estimated in 1911 as follows :—

Country.	Acreage.[1]
Ceylon	200,000
Malay States	400,000
Java, Sumatra, and Borneo	200,000
Southern India and Burma	35,000
German Colonies	45,000
Mexico, Brazil, Africa, and W. Indies	100,000
Total	980,000

Estimates as to the yield of rubber from plantation sources differ considerably, but there appears reason to believe that the average yield amounts to between 300 and 400 pounds per acre per annum.

As to the whole world production it would be very difficult to state a figure even approximately trustworthy, but it certainly appears to be steadily increasing, as might be expected from the increasing demand and the gradual extension of the plantations.

That the consumption is very large may be inferred from the statistics of the business done in the United Kingdom alone.

The following statement, based on the official figures of the Board of Trade, show that our imports of raw rubber during the year 1915 reached the record figure of 182,565,900 lb. The value was £20,225,060, which gives an average value at slightly over 2s. 2d. per lb. The great bulk of our importations is of British growth, and plantation production now exceeds wild forest production, which amounts to little more than one-fourth of the total. It is obvious that business during the year

[1] *India Rubber Journal*, 1911.

Fig. 127.—TAPPING HEVEA IN JAVA.

To face page 364.

FIG. 128.—YOUNG RUBBER TREES. STRAITS SETTLEMENTS.

FIG. 129.—RUBBER AND TEA. CEYLON.

To face page 365.

1915 has been seriously interfered with by the state of war in Europe, and for the same reason some of the statistics for this year are necessarily imperfect. The figures are, however, interesting as showing the steady growth of an industry which belongs entirely to modern times, and which is destined unquestionably to expand still further with the development of the motor and the consequent consumption of tyres.

These figures do not include gutta-percha.

RUBBER BY QUANTITY

IMPORTS OF RUBBER.		1913	1914	1915
From Dutch East Indies .	Centals of 100 lbs.	N.S.	N.S.	64,119
,, French West Africa . .	,,	22,610	6,290	42,552
,, Gold Coast . . .	,,	14,935	5,644	6,318
,, Other Countries in Africa .	,,	N.S.	N.S.	101,340
,, Peru	,,	29,133	15,523	15,582
,, Brazil	,,	363,595	277,433	286,391
,, British India . . .	,,	N.S.	N.S.	32,888
,, Straits Settlements and Dependencies, including Labuan	,,	338,313	473,599	660,532
,, Federated Malay States .	,,	221,304	219,991	288,803
,, Ceylon and Dependencies .	,,	150,182	209,693	286,097
,, Other Countries . .	,,	434,367	307,023	40,037
Total Imports . .	,,	1,574,439	1,515,196	1,825,659

RE-EXPORTS OF RUBBER.[1]		1913	1914	1915
To Russia . . .	Centals of 100 lbs.	142,326	168,156	259,061
,, Germany	,,	217,944	158,360	—
,, Belgium	,,	50,820	33,154	—
,, France	,,	118,908	110,573	152,097
,, United States of America .	,,	398,510	541,615	831,801
,, Other Countries . . .	,,	79,761	87,373	179,895
Total Re-Exports . .	,,	1,008,269	1,099,231	1,422,854

RUBBER BY VALUE

IMPORTS OF RUBBER.[1]	£	£	£
From Dutch East Indies . .	N.S.	N.S.	716,151
,, French West Africa . .	284,808	59,731	400,853
,, Gold Coast . . .	147,098	45,988	39,749
,, Other Countries in Africa .	N.S.	N.S.	915,608
,, Peru	445,681	186,078	178,271
,, Brazil	5,640,700	3,433,581	3,240,729
,, British India . . .	N.S.	N.S.	372,313
,, Straits Settlements and Dependencies, including Labuan .	5,299,206	5,248,734	7,384,830
,, Federated Malay States . .	3,532,173	2,512,500	3,340,071
,, Ceylon and Dependencies .	2,309,324	2,328,024	3,230,218
,, Other Countries . .	2,568,029	2,029,792	406,217
Total Imports . . .	20,524,019	15,844,428	20,225,060

[1] Prior to 1915 these figures include waste and reclaimed rubber as well as raw rubber.

Re-Exports of Rubber.[1]	£	£	£
To Russia	2,205,205	1,860,362	2,858,813
,, Germany	3,342,715	1,690,972	—
,, Belgium	789,151	364,856	—
,, France	1,876,506	1,294,656	1,772,100
,, United States of America	5,417,127	5,912,899	9,273,915
,, Other Countries	1,205,900	996,528	2,060,638
Total Re-Exports	14,836,604	12,120,273	15,965,496

TRADE IN RUBBER GOODS

	1913	1914	1915
Imports of Boots and Shoes, doz pairs	95,771	85,348	160,462
value	£119,921	£164,323	£264,260
Exports of ,, ,, doz. pairs	132,736	121,681	118,716
value	£138,006	£123,756	£139,340
Imports Waterproofed Apparel	£6,482	£8,456	£5,376
Exports ,, ,,	£1,021,393	£774,489	£525,234
Exports other than above and Tyres and Tubes	£1,656,246	£1,178,128	£1,061,540

It has already been mentioned that rubber is obtained from the milky juice or *latex* which exudes on wounding the bark of the tree. The age at which the process of tapping should commence is about four or five years, but this is dependent on various considerations, and differs somewhat according to the kind of tree and the climate and soil of the district, which affect the rate of growth. Tapping is a process which consists in scoring the bark by means of a gouge or some kind of knife with adjustable blade, of which a large number of varieties have been patented. In the plantations a vertical channel is often cut first and a collecting tin placed at the foot. A dozen or more oblique cuts are then provided to lead the juice into this channel. Various other systems of tapping are adopted in different countries. These are sufficiently indicated in the accompanying diagram (Fig. 130); in connection with which it may be noted that the half herring bone system is by far the most common.

The latex as it flows from the tree has a tendency to coagulate and to form clots or scrap which has to be dealt with separately. But the bulk of the liquid is conveyed as soon as possible to the factory, and after being strained, to remove impurities, it is mixed with a small quantity of acid, generally acetic acid. A clot soon forms which takes the shape of the containing vessel, and after washing the rubber is passed through rolls, and then dried.

[1] Prior to 1915 these figures include waste and reclaimed rubber as well as raw rubber.

The treatment of wild rubber on the Amazon is somewhat different. V-shaped incisions are made in the bark, and at the bottom of each cut a small collecting cup is placed. The latex, which contains about one-third of its weight of rubber, is then coagulated by exposing it to wood smoke which, of course, is accompanied by small quantities of acetic acid and vapour of creasote. In order to accomplish this the collector uses an earthen bottomless pot in which a smoky fire is made by igniting a pile of dry twigs, to which is added from time to time the nuts of a kind of palm abundant in the district. A long wooden paddle,

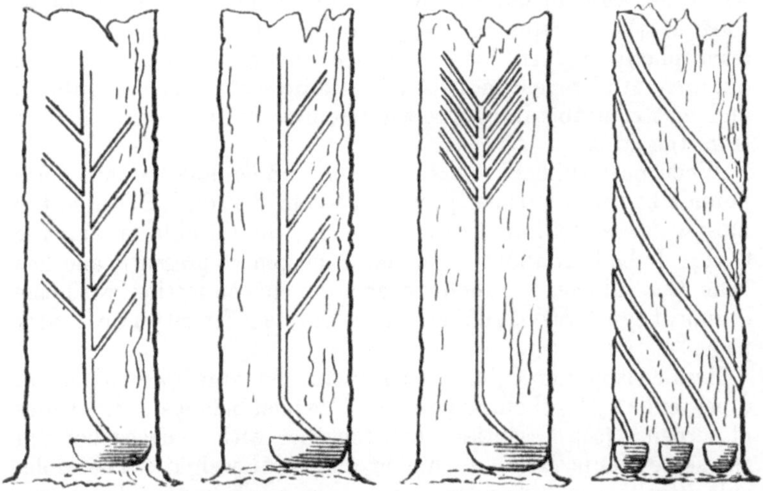

Herring bone. Half herring bone. Basal V system. Spiral.

FIG. 130. VARIOUS SYSTEMS OF TAPPING.

of which the blade is first smeared with wet clay to prevent the rubber from sticking, is then dipped in the latex and held in the smoke. A thin sheet of coagulated rubber is then almost immediately produced, and by alternately dipping in the milk and rotating the paddle over the fire, successive layers of rubber are deposited until a ball is produced of the required size, which in as much as a man can conveniently lift. The ball is then split by means of a moistened knife, and the rubber detached from the paddle. As the latex contains, beside rubber, a considerable quantity of albuminous matter, of which a portion is retained by the rubber, it is necessary to sterilise it, otherwise the impurity is liable to putrefy, and in its decomposition to affect the quality of the rubber. By the smoking process the rubber is preserved and the dark colour commonly exhibited by the loaves or blocks is accounted for.

COMPOSITION AND CONSTITUTION OF RUBBER

Rubber has long been known to consist essentially of a mixture of two or more hydrocarbons having the ultimate composition expressed by the formula $C_{10}H_{16}$. But rubber always contains larger or smaller amounts of substances containing oxygen, which for want of more knowledge are commonly called resins. Part of these resins probably result from the absorption of atmospheric oxygen by the rubber hydrocarbons. A small quantity of nitrogenous matter is also present in natural rubber, and is attributable in part to the retention of albuminous matter from the latex.

When pure rubber is heated it splits up completely into compounds having the same percentage composition. Of these the most volatile is a liquid called *isoprene*, the formula of which is C_5H_8. It boils at about 37° under atmospheric pressure, and has been the subject of much experiment in connection with the chemical synthesis of rubber. This will be referred to on a later page.

Beside isoprene rubber also yields a large proportion of dipentene $C_{10}H_{16}$ (boiling point 175°) which belongs to the series of terpenes (see Essential Oils), together with hydrocarbons of the same composition but higher molecular weight. The molecule of rubber is undoubtedly very large and complex, how large it is impossible as yet to say with certainty, but it has been suggested that the molecule consists of at least eight groups of the composition C_5H_8, that is that the molecular formula is $C_{40}H_{64}$. The difficult solubility of rubber and its colloidal character would be consistent with a still more complex formula.

Whatever the constitution of rubber may be it has one definite chemical characteristic; it is unsaturated. On this depends its power of entering into direct chemical combination with such elements as chlorine, bromine, and sulphur as well as with certain oxides of nitrogen. Its capacity for combination with sulphur and with sulphur chloride is the explanation of the important process known as "vulcanisation," upon which depend so many applications of rubber to practical purposes.

Vulcanisation is effected by two principal processes. In the one the rubber, made into a stiff semi-solid solution in coal-tar naphtha, is mixed with the requisite quantity of flowers of sulphur together with certain other materials, such as litharge (lead oxide), zinc oxide, magnesia, and antimony sulphide, some of which seem to accelerate combination. The mixture is exposed to a steam heat and is then usually spread by means of rollers on a cotton cloth, whereby a sheet is formed from which the majority of rubber articles are made. The cloth is finally hung in chambers heated by steam pipes, and the solvent naphtha employed at the beginning of the process dries off.

The other process, called cold vulcanisation, consists first in spreading a thin layer of rubber paste on cloth, and then by means of rollers passing the coated material through a trough containing a mixture of sulphur chloride, S_2Cl_2, and carbon bisulphide. In this case it is not merely the sulphur which is added on to the rubber molecule, but the chlorine as well. The formula of the compound produced is $(C_{10}H_{16})nS_2Cl_2$, but there is great difference of opinion as to the value of n in the formula, that is as to the number of molecules of sulphur chloride which are associated chemically with the rubber molecule, and whether the compound so formed is mechanically united with more rubber. In fact the exact nature of the vulcanised product of the cold curing process is still a subject for further investigation.

Several "substitutes" for rubber have long been used for incorporation with true rubber in order to cheapen the material. Of these the most interesting and important is a peculiar tough substance produced by the action of sulphur chloride on various vegetable oils. But several other additions to rubber are employed when toughness and tensile strength is not the most important quality looked for in the material. When, for example, rubber is required as an electrically insulatory material, as in coating cables, various resins, nitrocellulose, bitumen, and other substances are used.

A considerable quantity of rubber is reclaimed from vulcanised waste by heating it with an alkaline solution, and subsequently washing the desulphurised mass.

This, however, is not to be regarded as a treatise on the manufacture of rubber, and those who are interested in the scientific principles of the industry would consult with advantage such a work as Dr. Schidrowitz's *Rubber* (Methuen and Co.).

The cultivation of Hevea in the East is fully treated from the scientific and practical points of view in Mr. Herbert Wright's *Hevea Brasiliensis or Para Rubber* (MacLaren and Sons, London).

HISTORY OF SYNTHETIC RUBBER

In the course of researches in connection with the hydrocarbons called terpenes, which include turpentine oil, the author obtained from turpentine by the action of heat a peculiar very volatile liquid, called *isoprene,* which had previously been produced only by the destructive distillation of india-rubber. A peculiarity of this limpid liquid, which possesses a boiling point close to that of common ether, is that in contact with certain reagents, common hydrochloric acid among them, it is converted partly into rubber. The liquid which remained over, after the termination of this series of experiments, was preserved in well-closed bottles. Some few years later, in May, 1892, in a paper read by the author to the Philosophical Society of Birmingham, the following passage occurs: "I was surprised a few weeks ago at finding the contents of the bottles containing isoprene from turpentine entirely changed in appearance. In place of a limpid colourless liquid the bottles contained a dense syrup in which were floating several large masses of solid, of a yellowish colour. Upon examination this turned out to be india-rubber. . . . The artificial, like natural, rubber appears to consist of two substances, one of which is more soluble in benzene or carbon bisulphide than the other. A solution of the artificial rubber leaves on evaporation a residue which agrees in all characters with a similar preparation from Para rubber. The artificial rubber unites with sulphur in the same way as ordinary rubber, forming a tough elastic compound." At the time mentioned there was no means of further testing rubber chemically so as to establish the relation of synthetic to natural rubber, but the discovery many years later of ozonides of rubber by Professor Harries of Berlin, and their decomposition products, has supplied the means of testing the identity of rubbers from different sources. This test has been applied by Professor Perkin of Oxford to Tilden's original specimens and their true character as rubber has thus been established. The remains of the original specimens examined in 1892 and exhibited at the York meeting of the British Association in 1906 have been deposited in the Victoria and Albert Science Museum at South Kensington. As it has now been proved that true rubber can be made from the hydrocarbon isoprene for which the formula $CH_2 : C(CH_3)\cdot CH : CH_2$ was first proposed by Tilden, the question arises from what sources such a hydrocarbon can be manufactured on a large scale. It is scarcely necessary to point out that for the production of rubber turpentine is out of the question, on account of its cost and the comparatively small amount obtainable even supposing the whole world produce were available.

Since these investigations much research has been undertaken on the problem. One result is that it is now recognised that other hydrocarbons presenting the peculiarity of constitution exhibited by isoprene, namely the presence of two double linkages in the carbon chain, will also yield rubber-like substances. Another wholly unexpected observation has been made by Dr. F. E. Matthews. He found in 1910 that the hydrocarbon isoprene in contact with a small quantity of metallic sodium is converted in the course of a few hours or a few days, according to the temperature, into a mass of pure rubber. The process was of course patented. The same action of the metal was discovered soon afterwards and independently by Professor Harries. The great importance of this discovery of the sodium polymerisation process lies in the fact that it is not seriously interfered with by the presence of impurities, and that it does not require either a high temperature or any considerable consumption of time.

In what direction then are we to turn for a supply of a raw material from which isoprene or some similar hydrocarbon can be made ? The two requirements are that this raw material shall be cheap and procurable in indefinitely large quantities. The only substances fulfilling these conditions seem to be wood, starch or sugar, petroleum, and coal. To describe even very briefly the numerous attempts which have been made, in some cases with considerable success, to produce rubber from compounds originating in such materials would provide rather tedious reading. It will be sufficient to indicate briefly the general nature of the operations involved in two cases which appear among the most promising.

The first process is the subject of patents taken out by the Badische Anilin und Soda Fabrik, the famous colour-makers at Ludwigshafen. It starts with a fraction of petroleum spirit which consists essentially of a mixture of pentanes C_5H_{12}. These compounds are first exposed to the action of chlorine, and from the product is separated a mixture of mono-chlorinated derivatives $C_5H_{11}Cl$. These are brought into contact with heated lime by which the elements of hydrogen chloride are abstracted, and the result is a mixture of amylenes C_5H_{10}. These hydrocarbons which belong to the series of olefines are mixed with hydrogen chloride whereby one only, namely trimethylethylene, combines with the acid in the cold, the two others are separated and subjected to treatment by which they also are converted into trimethylethylene. The product of the union of this hydrocarbon with hydrogen chloride is the compound $CH_3{\cdot}C(CH_3)Cl{\cdot}CH_2{\cdot}CH_3$, to which it is not necessary to apply a name. This compound treated with chlorine yields two products

$$CH_3{\cdot}C(CH_3)Cl{\cdot}CHCl{\cdot}CH_3$$
$$\text{and } CH_3{\cdot}C(CH_3)Cl{\cdot}CH_2{\cdot}CH_2Cl$$

both of which when deprived of HCl by lime or soda yield isoprene

$$CH_2 : C(CH_3){\cdot}CH : CH_2.$$

There are several variations of the procedure which lead to the same result. In any case many operations are involved.

The second process for rubber synthesis is the property of the Synthetic Products Company. In this case isoprene is not the intermediate material aimed at. Starch in any cheap form is dissolved in boiling water which gelatinises it, and at the same time destroys other ferments accidentally present. A peculiar microbe discovered by Professor Fernbach of the Pasteur Institute is then added and a fermentation ensues which results in the production of a mixture of normal butyl alcohol and acetone. These are easily separated by distillation as their boiling points lie far apart. The butyl alcohol is converted by hydrogen chloride gas into butyl chloride, $CH_3{\cdot}CH_2{\cdot}CH_2{\cdot}CH_2Cl$, which is then acted on by chlorine gas with production of a mixture of dichlorides from which, by removing hydrogen chloride by passage over heated soda lime, the hydrocarbon butadiene $CH_2 : CH{-}CH : CH_2$ is produced. This compound is more volatile even than isoprene and has to be condensed to the liquid state by cooling, but like isoprene it undergoes condensation when kept in contact with sodium. The product is a rubber not identical with natural rubber, but one which is believed to be superior in some respects to that substance.

As in the production of isoprene the operations which lead to the production of butadiene are in practice open to various modifications. Further research is necessary to determine which of these methods and what details must be adopted to give the best results from the commercial point of view. It must be borne in mind in considering the fermentation process just outlined, that, in addition to the butyl alcohol required for making rubber, acetone is the other product. This liquid, hitherto obtained from acetic acid, a product of the destructive distillation of wood, is a solvent which has much increased in value since its employment in making cordite. This by-product then may become so profitable as to reduce the cost of butadiene rubber considerably, and so assist the synthetic in competition with the natural rubber.

A more recent patent taken out in Germany starts from acetone, which is first subjected to the action of fuming sulphuric acid, and ethylene gas is then passed into the liquid at a temperature of 100° to 110° C. Isoprene is said to be formed together with caoutchouc which is the product of its polymerisation. From 6 parts of acetone, 5 parts of raw caoutchouc, with $1\frac{1}{2}$ parts of isoprene, and other volatile products are stated to have been produced. Homologues of acetone, such as diethyl-ketone, are included in the patent (*India-rubber Journal*, Nov. 6th, 1915, p. 12). Many of these attempts are not likely to be successful from the commercial point of view, but the frequent recourse to patent protection is an indication that chemists are still busy with the problem.

The production of synthetic alizarin and indigo, and the influence of the resulting manufacture of these dyes on the cultivation of the madder and indigo plants respectively, have been discussed in many quarters as indicating the possible fate of the rubber plantations which, during the last twenty years, have extended over very large areas of land in the East. The case of rubber, however, appears to present one feature in which it differs from the position of madder and indigo. For these two materials the demand though very large is limited, whereas the uses of rubber multiply in the imagination of anyone who seriously considers the question.

Should the many difficulties at present attending the several synthetic processes be successfully overcome, the first effect would probably be a fall in the price of rubber generally, but it would then find applications on a scale much greater than anything previously known, as for example in paving the streets, and the extra demand would for a time at least tend to raise the price again. At any rate there seems no reason at the present stage of the researches which are going forward for rubber planters to entertain alarm. The change, if it came about, would not take place in a moment, and there would be ample time for the land now occupied with rubber trees to revert to its primitive use in the provision of food.

CHAPTER XXVI

EXPLOSIVES

"... it was great pity, so it was,
This villainous saltpetre should be digg'd
Out of the bowels of the harmless earth,
Which many a good tall fellow had destroy'd
So cowardly; and, but for these vile guns
He would himself have been a soldier."

When Shakespeare put these words into the mouth of Hotspur the only use for gunpowder was in the practice of war, and for purposes of destruction such as was contemplated in the Gunpowder Plot of 1605. But though at the time of writing this book the greater part of Europe is devastated and millions of men are exposed to destruction by the wholesale use of explosives in war, it must not be forgotten that these agents have been among the most powerful auxiliaries in the arts of peace. It is only necessary to consider how many roads, railways, tunnels, and water works have been rendered possible by the use of dynamite and other blasting materials to perceive that explosives have a civilising mission of their own, and probably next to steam have done more to facilitate inter-communication between different countries than any other of the works of man's invention.

The chemist of the twentieth century is acquainted with a large number of substances which when heated or struck or in some cases even merely shaken explode, but the great majority of them are useless for practical purposes, being too unstable to be handled or carried about without great danger to the person. By an explosion the chemist understands the sudden production of a relatively large volume of a gas or gases from a solid, liquid, or mixture of gases. And as such changes are almost always attended by the production of much heat, the hot gases formed are still further expanded. For the moment the last case must be postponed from consideration, but the reader will easily understand what is referred to by thinking of the disastrous effect in a coal-pit when a mixture of air with inflammable gas from the coal, called *fire-damp*, comes into contact with a flame. The resulting explosion which, under such circumstances, does nothing but mischief, can in another form be turned to useful practical account when under control in the gas-engine or internal combustion engine of the motor.

But although explosive substances are familiar in the chemical laboratory, and have multiplied among the products of modern chemical research, it is curious to note that nearly all the explosives employed as propellants or for blasting purposes are produced more or less directly by the use of the "villainous saltpetre" so long an ingredient in old-fashioned black gunpowder. The object in all cases is to introduce into a mixture or compound containing the combustible elements, carbon and hydrogen, so large a quantity of oxygen that the product will burn without the assistance of atmospheric air.

This is effected in the case of gunpowder through the agency of the nitre or saltpetre which supplies oxygen to the sulphur and charcoal with which it is mixed. Or it may be by bringing cotton or glycerine or phenol or some other compound of this kind into contact with nitric acid. An interchange is then effected whereby a portion of the hydrogen of the original substance is removed in the form of water and the group of atoms, NO_2, characteristic of the nitrates is introduced. When the nitrated compound is fired the oxygen combines with carbon forming gaseous oxides of carbon, and with the hydrogen forming water, which is of course liberated in the form of steam, while the nitrogen is set free in the state of gas and thus contributes to the total volume of gas formed in the act of explosion.

This chapter must be devoted to an account of the chemical composition and action of the modern explosives, some of them of quite recent introduction, but to understand why some of the changes which have taken place of late years have been introduced, it is necessary in passing to glance at the changes which have taken place in the construction of military and naval guns.

At the time of the Crimean War the largest guns ashore or

afloat were the 68 pounders with smooth bores. The idea of rifling the gun for the purpose of giving the projectile the spin which increases greatly its accuracy of fire had not at this time been actually adopted in practice. With this very important change two names will always be connected, the late Lord Armstrong (died 1900) and the late Sir Andrew Noble (died 1915), who for some forty years were associated together in the great Elswick Ordnance Works near Newcastle-on-Tyne. To the former we owe the rifled breech-loading gun with wire-wound cylinder, to the latter the invention of the chronoscope, by which minute fractions of time may be measured, beside famous experiments on the pressures attained in large guns.

Up to about 1886 black gunpowder had been used, but as it had been found that with increased length of the gun the pressure on the breech became injurious to the gun without giving the desired velocity to the projectile, many modifications were tried in the size of the grain, and in the cubes, prisms or perforated slabs in which form the powder was used. The old powder, however, had one inseparable defect, namely, the large quantity of smoke produced in firing. This arises from the fact that black gunpowder is composed of nitre, charcoal, and sulphur in the proportions on the average of 75 : 15 : 10 per cent respectively. Hence when burnt the potassium of the nitre is converted into a mixture of potassium carbonate, potassium sulphate, with a small quantity of potassium sulphide, all of which are solids, and being dispersed in fine powder give rise to clouds of smoke. At the time referred to the service powders used by the various European Powers had the composition shown in the following table :—

Country.	Nitre.	Charcoal.	Sulphur.
England Black Powder	75	15	10
,, Brown ,,	79	18	3
Sweden	75	15	10
Russia	75	15	10
Prussia	74	16	10
Saxony	74	16	10
United States	76	14	10
Austria	75·5	14·5	10
France	75	12·5	12·5

According to Thorpe's *Dictionary of Applied Chemistry* Chinese gunpowder contained of nitre 61·5, charcoal 23, and sulphur 15·5 parts per cent. This departure from the type, which has been established by modern scientific methods of manufacture, is interesting when the tradition is recalled which attributed the invention of gunpowder to the Chinese.[1]

Changes in the guns then demanded changes in the rate of combustion of the powder used in them, while the conditions of modern warfare required a propellant which should be practically smokeless. It seemed useless to construct quick-firing guns and machine guns capable of delivering a shower of bullets if after the first discharge or two all view of the enemy in front of the guns became impossible. Gun-cotton, which is the essential basis of all modern propellants, differs from the old powder in yielding only gaseous products in its explosion, without any solid and hence without smoke. There is also an important difference between the two, in the fact that the old powder is merely a mechanical mixture of solid ingredients, the particles of which, under a microscope, can be seen lying side by side but quite distinct from one another, while gun-cotton is a chemical compound. In the former, therefore, the oxygen required to combine with the sulphur and with the carbon of the charcoal has to be liberated first from the particles of the nitrate and then to attack separately the particles of the combustible sulphur and carbon.

In gun-cotton and similar substances, each molecule of the compound contains within itself the elements which are to combine together to form the gaseous products of the explosion. This will be understood by reference to the equation given below.

Cotton consists of the hairs from the seed of the cotton plant (*Gossypium herbaceum* and other species, N.O. *Malvaceæ*). When looked at with a microscope they are seen to consist of long flattened twisted tubes of translucent substance. This

[1] The invention of gunpowder is by the English attributed to Roger Bacon, who was born in 1214 Others suppose a certain monk, of whom nothing positive is known, but who is supposed to have lived in the early part of the fourteenth century, to have been the inventor. He is commonly spoken of as Berthold Schwarz, a purely imaginary name.

Gunpowder and cannon were known to have been used in England in 1344, in France in 1338, and the Oxford MS. "De officiis regum," dated 1325, gives an illustration of a gun. The invention of gunpowder must therefore be placed at an earlier date.

Those who are interested in the history of the subject should consult *Monumenta Pulveris Pyrii*, by the late Oscar Guttmann, 1906.

substance is called *cellulose*, it has the composition expressed by the formula $(C_6H_{10}O_5)n$, and it forms the fundamental material of vegetable tissues in general. Clean cotton consists of almost pure cellulose, and when ignited it burns away leaving only a minute quantity of mineral matter in the form of ash.

When cotton is immersed in strong nitric acid an interchange takes place which may be expressed by the following equation :—

$$(C_6H_{10}O_5)_2 + 3HNO_3 = [C_6H_7O_2(NO_3)_3]_2 + 3H_2O$$

in which it is obvious that the product is a nitrate, and its formation is comparable with the production of a nitrate when caustic potash is mixed with nitric acid. Water is in both cases formed simultaneously :—

$$KHO + HNO_3 = KNO_3 + H_2O.$$

In the case of cellulose three stages of nitration are possible, the products being represented by formulæ, thus :—

$$[C_6H_9O_4(NO_3)]_2 \qquad [C_6H_8O_3(NO_3)_2]_2 \qquad [C_6H_7O_2(NO_3)_3]_2.$$

It has long been known that when starch, paper, cotton fibre, or other vegetable material is soaked in very strong nitric acid and is subsequently washed in water and dried, the cotton or other material is scarcely changed in appearance, but it is found to have increased in weight, 1 part of cotton giving, according to the theory explained above, 1·8 parts of nitrated cotton. This material is extremely inflammable, and on contact with a flame disappears instantaneously with a bright flash. The Swiss chemist Schönbein, so long ago as 1845, proposed to use this product as a substitute for gunpowder. It was, however, many years before the manufacture of gun-cotton could be carried on without danger of explosion, and before the product could be obtained in a condition in which it could be stored and used for any purpose with reasonable safety. A long series of experiments, conducted first by the Austrian General von Lenk, and later by Sir Frederick Abel in this country, led to the discovery of the conditions necessary for this object, the first essential being the removal of the last traces of acid from the nitrated cotton.

At the present day gun-cotton as well as nitroglycerine, to be described later, is manufactured in large quantity in many countries in which the regulations controlling the operations vary. In the United Kingdom the Explosives Department of the Home Office prescribes the conditions which must be obeyed.

The following account of the manufacture of gun-cotton is chiefly taken from a lecture given by Mr. William Macnab before the Institute of Chemistry of Great Britain and Ireland in February, 1914.

In laying out explosives works it is necessary to distinguish the danger area from the non-danger area. In the latter, boilers, engines, acid stores, and other departments may be arranged in any manner found to be most convenient, but in the former where the manufacture of the explosive is carried out the case is quite different. " The object of the restrictions is to allow only limited quantities of explosive material and a limited number of work-people in one building at a time, and further to place the different buildings at such distances from each other, or surround them by protecting earth mounds (Fig. 131), that in the event of an explosion the effect is localised as much as possible, and the explosives in the adjacent buildings are not 'set off.'" Special precautions are taken to prevent the accumulation of dusty explosive matter, and scrupulous cleanliness is enforced. No naked iron or steel is allowed where the more explosive materials are treated; the workers have to wear shoes containing no iron or steel nails; and in order to prevent the introduction of grit from the outside those entering the building temporarily have to slip on large shoes which are kept at each building specially for this purpose. Everyone on entering an explosive works has to give up any matches he may have in his possession; the work-people have to wear special outer clothing without pockets; and women have to fix their hair without pins which might possibly fall in among the explosives with which they are working.

The lighting of the buildings is nearly always electrical, and where motive power is required, it is usually supplied by electric motors placed outside the building.

It is not permissible to use a house for a different operation from that for which it is licensed without special authorisation. Serious penalties follow the breach of the terms of the licence under which the factory is allowed to work, and surprise visits from the Inspectors of Explosives help to maintain a good state of discipline.

The manufacture of gun-cotton and the other forms of nitro-cellulose is carried out in the first stages in the non-danger part of the factory. The raw material is cotton waste, which is specially prepared for the explosive manufacturer. First it is hand picked in order to remove all foreign matter as much as possible, and it it amazing to see how much rubbish in the form of pieces of wire, wood, nails, etc., is thus removed. Next it is teased and dried, because cotton ordinarily contains about 10 per cent of moisture and this water would needlessly dilute the nitrating acids. The photograph (Fig. 133) shows a drying plant in use at Waltham Abbey. Here it is exposed to a temperature of about 80° C. for twenty minutes. It is then weighed up, according to the older method introduced by Sir Frederick Abel, into lots of $1\frac{1}{4}$ lb. called a charge, and is kept dry in an air-tight box till it is dipped.

The acids used consist of a mixture of 1 part by weight of strong nitric acid of specific gravity 1·5, with 3 parts by weight of strong sulphuric acid of specific gravity 1·84. Mixing the acids is attended by evolution of heat and the mixture is allowed to become completely cool before it is run into the cast-iron dipping tank.

The charges of cotton are immersed in the acid for a few minutes, then placed on a grating and the excess of acid squeezed out. The partially changed cotton, still saturated with acid, is placed in an earthenware covered pot standing in water, and left for about twelve hours (Fig. 134). The nitration is then complete, and the contents of the pots are lifted out by tongs and placed in a centrifugal machine, where the excess of acid is wrung out. The gun-cotton is then placed in a tank full of running water till the water no longer answers to a test for acid.

To remove the last traces of acid the cotton requires to be boiled with water repeatedly. It is then reduced to pulp by means of a machine similar in construction to the machines used by paper-makers. It is then in a very fine state of division, and, suspended in water, is passed by a pipe into the " poaching " machine, where paddles keep the fine pulp agitated with water and thoroughly wash every portion of it. After some hours a small quantity of lime-water, whiting, and caustic soda is added so as to leave the cotton pulp slightly alkaline. It is then drawn off by means of a vacuum pump, and the pulp strained off in measured quantities into moulds, where pressure is applied sufficient to reduce the substance to the condition of a solid cake hard enough to bear handling. Finally, the moulded cotton is submitted to hydraulic pressure amounting to about five tons

Fig. 131.—MOUNDED HOUSE. COTTON POWDER WORKS.

Fig. 132.—BUILDINGS AND PIPE CONNECTIONS. COTTON POWDER WORKS.

To face page 380.

FIG. 133.—DRYING MACHINE. WALTHAM ABBEY.

FIG. 134.—ABEL NITRATION PROCESS. DIPPING PANS. WALTHAM ABBEY.

To face page 381.

on the square inch, which leaves the cake so hard that it does not yield perceptibly to pressure by the finger.

Newer methods of nitration have been introduced by which a larger quantity of cotton can be immersed in the acids at one time.

Centrifugal machines have been constructed which can be filled with the acids and a much larger weight of cotton, generally about 17 lbs., can be immersed. When the nitration is complete the acid can be run off and the cotton drained by setting the machine in motion.

Another method employed at the Royal Factory, Waltham Abbey, is known as the displacement process. The plant consists of shallow earthenware circular pans grouped together in sets of four. They are provided with perforated false bottoms, and the bottom of each pan is connected with a pipe by which the nitrating acid can be supplied, and a pipe by which the spent acid can be drawn off. These pans will each take a charge of 20 lb. of dry cotton.

Hoods connected with an exhaust fan draw off the fumes from the acids, and these hoods are made of aluminium, a metal which is practically unacted on by nitric acid. When all the cotton is immersed perforated earthenware plates are laid on top of the cotton to keep it under the acid, and a thin layer of water is cautiously run over the surface of the acid. This prevents the escape of acid fumes and allows of the removal of the hoods. After two and a half hours the nitration is complete ; the spent acid can be drawn off, and an equivalent quantity of water run into each pan. In this way the spent acid is displaced much more completely than by the older methods.

After draining off the water from the pans the gun-cotton is ready for the processes of purification already described.

Up to this point the nitrated cotton has been treated as non-explosive, but in order to dry it, it is removed to one of the stoves in the danger area. Dry gun-cotton is one of the most dangerous explosives, as when dry and warm it is very liable to explode by friction, and the greatest care has to be exercised in handling it.

In the production of gun-cotton the composition of the acid mixture is of the utmost importance, and if the sulphuric acid present is deficient in amount, or the proportion of water formed in the process is allowed to exceed a certain amount the nitration does not reach the maximum. Nitrocellulose having the composition expressed by the formula given above contains just over 14 per cent of nitrogen. Gun-cotton, however, usually contains somewhat less than this percentage, namely, about 13·3 per cent, owing probably to the presence of small quantities of one of the lower nitrates, the formula of which has already been given.

Generally speaking the lower nitrates are soluble in a mixture of ether and alcohol, while gun-cotton is not dissolved by this liquid.

The solution of these lower nitrates in ether-alcohol constitutes "collodion." It must be remembered that cotton is not strictly speaking a definite chemical substance, and it varies somewhat in physical state, and hence that cottons from different sources, under the same conditions in a bath of the same composition, while yielding nitrocellulose containing the same percentage of nitrogen, may vary considerably in solubility. In the early days a high degree of nitration, say 12·8 per cent of nitrogen or upwards, was generally associated with insolubility in ether alcohol, while lower content of nitrogen corresponded with greater solubility. With greater experience, however, it is now possible to produce nitrocellulose with a high percentage of nitrogen and complete solubility in ether-alcohol.

Gun-cotton requires a lower temperature than gunpowder for its ignition. The rate at which it burns depends on the mode of ignition and the conditions under which it is fired. A mass of loose gun-cotton may be ignited on the open hand without burning the skin or producing more than a momentary sensation of warmth, while the same cotton lightly twisted would produce a burn, and if confined in any sort of strong envelope would explode. The difference consists in the rate at which decomposition is transmitted through the mass, and the discovery that the explosion of a detonating fuse containing fulminate of mercury or some similar compound in contact with a mass of gun-cotton would cause it also to explode was a step of great practical importance.

Nitroglycerine, a compound similar in constitution to nitrocellulose, both being nitrates, was discovered by Sobrero, an Italian chemist, in 1847. Though its explosive properties were known it was regarded as dangerous, and was not generally used as a blasting agent till after 1867 when Alfred Nobel discovered a method of rendering it portable and less dangerous by incor-

Fig. 135.—ABEL NITRATION PROCESS. WALTHAM ABBEY.

Fig. 136.—BEATING ENGINES AND POACHER. WALTHAM ABBEY.

To face page 382.

FIG. 137.—MOULDING MACHINE. WALTHAM ABBEY.

FIG. 138.—DISPLACEMENT PROCESS. WALTHAM ABBEY.

To face page 383.

porating the liquid with a sufficient quantity of a fine silicious earth, called kieselguhr. The product is dynamite, which is familiar enough by name to the public.

Nitroglycerine is produced very simply by the interaction of a mixture of nitric and sulphuric acid with pure glycerine.

Glycerine is the secondary product obtained in boiling fat or oil with caustic alkali for the purpose of producing soap. But a large quantity is also produced by distilling fats in super-heated steam, when the fatty acid and glycerine are obtained, and it is only necessary to evaporate the watery part of the distillate to obtain the glycerine.

Glycerine, or glycerol as it is called in systematic chemical language, is a familiar colourless syrupy liquid, with a sweet taste. It mixes with water in all proportions, and when mixed with nitric acid it is converted into the nitrate, or nitroglycerine, at the same time that water is produced :

$$C_3H_5(HO)_3+3HNO_3=C_3H_5(NO_3)_3+3H_2O.$$

While formerly only small quantities at one time of glycerine were acted on by the acids, a charge of 1400 lbs. of glycerine may be now used in one operation in the apparatus called a nitrator-separator. In the modern practice a mixture of strong nitric acid with sulphuric acid is used, to which is added a certain amount of anhydrous sulphuric acid in the form of what is called *oleum*, which combines with a larger proportion of water, with the result that the yield of nitroglycerine is not far short of the theoretically possible amount. From the formulæ 100 parts of glycerine should yield 246·7 parts of the nitrate, while in practice upwards of 230 parts are obtained.

" The nitrator separator is a cylindrical leaden vessel with a coned top ; inside are placed leaden coils, through which cooling water circulates, and pipes through which compressed air is blown to mix the contents. The glycerine is introduced in the form of a fine spray under the acid by means of a special injector worked also by compressed air. Long thermometers passing through the top of the nitrator-separator enable the temperature to be watched, and it is the business of the man in charge of the operation to see that the temperature does not rise beyond a certain point, generally 28° C. By reducing the flow of the glycerine and by increasing the agitation with the air any undue tendency to rise can usually be checked.

"Should, however, the temperature continue to rise and pass the danger mark then a large cock in the bottom of the nitrator is opened, and the contents are rapidly discharged into a large tank, containing water, outside the building, where the charge is 'drowned,' and thereby the danger avoided of serious decomposition and probable explosion.

"When everything goes right the nitration of the charge is usually completed in about one hour, the agitation with the air is discontinued, and the separation of the nitroglycerine from the acids takes place; being lighter it comes to the top. A pipe in which a glass window is fitted leads from the top of the nitrator-separator to a pre-washing tank; by allowing waste acid from a previous operation to enter at the bottom the nitroglycerine is forced over into the washing tank, and the flow of acid is stopped whenever all the nitroglycerine has passed into the washing tank, which can be observed through the window." In the washing tank the nitroglycerine is stirred up repeatedly with fresh water, then with a solution of sodium carbonate, and finally with water. After this it is filtered to remove traces of water or impurities.

Nitroglycerine is a colourless oil of specific gravity 1·6, and therefore sinks in water in which it is insoluble. It has a sweetish taste and is poisonous. In minute doses it is used in medicine. When a lighted match is applied it burns quietly away, but it detonates violently when struck on an anvil by a hammer or by sudden heating to 257° C. Nitroglycerine becomes solid when exposed to frost and in use it requires to be thawed, an operation attended by considerable risk.

When nitroglycerine is exploded it yields a mixture of carbon dioxide and nitrogen with 4 per cent of free oxygen, whereas when nitrocellulose is fired the carbon dioxide and nitrogen are accompanied by carbon monoxide and a considerable quantity of free hydrogen. In the latter case the relative proportions of these gases vary with the pressure developed in the space in which explosion occurs. It appears that even when oxygen is present in excess, oxides of nitrogen are never formed in a normal explosion. Nitrous fumes are however formed when one of these high explosives burns freely without explosion.

In 1875 it was discovered by Alfred Nobel that when a low grade of gun-cotton and nitroglycerine are mixed together the cotton loses its fibrous or cellular structure and becomes gela-

FIG. 139.—NITRATOR-SEPARATORS.

FIG. 140.—CORDITE MIXING MACHINE.

To face page 384.

Fig. 141.—CORDITE PRESS.

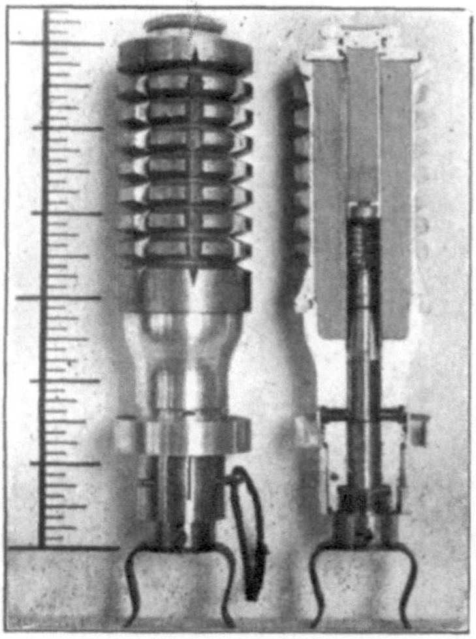

Fig. 142.—MARTIN HALE'S GRENADE.

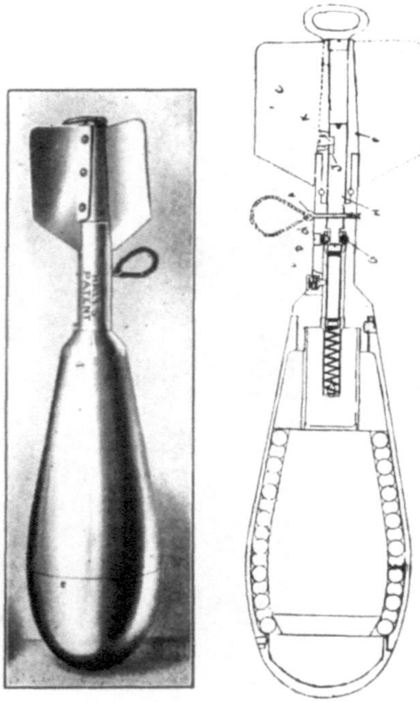

Fig. 143.—MARTIN HALE'S BOMB.

To face page 385.

tinised. In the product each constituent has its explosive properties modified, and the mass becomes better suited to blasting purposes than either ingredient separately. This substance has been largely used under the name "blasting gelatine," and it is otherwise interesting as the forerunner of the various mixtures which have been the subject of experiment and which have resulted in the production of the chief military propellant *cordite*. It was discovered that not only could the lower nitrocelluloses be gelatinised by nitroglycerine, but that the most highly nitrated cotton could be blended with nitroglycerine if the mixture was treated with a common solvent such as acetone.

To manufacture cordite the nitroglycerine is poured on to the gun-cotton contained in rubber bags and hand-mixed. The paste produced is then transferred to a large Pfleiderer mixing machine, similar to the machine used in some bakeries for mixing dough, and the requisite quantity of acetone added. After working the mixer for some time, 5 per cent of vaseline is added to increase the stability of the product and lubricate the gun. When gelatinisation is complete the mass is pressed through a die of the requisite size, and the *cord* which is thus formed wound on a reel, or in the case of the thicker sizes it is cut into suitable lengths. The cordite is then dried slowly to drive off the last traces of acetone. In the case of the larger sticks, containing the smaller quantity of nitroglycerine, 30 per cent, this drying takes about two months.

It is interesting, says Mr. Macnab, to note the accuracy which has been attained in this manufacture. For rifles, for instance, the velocity prescribed is 2380 foot seconds, with a *plus* or *minus* of only 40 feet, and a pressure of 19·5 tons, with a maximum of 20 tons per square inch; for larger guns it may be 2500 foot seconds + 15 foot seconds, and the pressure must not exceed 19 tons per square inch.

In July last (1915) Professor Vivian Lewes[1] of the Royal Naval College, Greenwich, in some lectures delivered before the Royal Society of Arts gave some interesting facts concerning the explosives used in the European war, from which the following condensed account is taken. The shells used in big guns and field artillery may be divided into two main classes, namely shrapnel and high explosive shells. The shrapnel shell, named after its inventor, is a hollow cylindrical projectile packed with

[1] Professor Lewes, unfortunately, died on October 23rd, 1915.

bullets, at the base of which is a bursting charge, which may be gunpowder or a high explosive, while in the nose of the shell is arranged the time fuse connected by a tube with the bursting charge. This can be so regulated that the shell bursts in the air at any desired point. Shrapnel, however effective against troops in the field, does but little damage to earth works, wire entanglements, and other defences. Hence for the latter purpose high explosive shells are required. These consist of forged steel with comparatively thin walls and a heavy bursting charge. The explosive with which such shells are charged is usually one of the products of nitration obtained by acting on one or other of the constituents of coal-tar (see Chapter XX, p. 310) with strong nitric acid.

Phenol or carbolic acid mixed first with an equal weight of strong sulphuric acid and the compound introduced gradually into three times its weight of strong nitric acid gives trinitrophenol or picric acid. This is a lemon yellow crystalline substance which has long been used as a dye for silk and wool. It melts at 122°·5 C., and is a moderately strong acid, forming a variety of salts with bases.

Many of the picrates explode when heated or struck, but picric acid burns quietly. When the fused acid is supplied with a detonator it explodes violently, and it has been largely used under the name lyddite, or melinite, for charging shells. Experience in the South African War showed that lyddite shells are, however, somewhat erratic.

Trinitrotoluene, T.N.T., is found to be more trustworthy, and though its explosive force is somewhat less than that of picric acid it is preferred on account of its stability, and being not an acid but perfectly neutral it is not liable to attack the surface of metals.

Toluene is a colourless liquid which by the action of strong nitric acid is converted successively into three nitro-compounds :

C_7H_8	toluene
$C_7H_7NO_2$	mononitrotoluene
$C_7H_6(NO_2)_2$	dinitrotoluene
$C_7H_5(NO_2)_3$	trinitrotoluene or T.N.T.

Trinitrotoluene is a yellowish crystalline powder with a melting point about 79° C. When detonated by mercuric fulminate it explodes with great violence giving a quantity of black smoke, whence some of the names—Black Maria or Coal Box—given by the soldiers to shells of this kind.

T.N.T. is sometimes mixed with other substances, especially with an oxidising compound such as ammonium nitrate, together with a little aluminium powder and a trace of charcoal, the mixture being known as ammonal.

Other constituents of coal-tar yield explosive compounds under the action of nitric acid.

Dinitrobenzene, for example, enters into the composition of the mining explosives roburite and bellite. Trinitrocresol has been used in place of picric acid under the name ecrasite, but it shares the disadvantages of picric acid.

Cheddite is a name given to a permitted explosive containing potassium chlorate mixed with mononitronaphthalene, dinitro-toluene, and a little castor oil. Another variety of cheddite contains ammonium perchlorate.

Probably the most powerful explosive known is tetranitro aniline, and another similar compound tetranitromethyl aniline, known as "tetryl," is already used for detonators in place of mercuric fulminate. Another compound which has recently found application as a detonator is lead hydrazoate or triazide, PbN_6, derived from hydrazoic acid or azoimide HN_3. The acid itself when in the pure anhydrous state and some of its organic derivatives are among the most dangerously explosible compounds known, as they sometimes explode violently without obvious cause. But several of the metallic salts, such as the lead salt mentioned above, and the barium salt, are fairly stable and can be manipulated without risk, if proper precautions are taken.

In blasting operations gunpowder and detonators are fired by a time fuse or electrically. The time fuse is a case containing gunpowder which is made to burn at a known rate, generally 2 feet per minute. The instantaneous fuse which burns at the rate of 100 to 300 feet per second affords the means of firing many charges simultaneously.

Of the bombs which have come into use in warfare with the development of airships and aeroplanes there are several varieties. The British air service during the war is understood to have made use of the bomb designed by Marten Hale, Fig. 143. This is an ingenious arrangement which has the great advantage that it can be handled and transported quite safely. In the neck of the bomb is a propeller which, when free and falling through the air, spins round and releases the detonator so that on impact it flies forward into the bursting charge and strikes a firing needle which causes the explosion of the bomb. Before being dropped a pin which holds the vanes is withdrawn and a fall of 200 feet suffices to set the vanes spinning as described.

The incendiary bombs used by the Germans consist of a shell wound round with tarred rope and containing a quantity of resin and other inflammable matter, in the midst of which is a charge of "thermit" (p. 262) with usually a quantity of red phosphorus at the bottom. In thermit advantage is taken of the very high temperature produced by the combination of metallic aluminium with oxygen. A mixture of fine powder of aluminium with oxide of iron was introduced under this name about 1898 for the purpose of welding together steel rails, repairing castings, or heating iron bolts white hot. The mixture being packed round the object to be heated is ignited by means of a piece of magnesium ribbon which can be lighted by a match. The iron in the oxide is reduced to the metallic state and remains when the action is over as a fused mass.

Enough has now been written to show the reader the general character of the chemical mixtures and compounds employed for military and naval use and for the peaceful purposes of the miner. But the subject is a very extensive one, and those who desire more technical information can only be advised to read the article on Explosives in Thorpe's *Dictionary of Applied Chemistry*.

An interesting application of explosives to the purposes of agriculture has attracted some attention during very recent years, especially on the other side of the Atlantic. In new countries land has often to be cleared of wood and sometimes of masses of rock before it can be brought into cultivation. In order to get rid of trees it has been the custom in past times to burn them and leave the stumps to rot, before attempting their removal. This necessarily occupies a good many years, and the work is difficult and laborious.

As soon as modern explosives became available the idea of blowing up such obstructions naturally arose and has been put into operation on a considerable scale. But latterly the use of dynamite has been resorted to for the purpose of preparing holes for planting fruit trees and for loosening the soil between trees in orchards. As with every newly introduced practice there has been evidence of some degree of exaggeration in the reports which have appeared in the press concerning the advantages of soil explosions.

There can be no doubt that the aeration of the soil, the breaking up of the subsoil, especially when hard, the destruction of vermin, and the saving of labour are advantages generally recognised. The two questions in respect to soil improvement by explosion which must be considered are first whether it is effectual in all cases, and secondly does it pay? There seem on both these points to be as yet a lack of unanimity, which perhaps is due to want of experience, as the method is so new. The experiments on limes, bananas, and other crops in the West Indian Islands as reported in the *Agricultural News* published by the Imperial Department of Agriculture, Barbados (March 11, 1916), show that much further experience is necessary before a definite conclusion can be reached, as it appears that in these islands and for the crops referred to the results obtained have not been encouraging.

The phenomena of combustion and explosion in gases have an interest both for the scientific man and for the coal miner, exposed as he is in the majority of pits to imminent risk in his daily work.

During the last forty years great advances have been made in the theory of gaseous explosion, and in a knowledge of the rate of transmission of an explosion wave. The first steps in this direction were taken by the famous French chemist, M. Berthelot. At the time of the siege of Paris in 1870, Berthelot, then Professor in the Collège de France, became President of the Scientific Committee of National Defence. The superintendence of the manufacture of explosives to be used against the enemy naturally led him, after the war, to turn his attention to the systematic investigation of the phenomena of explosions. In the result he was able to connect the maximum velocity of the flame in a mixture of gases with the mean velocity of the molecules, according to the kinetic theory of gases. A long series of researches on the propagation of flame through mixtures of gases and on cognate subjects was begun by Messieurs Mallard and Le Chatelier in 1879, and the work of these distinguished French investigators is still frequently referred to.

Another very important discovery was made in 1880 by Mr. Harold B. Dixon, a few years later Professor of Chemistry in the

University of Manchester. Dixon found that carbon monoxide mixed with oxygen, when dried as perfectly as possible, by long contact with phosphoric oxide, does not explode when an electric spark is passed through the gas. The admission of a minute trace of water vapour at once restores to the mixture its inflammability. This discovery has been very fruitful in the way of discussion, and a hypothesis put forward soon afterwards to the effect that chemical combination between two substances was impossible without the presence of a small quantity of a third substance met with a good deal of favour. This hypothesis seemed to be further supported by discoveries of a similar kind made a few years later by Dr. H. Brereton Baker, now Professor in the Imperial College of Science and Technology at South Kensington. Dr. Baker's experiments showed that carbon, sulphur, and even phosphorus, when carefully dried, refuse to burn in oxygen when heated above the temperature at which they usually ignite. He also found that ammonia mixed with hydrogen chloride, and nitric oxide with oxygen are indifferent when the gases are well dried. Whether in all cases a third substance is essential to the act of chemical combination must, however, be still regarded as an open question, notwithstanding the interesting suggestiveness of the experiments referred to. Much has yet to be learnt as to the constitution of gases and the real nature of chemical action, especially since the doctrine concerning electrons and their functions has become generally accepted (see pp. 118 and 212).

Notwithstanding the greatly increased knowledge in our time about the properties of inflammable gases and of the conditions prevailing in coal pits, it is, unhappily, true that disastrous explosions continue to occur, in which many lives are lost, as they were before the invention of the safety lamp, in 1817, by Sir Humphry Davy. This fact is, of course, no ground for argument against the utility of the safety lamp.

The explosions which occur are due either to abuse of the lamp, to gross neglect of rules by miners, to blown-out shots, or some other cause. Among the sources of danger not recognised a few years ago is the accumulation of fine coal-dust in many workings.

Attention was first called to the subject by Mr. William Galloway so long ago as 1876, and much discussion and experimentation has been carried on since that time. The presence of fine coal-dust suspended in the air of a mine has long been known to add to the danger of explosions when they occur from presence of fire-damp, but it has only been recognised within recent years that dust alone, diffused through air, forms an explosive mixture through which flame is propagated, when once started, with the violence characteristic of gas explosion.

In France and in England large scale experiments have been carried out within the last few years which have supplied very valuable information. The English experiments at Altofts have been provided for by the Mining Association of Great Britain, and have been described by Professor Dixon in his Presidential Address to the Chemical Society (London) in 1911 in the following passage :—

" An iron gallery 600 feet long and $7\frac{1}{2}$ feet in diameter was constructed of cylindrical boilers bolted together. Inside a tram-line on a concrete floor, with props and cross-timbers placed at 9 feet intervals, made a travelling road, comparable with the main haulage road of a mine. Shelves fastened to the sides provided ledges for holding dust, and the flame of a blown-out shot was reproduced by firing a stemmed gunpowder charge from a cannon. Just before firing a current of air was drawn into the main gallery by a fan placed at the end of a 'return' gallery. By this means a pure coal-dust explosion, extending over several hundreds of feet, could be obtained, and the propagation of the flame and pressure studied."

The reports of the French Coal-Dust Experiments conducted at the Liévin Experimental Station, near Lens, in 1907–10, have been published in English by the *Colliery Guardian*. Experiments were made similar to those described, and with similar results. The principal gallery, constructed originally only 71 yards long, was extended till in 1910 it was 328 yards long, with an internal height of 6 feet.

The fact thus established is consistent with what is known of other dust explosions, as in flour mills, where there can be no question of the existence of inflammable gas in the atmosphere. The initiation of the flame does not apparently depend on the production of gas from the dust by a preliminary process of distillation, and Dr. R. V. Wheeler, who has been in charge of the laboratory at Altofts, has been able to show that an explosion is propagated through a cloud of *charcoal* dust in air.

With the object of limiting the risk of explosions in coal mines, whether originating from gas or dust, the explosives to be used in fiery or dusty mines have to pass a Government test. A testing gallery has been erected by the Home Office at Rotherham, and there the effects of various explosives on an explosive mixture of gas and air are carried out. A charge is fired from a gun with a 2-inch bore, which represents a bore-hole, into the cylinder containing an explosive mixture of gas and air, or air in the presence of coal-dust laid along the cylinder. Shots are then fired electrically till the largest charge is found, which can be fired without igniting the mixture. Further shots are then fired till five shots of the same weight have been fired without igniting the mixture.

Strictly speaking there is no such thing as a perfectly safe explosive; under certain unfavourable conditions they will all ignite gas or coal-dust, but the "permitted test" does enable the various explosives to be sorted into grades of safety, and only those which have shown themselves to be the safest are allowed to be used (Macnab).

In consequence of the extensive manufacture and use of explosives in modern times it has been necessary, in all civilised countries, to regulate by legislation the conditions under which they may be made, stored, and distributed. Many of the enactments are self-evident in their application: buildings for the factory must be licensed, stores in mines and quarries must be registered, explosives must be properly packed, and imports from abroad require special licences. Inspectors are also appointed whose business it is to make surprise visits for the purpose of observing that the regulations for the safety of workpeople and all the conditions of the licences are duly carried out.

Explosives of any new composition require to pass a strict examination before they are authorised, and all must be in a condition which indicates reasonable safety when kept and freedom from serious danger from friction or blows when packed or in transit.

Explosives differ considerably in stability, some being liable to slow decomposition which in course of time may assume a dangerous character. This is especially true of the nitric "esters," that is so-called nitrocelluloses and nitroglycerin, and more especially if the temperature is somewhat elevated as in tropical countries, in the holds of ships, and especially in positions where the temperature may be raised in consequence of the position of boilers, or of cargo like coal, which may undergo chemical change and therefore possible spontaneous heating.

When nitrocellulose commences to decompose from the presence of minute traces of acid, a mixture of oxides of nitrogen is given off among which nitrogen peroxide is recognisable by its orange-brown colour. A test therefore is based on the heating of the material in a long narrow test tube to 135° C. and noting the lapse of time before the first faint yellow colour is seen in the air contained in the tube. A more delicate test consists in heating in a closed test tube a quantity of the cotton to a prescribed temperature, while a piece of paper impregnated with a mixture of starch with an iodide and moistened at the end with glycerine is suspended in the tube. The number of minutes which elapses before the paper becomes discoloured serves to indicate the quality of the explosive according to its class.

CHAPTER XXVII

FIXATION OF ATMOSPHERIC NITROGEN

At the meeting of the British Association for the Advancement of Science held at Bristol in 1898, Sir William Crookes in his address as president drew attention to what he called the "Wheat Problem." In the course of his discussion of the facts he produced something approaching a serious sensation by the statement that "England and all civilised nations stand in deadly peril of not having enough to eat. As mouths multiply, food resources dwindle. Land is a limited quantity, and the land that will grow wheat is absolutely dependent on difficult and capricious natural phenomena."

It is true that he added to this alarming view, "I hope to point a way out of the colossal dilemma. It is the chemist who must come to the rescue of the threatened communities. It is through the laboratory that starvation may ultimately be turned into plenty."

Fortunately for public peace of mind some relief from anxiety was provided a few months later in a letter addressed to the *Times* on December 2nd, 1898, by Sir John Bennett Lawes and Sir J. Henry Gilbert, the famous experimental agriculturists of Rothamsted, England. They said: "To sum up the world's wheat supply it may be said that whilst wheat is capable of producing very large crops under favourable conditions as to soil, climate and manuring, it possesses a remarkable power of obtaining food from a poor soil. It can stand a considerable amount of frost, and it can thrive over an immense area of the world's surface. Although endorsing all that Sir William Crookes says as to the importance of wheat as a food, we cannot adopt his desponding views in regard to the future supplies of it. That we may have considerable fluctuations in produce and in price, the result of war, or of the vicissitudes of the seasons in different countries, is very probable; but we believe that there will always be a sufficient supply forthcoming, for those who will find the money to purchase it at a remunerative price."

To this assurance from so respectable an authority may be added a few considerations arising out of the progress which has been made by agriculture in the years which have elapsed since their words were written. The production of wheat for the whole world has increased very largely for several reasons.

Wheat is now grown over large areas not counted on in 1898, including Australia, India, Egypt, South America, while it has increased enormously in Canada and considerably in the Russian Empire and the United States.

Wheat breeders have also succeeded in raising varieties more suitable to local conditions than the older sorts, and therefore in improving yields, while better rotations and more manure are now used than formerly.

Crookes' remedy for shortness of wheat supply was the production and application to the land of much larger amounts of nitrogen in the form of nitrate. In reviewing the world's annual wheat crop, and the known results of applying nitrate of soda to the experimental plots at Rothamsted, he calculated that to raise the 12·7 bushels per acre, which was the average yield of wheat of the world's crop, to 20 bushels, it would require 12 million tons of nitrate annually to be distributed in varying amounts over the wheat-growing countries of the world, in addition to the 1¼ million tons already absorbed by various crops. But though Lawes and Gilbert would regard a cheap and liberal supply of nitrate as a very great boon to the agricultural world, they thought it very doubtful whether an average of 20 bushels per acre would be obtained year after year the world over by the annual application of 12 million tons of nitrate. They pointed out that if nitrate were used alone the available minerals such as potash and phosphate would soon show a deficiency.

There can be no doubt as to the benefit derived from the use of nitrogenous manures, but there can also be no doubt that in a comparatively few years the supplies of natural nitrate are certain to be exhausted. This substance occurs in the rainless district in the northern provinces of Chile between the Andes and the coast. In recent times of peace the exports of nitrate from Chile are stated to have risen from the estimated 1,200,000 tons in 1898 to more than double that amount per annum. Into the United Kingdom alone the imports, according to the Board of Trade returns, have been as follows :—

Year.	Tons.	Value.
1909	90,207	£ 860,860
1910	126,498	1,161,127
1911	128,487	1,189,019
1912	123,580	1,274,752
1913	140,926	1,490,669

figures which show a gradual increase of price. The consumption of nitrates during the war must be enormous, and probably exceeds the total consumption for agricultural purposes.

Looking to other available sources of combined nitrogen the next in importance is the ammonia derived from the nitrogen of coal and employed in the form of sulphate.

Animal manures come next, and in the form of the dung of animals, fed on pasture, a certain amount of the nitrogen derived from the grasses and other herbage is transferred to the arable, including wheat lands. It is impossible in this connection to avoid deploring the sewage system which is so generally prevalent in towns and cities, for by this means practically the whole of the nitrogen from the food of the human population is irrecoverably wasted. A simple calculation will show how very great is the waste. Assuming that in round numbers there are 30 million adults and 15 million children in the United Kingdom, and that each adult excretes 1 ounce of urea and a child $\frac{1}{2}$ ounce of urea in a day, these figures correspond to 381,790 tons of urea in the year. This quantity of urea contains the same amount of nitrogen as 839,942 tons of ammonium sulphate or 1,081,744 tons of sodium nitrate. A small quantity of this nitrogen passes direct to the soil and a small quantity to sewage farms, but the saving is practically insignificant. The rest, with the phosphates, is discharged into the sea.

In all pasture land a certain amount of fixation of atmospheric nitrogen is always going on through the agency of bacteria. And it is fortunate that this is so, for without the secret, obscure operations of such tiny things as the *azotobacter*, *clostridium*, and a few other organisms, the necessary stimulant would be missing from large parts of the earth's surface. The albuminous matters thus stored up in the clovers and other leguminous plants yield up their nitrogen again by decay, ammonia passing into the soil and becoming the food of another generation. Here it may perhaps be as well to warn the reader against confusing the action of the bacteria which bring atmospheric nitrogen into chemical combination producing protein substances in the plant, with the action of those other properly called nitrifying organisms by which ammonia is converted first by one microbe into nitrite, and then by another into nitrate. These operations are of great physiological importance as bringing the nitrogen into an assimilable condition, but they add nothing to the soil. The farmer then must look for cheap nitrogen in the form of ammonium sulphate or a nitrate to the chemist, in accordance with the indication of Sir William Crookes in 1898. The supply has begun, but at present the synthetical nitrogenous manures play no great part in agriculture. Their chief function at present seems to be to keep down the price of ammonium sulphate and sodium nitrate. In a few years, however, it seems probable that the demand for nitrates in other directions will increase to such an extent as to render necessary a greatly increased production of the artificial compound. We may now turn to the processes by which atmospheric nitrogen is being brought into forms of practical utility.

Up to comparatively recent times gaseous nitrogen was described in chemical text-books as a very sluggish substance, and was often stated, quite improperly, to be incapable of entering into chemical combination with other elements by any direct method. It constitutes four-fifths of atmospheric air, and is usually said to serve the purpose of diluting the oxygen of the air, which would otherwise be too stimulant for the health of the animals which live in it. There is a certain fallacy implied in this statement. If an atmosphere of pure oxygen had been provided as the outcome of the chemical changes which attended the early history of the planet there can be no doubt that the animal organism would have adapted itself to it. It is pretty certain that the composition of the earth's atmosphere has changed considerably since life appeared on the globe, and the physiological processes going on in the present animal and vegetable inhabitants of the earth are the result of adaptation. It is true that atmospheric nitrogen appears to take no direct part in the animal economy. But it is now known that certain plants by a mechanism of their own, namely the nodules swarming with bacteria which are formed on the rootlets of plants of the natural order *Leguminosæ*, the bean and pea tribe, have the power of taking in the nitrogen of the air and using it in building up some of the albuminous or protein compounds contained in their tissues. Plants, however, usually derive their nitrogen from chemical compounds which are formed in minute quantity in the atmosphere.

Of these probably the ammonia results from products of decay of animal and vegetable remains escaping into the air. But the oxides of nitrogen which are formed, and which come down to the soil in the form of nitrous or nitric acid with the rain, are undoubtedly produced by electric discharges taking place through the atmosphere, perhaps more or less at all times, but especially during thunderstorms. This production of nitrate by electricity can not only be demonstrated in a few minutes on the lecture table, but has become the basis of a most important manufacture. It has also been long known that certain metals when heated in nitrogen gas combine with it forming a class of compounds called *nitrides*.

The companion element in the atmosphere, oxygen, enters directly into a considerable number of combinations giving rise, for instance, to the rusting of moist iron, and as everyone knows it is taken up in the lungs of animals and changes the venous into arterial blood. It is, however, a comparatively inactive gas unless its temperature is raised.

It would be useless, for example, to expect the ordinary materials of fuel to burn in the air unless at some point it is heated by the application of a flame. But by a silent electrical discharge at the common temperature oxygen can be converted into the very active substance known as ozone. This attacks all sorts of substances which, under the same circumstances, would be indifferent to common oxygen. Ozone consists of the same matter as oxygen but in a condensed state. Its molecule consists of three atoms, O_3, while the molecule of common oxygen consists of two atoms, expressed as O_2. In most cases the activity of ozone results from the instability of its molecule, and the tendency to part with one of the atoms while common oxygen is regenerated.

An active modification of nitrogen has now been obtained, but its activity is apparently due to an entirely different condition, for it is not nitrogen in a condensed state. The new gas apparently consists of free nitrogen atoms (see p. 121).

Independently of natural agencies already referred to the fixation of atmospheric nitrogen may be effected in a variety of ways, and the products may take the following forms :—

1. *Ammonia.* The synthetical formation of ammonia by Haber's process has been already described under *Catalysis* (p. 201).
2. *Metallic nitride.*
3. *Cyanide.*
4. *Cyanamide.*
5. *Nitric oxide* leading to nitrate.

The formation of nitrides by heating the metals of the alkalis or alkaline earths in contact with nitrogen gas has long been recognised. A mixture of magnesium with quicklime was used for absorbing nitrogen in the Rayleigh-Ramsay experiments on the isolation of argon from air. Boron, silicon, titanium, and some other elements also combine with nitrogen at a red heat. The compounds which result are decomposed by water or steam with evolution of ammonia, but the production of such compounds on a large scale is for the present impracticable, and such a process of getting nitrogen from the atmosphere has as yet no technical value.

The only process in which a nitride is concerned which has received serious attention from the industrial world is that which is known as the Serpek process. This is based on the production of an aluminium nitride and its subsequent decomposition by water or alkali with formation of ammonia and alumina.

The process has been tested in France by the Société Générale des Nitrures, which has acquired the patents, and by the Badische Anilin und Soda Fabrik at Ludwigshafen.

Bauxite, which is natural alumina containing a good deal of iron, is mixed with coke and is heated to the requisite temperature by electricity. The favourable temperature is said to lie between 1800° and 1900° C., a higher temperature causing decomposition of the preformed nitride. The presence of some metallic oxides is said to be favourable to the formation of the nitride, and therefore low-grade bauxite can be used, and in preference to pure alumina. The gas employed is producer gas, which consists roughly of one-third of its volume of carbonic oxide with two-thirds of its volume of nitrogen. The simplest expression for the chemical change is shown by the following equation :—

$$Al_2O_3+3C+N_2=2AlN+3CO,$$

but there seems to be some difference of opinion as to the manner in which the change is effected and the composition of the nitride. The aluminium nitride can be decomposed by water or solution of caustic soda :—

$$AlN+3H_2O=Al(OH)_3+NH_3.$$

When alkali is used the alumina is dissolved out leaving the iron behind, and a very pure alumina may thus be separated specially suitable for use in the electrolytic process by which the metal is usually obtained. The aluminium nitride may also be decomposed by a limited quantity of acid, sufficient to fix the ammonia, while leaving the alumina insoluble. If sulphuric acid is used soluble ammonium sulphate is formed as follows :—

$$2AlN+H_2SO_4+6H_2O=2Al(OH_3)+(NH_4)_2SO_4.$$

The formation of cyanides when carbon and nitrogen are heated together to a high temperature in contact with hydrogen or water vapour or with alkaline salts or silicates has long been familiar. Thus it is known that hydrocyanic acid is formed around the electric arc taken in ordinary moist air, and Bunsen and Playfair in 1845 found cyanogen in the gases from a blast furnace at Alfreton. The white incrustation which is often seen at the joints of the iron furnaces consists chiefly of potassium cyanide, and at one time it was supposed that this salt played an important part in the reduction of iron ores. The alkali metal is obviously derived from the ash of the coal and from a small quantity of alkaline salt contained in the ore. Upon the recognition of these facts attempts were made to establish a process for the synthetical production of cyanides by passing nitrogen over a strongly heated mixture of coal and potash. The high temperature required is, however, a great disadvantage, and the process in this form was soon abandoned.

The production of cyanamide and of basic calcium nitrate are processes which have assumed great commercial importance within the last few years. They both involve the use of electricity and the necessary power can be obtained most economically in countries provided by nature with elevated water supply on a liberal scale. Hence some of the largest works for the utilisation of water power have been established in Norway, Sweden, and Switzerland, as well as at Niagara and on the Pacific coast of North America.

The works of the Alby United Carbide Company for the manufacture of calcium carbide, and its conversion into cyanamide at the works of the North-Western Cyanamide Company, Limited, situated at Odda on the Sondre Fiord in Norway, may be taken as representative.

We may glance first at the sources of power utilised in the hydro-electric installation on which depend all the electro-chemical operations. The source of hydraulic power for the turbines is the river Tysse which flows into the Sondre Fiord at a point six kilometres from Odda. This village has long been a favourite resort of tourists on account of the magnificent waterfall scenery of the near neighbourhood. The Skjeggedalsfos is perhaps the most famous of these falls, and from the Ringedalsvand, or lake, just below, the Tysse river falls 436 metres (1430 feet) in a distance of about 6 kilometres (3¾ miles). The Ringedalsvand is the great collecting basin of the locality, the area from which its water is received being about 380 square kilometres.

The water from the lake is brought in pipes which for the upper 90 metres (295 feet) are of cast iron from 7 to 9 millimetres (0·28 to 0·35 inch) thick. The lower parts are of steel plates in 20 feet lengths, the metal being from 10 to 25 millimetres (0·39 to 0·98 inch) thick. The power station is situated on the shore of the lake near the mouth of the river about 3¾ miles from Odda. Here are placed the turbines and dynamos by which the power is generated and distributed. The installation was constructed to give 23,000 electrical horse-power, while the available water makes it possible to obtain 75,000 to 80,000 horse-power if required.[1]

[1] A detailed description of the whole installation is given in *Engineering*, vol. 87 (1909).

FIG. 144.—PIPES CARRYING WATER TO THE RJUKAN POWER STATION.

To face page 400.

Fig. 145.—NITROLIME FURNACES AT ODDA, NORWAY. NORTH-WESTERN CYANAMIDE COMPANY, LTD.

To face page 401.

A still larger water-power installation established in Southern Norway derives its water supply from three lakes, Maarvand, Mösvand, and Tinnsjö. The two former drain through the river Maan into the upper end of Tinnsjö. The last-named lake has a capacity of 168 million cubic metres, and is 190 metres above the sea. The principal power houses are at Rjukan, below Mösvand and above lake Tinnsjö, the water being brought in ten steel tubes. Two other power houses lower down on the same stream at Lienfos and Svaelgfos supply the nitrate works at Notodden. When these and other factories contemplated in this district are completed about 540,000 horse-power will be employed.

Nitrolime or Calcium Cyanamide

The production of this substance depends on the combination of calcium carbide with nitrogen gas under the influence of a moderately high temperature, produced by an electric current passing through carbon resistances embedded in the mass.

$$CaC_2+N_2=CaNCN+C.$$

The first stage in the series of operations is the production of calcium carbide. At Odda this is the business of the Alby United Carbide Factories, Limited. The materials employed are *lime* made from Norwegian limestone as free as possible from impurities, and *carbon* in the form of Welsh anthracite. When these materials, mixed together in proper proportions, are heated in electric furnaces to a temperature approaching 3000° C. (5432° F.), a reaction ensues in which, after expulsion of a small quantity of gas given off by the anthracite, the carbon unites with the oxygen of the lime forming carbonic oxide gas, and with the calcium, forming calcium carbide, which, with the small quantity (3 p.c.) of ash left by the coal and a small quantity of coke, constitutes the solid residue :—

$$CaO+3C=CaC_2+CO.$$

The lime required is made in five kilns each of 30 tons capacity heated by producer gas. Four of these are sufficient to supply the twelve electric furnaces in which the carbide is produced, so that one of the kilns is a standby.

The separation of the requisite nitrogen from the atmosphere by the Linde liquefaction process has already been described (see Nitrogen, p. 251), and we may, therefore, now proceed to examine the operations involved in the production of cyanamide.

The first step is to reduce the carbide to powder, and to avoid access of moisture, the grinding is performed in air-tight apparatus. This grinding is associated with a system of screening, and separation of the material into several different grades of fineness according to the size of the perforations through which it passes. The powder exposed to contact with nitrogen at a temperature of 800° to 900° C. absorbs the gas, and the process is attended by the evolution of much heat. The heat liberated is, however, not sufficient to make the process automatically continuous. At Odda the furnaces in which the combination is effected are drum-shaped portable vessels and are brought into position, when charged with carbide, by means of an overhead electric traveller, by which also they are removed when absorption of nitrogen is complete.

After being placed in position, seven in a row, the necessary connections are made for admitting the nitrogen under pressure, and the supply of electric current which is passed through a carbon rod placed in the centre of each. In the view of the furnace house shown there are 196 furnaces, each producing about 1 ton of nitrolime per week, the total output being, therefore, about 10,000 tons per annum.

The nitrolime of commerce consists of nearly two-thirds of its weight of calcium cyanamide, $CaN{\cdot}CN$, corresponding to a total content of nitrogen equal to 20 to 22 per cent.

The use of nitrolime as a manure is explained by the ultimate conversion of the whole of this nitrogen when in contact with water into ammonia. Nitrolime also contains about 20 per cent of free lime, 7 to 8 per cent of silica, alumina, and iron as impurities, and 14 per cent of carbon, which is in the form of graphite.

By the slow action of atmospheric air, which contains carbon dioxide and moisture, cyanamide is converted into urea, and hence old specimens of nitrolime may contain appreciable quantities of this substance.

It is probable that in the soil a change of this kind precedes the final elimination of the nitrogen in the form of ammonia. First carbon dioxide and water vapour would produce calcium carbonate and cyanamide :—

$$CaNCN+CO_2+H_2O=CaCO_3+H_2NCN.$$

The cyanamide by uniting with water forms urea :—

$$H_2N{\cdot}CN+H_2O=(H_2N)_2CO$$

and the latter passes into ammonium carbonate :—

$$(H_2N)_2CO+2H_2O=(H_4N)_2CO_3.$$

Other uses, however, have been found for nitrolime, and it is probable that still further applications may be made of its combined nitrogen. Thus by melting with proper alkaline salts alkaline cyanides may be made, and the manufacture has been established at Spandau near Berlin. The consumption of cyanides for the extraction of gold from quartz is very large, and hence the cheapest possible production of these salts is very desirable. The simplest form of the reaction is represented by the following equation :—

$$CaNCN+C \rightleftarrows Ca(CN)_2$$

in which the carbon already present in the nitrolime is concerned, and no addition is necessary except some fusible material to act as a flux. The process is, however, at present a secret.

Cyanamide treated with super-heated steam gives off all its nitrogen in the form of ammonia, and inasmuch as this process leaves all the lime and one-third of the required carbon in the form suitable for the manufacture of carbide, it is probable that this may prove to be an economical method for the manufacture of ammonia, and hence of such salts as ammonium sulphate, which is so largely used in agriculture.

Nitrates from Atmospheric Air

So long ago as 1781 and the few following years experiments placed on record by the Hon. Henry Cavendish show that he had established the fundamental facts which more than one hundred and twenty years later have found practical application on a large scale. He found that when hydrogen and air are exploded together the water which is formed is always accompanied with a small quantity of nitric or nitrous acid. He also showed[1] that when common air mixed with some oxygen is exposed to electric sparks the mixture is wholly absorbed by an alkaline solution with production of common nitre. (See passage quoted in connection with discovery of Argon, p. 136.)

Attempts to utilise these facts for manufacturing purposes led to no practical result till in 1897 Lord Rayleigh gave an account of his " Observations on the Oxidation of Nitrogen Gas " in the Transactions of the Chemical Society. The experiments described

[1] *Philosophical Transactions of the Royal Society*, vol. 75 (1785).

were made in connection with the isolation of argon from air by removal of the nitrogen, and there can be no doubt that their publication gave encouragement to the idea that it would be possible to utilise the process for the production of nitric acid and nitrates. In these experiments a mixture of oxygen and atmospheric air supplied to an inverted glass globe of 50 litres capacity containing a solution of caustic soda was exposed to a kind of electric flame. Some difficulties were encountèred in producing a steady flame, but a definite relation was established between the electric energy consumed and the amount of nitrate produced. Passing over the work of the numerous experimenters attracted to the subject in all the civilised countries of the world, it may be noted as an interesting fact that the first practical application of the principle of electric combination of atmospheric gases to the production of nitrates was made by two Englishmen, McDougall and Howles, in 1899. Their works were not commercially successful, and another attempt was made by Bradley and Lovejoy in 1902 in connection with the use of water from Niagara, but it lasted only two years.

The first practical success was achieved by Dr. Samuel Eyde, Engineer of Christiania, in association with Professor Kristian Birkeland,[1] who established the now famous works at Notodden below lake Tinnsjö. The French Company which found the greater part of the capital has also joined forces with the Badische Anilin und Soda Fabrik which built works at Christiansand in Norway, and two other German firms. The result has been the establishment of two new Norwegian companies, of which one, with a capital of 16 million krone (£900,000), undertakes the provision of water power from the Norwegian falls, and the other, with a capital of 18 million krone (about £1,000,000), is concerned in the application of this power to the manufacture of lime saltpetre.

There appears to be a little difference of opinion as to the exact part played by the electric arc in causing the combination of nitrogen and oxygen. Sir William Crookes and Lord Rayleigh appear to hold the view that it is a question only of temperature. Nitric oxide is an endothermic compound, and its production according to the equation :—

$$N_2+O_2 \rightleftarrows 2NO$$

[1] Professor of Physics in the University of Christiania, Norway.

is reversible. It never proceeds beyond the formation of a very small amount of nitric oxide, when a state of equilibrium is produced corresponding to the temperature. It is essential therefore to use as high a temperature as possible to promote combination and to cool the gases as quickly as possible to avoid the dissociation of the nitric oxide. The temperature in the electric flame is said to exceed 3000° C., and the escaping gases are brought down to between 800° and 1000° C. As the gases in the atmosphere occur in the ratio oxygen 21 to nitrogen 79 by volume, and the theoretical proportions required are equal volumes, atmospheric air does not provide by itself the most favourable material. It is, of course, known from experiments on a small scale that the addition of oxygen to the air to be exposed to the electric flame greatly enhances the quantity of nitric oxide produced, but the cost of such addition seems to be, for manufacturing purposes at present, prohibitive. Some physicists seem to be of opinion that the action of the arc in causing combination between nitrogen and oxygen is not wholly thermal, and since it has been shown that union occurs under the action of the silent electric discharge, which is not hot, it is supposed that the result is brought about more directly by the electric discharge at temperatures below that at which nitric oxide becomes unstable.

The flame of the electric arc used in the Birkeland and Eyde furnaces is formed between two copper electrodes, which are close together and established in a highly magnetic field produced between the poles of a powerful electro-magnet. The electrodes are made of thick copper tubing through which a stream of water passes for cooling purposes. The chamber in which the flame burns is circular, only a few centimetres in width and about three metres in diameter.

The production of the flame and its appearance as a disc has been explained as follows by Birkeland in a lecture to the Faraday Society in London, 1906 :—

" At the terminals of the closely adjacent electrodes a short arc is formed, thus establishing an easily movable and ductile current-conductor in a strong and extensive magnetic field—i.e. 4000–5000 lines of force per square centimetre in the centre. The arc thus formed then moves in the direction perpendicular to the lines of force, at first with enormous velocity, which subsequently diminishes ; and the extremities of the arc retire from the terminals of the electrodes. While the length of the arc increases its electric resistance also increases, so that the tension is heightened until it becomes sufficient to create a new arc at the points of the electrodes. The resistance of this short arc is very small, and the tension of the electrodes sinks suddenly, with the consequence that the outer long arc is extinguished. It is assumed that while this is taking place the strength of the current is regulated by an inductive resistance in series with the flame. In an alternating current all the arcs with a positive direction of current run one way, while all with a negative direction run the opposite way, presupposing the magnetising being effected by direct currents. In this way a complete luminous circular disc is presented to the eye. When the flame is burning it emits a loud noise, from which alone an impression may be obtained of the number of arcs per second formed in the flames which, however, may be more minutely investigated by means of an oscillograph."

The interior of the furnaces is lined with fire-clay brick, through the walls of which the air is admitted to the flame. The air is brought into each furnace by aid of centrifugal fans which drive it through tubes from the basement. The nitrous gases formed escape through a channel made along the casing of the furnace. This air at a temperature of about 1000° C. is then conducted to the steam boilers where its heat is utilised, and afterwards through a number of aluminium pipes cooled externally by water. The gases then pass into vertical iron cylinders lined with acid-proof stone, where the oxidation of the nitric oxide to nitric peroxide by the oxygen present is completed. The mixed gases then are driven into the absorption towers.

These towers are tall stone structures, of which the height, approximately 65 feet,[1] may be judged by the figures of two men visible in the adjoining illustration. They are filled with broken quartz over which water is continually trickling. The gases enter at the base of the first tower, and by means of a large earthenware pipe pass to the top of the second, through which they pass downwards to the bottom of the third tower. They then pass through two wooden towers in which they are washed by a solution of carbonate of soda. The liquid collected in the first series is pumped backward, the product of the third into the second and from the second into the first, and thus the

[1] Inside measure, 6 × 20 metres.

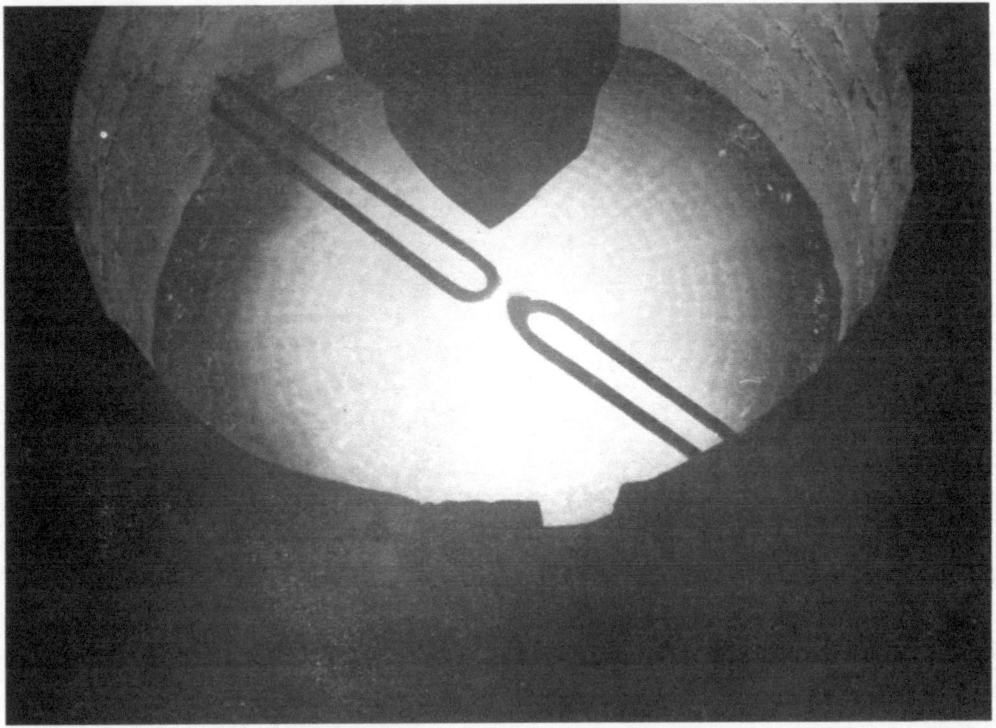

Fig. 146.—BIRKELAND AND EYDE ARC FLAME.

To face page 406.

FIG. 147.—POWER HOUSE AT SVAELGFOS, SUPPLYING NOTODDEN FURNACES.

FIG. 148.—NOTODDEN FURNACE HOUSE (ONE HALF)

To face page 407.

nitric acid formed is concentrated, so that the first tower yields an acid containing 50 per cent of nitric acid. The soda towers yield nitrite of soda.

To carry the gases forward each row of towers is provided with centrifugal fans made of aluminium which is not attacked by nitric acid.

The nitric acid collected from the towers is stored in granite tanks, neutralised by calcium carbonate, and after the addition of a small quantity of lime the solution is evaporated[1] and the residue, after solidification and granulation in a mill, is run into iron drums, in which it passes into commerce under the name of " Norwegian Saltpetre " or " Air Saltpetre."

The Birkeland-Eyde construction of furnaces is not the only system which has come into use. A few years later Dr. Schönherr, with the electrical engineer Hessberger of the Badische Company, perfected an electric furnace for the oxidation of atmospheric nitrogen based on different principles. The electro-magnet is done away with, and in place of the great disc of electric flame a long slender arc is formed in the axis of a narrow iron tube through which the current of air passes. The iron tube contains at one end an insulated electrode and itself forms the second electrode. The furnace is built up of iron plates 7 metres in height, and the air current passes into the internal reaction chamber through a number of tangential openings or slits arranged in several horizontal rows in the sides. As a rapid cooling of the gases after exposure to the flame is of importance the upper third of the tube has a water jacket through which the gases pass, and reversal of the combination is prevented to a notable degree.

The Schönherr furnaces at Christiansand, shown in the illustration (Fig. 150), take 600 horse-power, but larger furnaces have since been built.

According to Dr. Eyde the yield is practically the same as that obtained with the Birkeland-Eyde furnace.

At Notodden the furnaces are from 1000 to 3000 kilowatt capacity, and are of the Birkeland-Eyde type. At Rjukan there are furnaces of the Birkeland-Eyde system of 3000 kw. capacity as well as furnaces of the Schönherr type all of 1000 kws.

[1] All the necessary evaporations are effected by the waste heat of the electric furnaces, without the consumption of coal. This is a very important consideration from the industrial point of view.

The Notodden factories produce not only nitrate of lime and nitrite of soda, but have taken up the manufacture of concentrated nitric acid and nitrate of ammonia, and since nitric acid is the basis or starting point for the production of many other substances the development of industry in various directions may be expected in the future.

The following figures taken from Dr. Eyde's lecture to the Eighth International Congress of Applied Chemistry at New York in 1912, are interesting as giving some idea of the progress of the industry, and the important influence it has had on the prosperity of the country in which it was first established.

Dates.	Factories.	H.P. utilised.	Employees.	Workmen.
July, 1903.	Frognerkilens Febrik	25	2	2
Oct., 1903.	Ankerlökken	150	4	10
Sept., 1904.	Vasmoen and Arendal	1,000	6	20
May, 1905.	Notodden	2,500	4	35
May, 1907.	Notodden and Svaelgfos	42,500	12	403
Nov., 1911.	Notodden, Svaelgfos, Lienfos, and Rjukan I	200,000	143	1,340

A few years ago at Saaheim as well as at Notodden the population of the villages was limited to a small number of poor farmers. While Saaheim had only 50 inhabitants it has now between 5000 and 6000 citizens, and at Notodden, while there were formerly about 500 people, there are now upwards of 5000.

The economic aspect of this industry is of the utmost importance to the whole world for reasons sufficiently set forth at the beginning of the chapter. Hence it is probable that efforts to deal with the problem relating to the fixation of atmospheric nitrogen already begun in countries other than those already mentioned will be extended.

So far as the British Isles are concerned but little water power is available. There is, however, accessible water power in many of the British possessions, in Canada and in South Africa for example. There are also other sources of power in cheap producer gas from coal or peat and from mineral oil, and the more economical and efficient use of these materials is a subject on which there is still room for much scientific research.

Fig. 149.—NEW ABSORPTION TOWERS AT NOTODDEN.

To face page 408.

FIG. 150.—EXPERIMENTAL FURNACES AT CHRISTIANSAND.

To face page 409.

PART IV

MODERN PROGRESS IN ORGANIC CHEMISTRY

CHAPTER XXVIII

SUGAR

THE sugar cane must have been known from very ancient times as it is mentioned in Isaiah (Chap. 43, v. 24) and Jeremiah (Chap. 6, v. 21), but the production of sugar as an article of food probably belongs to a much later period. It is mentioned in some Greek and Roman writers as used in medicine. The cultivation of the cane seems to have come from the East, and the extraction of sugar practised in early Christian times then passed into Sicily and later into Spain and Portugal. About the fifteenth century the manufacture was established in the West Indies and in Brazil. Hawkins brought sugar to England from San Domingo in 1563, and about this time English planters were prospering in Barbadoes.

It is a little difficult to realise the conditions under which our remote forefathers before this time arranged a dietary with no direct sweetening agent except honey. The sugar which is now consumed in such large quantities is a remarkable example of a practically pure chemical compound which serves as an important food-stuff, and is very rapidly assimilated when taken into the stomach.

Common sugar exists in the juices of a great many plants and fruits, but the chief supplies of the world are derived from two sources, viz. the sugar cane, which is a tropical plant, and the sugar beet, which is cultivated only in temperate climates. Small quantities of sugar for local purposes are obtained from several species of palm in India and the East, and pass into commerce under the name joggery. Another source of sugar is the sugar maple (Acer saccharinum) which grows in some abundance in the northern states of America and in Lower Canada.

To give an idea of the large quantities of sugar used as food the following figures taken from the Annual Statement of the Board of Trade for the year 1913 show the extent of the imports into the United Kingdom from foreign and colonial sources. These do not include molasses, invert sugar, glucose, jams, or confectionery, but merely the substance sugar itself as we are accustomed to see it in loaf, crystal, or powder on our tables, or in the kitchen.

The countries from which the United Kingdom receives the largest supplies of refined sugar are Germany, Netherlands, Belgium, and Austria-Hungary. From this it may be inferred that the countries in which sugar is manufactured from the cane do not occupy themselves much with the process of refining and producing loaves and lump. This goes on in Europe, and the refined white sugar which appears on the table is very commonly a mixture of sugar from the cane and the beet, from which latter plant the whole of European-grown sugar is derived. This fact should be a corrective of the still surviving prejudice against beet sugar with which some housekeepers are haunted. Forty or more years ago there was substantial foundation for the prevalent objection to beet sugar which in those days retained, in the shops, a peculiar evil smell and an appreciable quantity of potassium chloride, by which its sweetening power was sensibly diminished. With the removal of these impurities by the processes of refining, sugar from whatever source it is derived has, like every other definite chemical compound, specific properties of its own by which it is always distinguished from every other chemical compound. Hence until it has been shown that sugar derived from beet differs from sugar obtained from the cane in crystalline form, in its action on polarised light, in solubility in water and in sweetening power as determined by an exact quantitative comparison, in which correctly weighed quantities are compared together, there is no justification for the suspicion that they are not identical.

Imports of sugar into the United Kingdom :—

	1913. Cwt.	Value.
Refined		
Total	18,450,897	£12,351,086
	Foreign and Colonial.	
Unrefined.		
Beet	13,542,112	£6,781,240
	Chiefly from Germany.	
Cane, etc.	7,392,181	£3,934,295
	Chiefly from Cuba and West Indian Islands.	

Sugar is obtained from sugar canes by passing them between rollers whereby the juice is squeezed out. The latter then receives the addition of a very small quantity of lime and is heated to boiling to coagulate albuminous matters. After skimming, the clarified liquid is run into a vacuum pan, where it is boiled down, under reduced pressure, and therefore at a lower temperature, till it becomes concentrated enough to concrete into a crystalline mass on cooling. This mass is drained in perforated casks or centrifugal machines whereby the uncrystallisable treacle is removed. Sugar obtained thus has more or less brownish colour, owing to changes which have been produced in some of the juice during evaporation. To obtain it in the form of white crystals or loaves the raw sugar is dissolved in water, and the syrup allowed to slowly percolate through a bed of bone charcoal, 30–40 feet thick, where the colouring matter is retained and a colourless syrup runs through. From the latter by concentration in vacuum pans separate crystals may be produced, or by standing in a frame a solid crystalline loaf may be formed.

Sugar beet is cultivated in nearly all the Continental countries, but at present in England to an extent which can only be spoken of as experimental. A factory has been established in Norfolk and good reports of successful results have been issued, but it is to be hoped that ere long the cultivation of this important crop will extend to other parts of the country. The difficulties which have stood in the way of progress have been twofold; on the one hand the capital required for establishing a factory with its expensive special apparatus, is very considerable, and the operations are intermittent, as they extend over only about three months in the year; on the other hand the factory would be useless without an assured supply of roots every season, and the English farmers have yet to learn the management of the crop, and too often expect to be guaranteed against possible loss in embarking on an unfamiliar venture.

Beet sugar is extracted by washing the roots and then cutting them into thin slices which are systematically exhausted of their soluble matters by immersion in successive portions of hot water, so arranged that the already partially charged liquid is used for treating the fresh roots and thus a strong solution is obtained. This is finally drawn off, heated with a little lime, which neutralises acid and coagulates albumin, and the clear solution concentrated in vacuum pans. In the operation of extraction it should be understood that the process is essentially based on liquid diffusion in which sugar and the salts present pass as "crystalloids" through the walls of the cells of the beet-tissue, while the gummy and albuminous matters being "colloidal" remain for the most part behind in the pulp. Mere expression of the juice would not, therefore, lead to satisfactory results, as the fluid would in that way be loaded with uncrystallisable matters which would be difficult to remove.

Beet contains about 15 per cent of sugar, which even rises under favourable circumstances to as much as 18 per cent. A small quantity remains in the waste pulp.

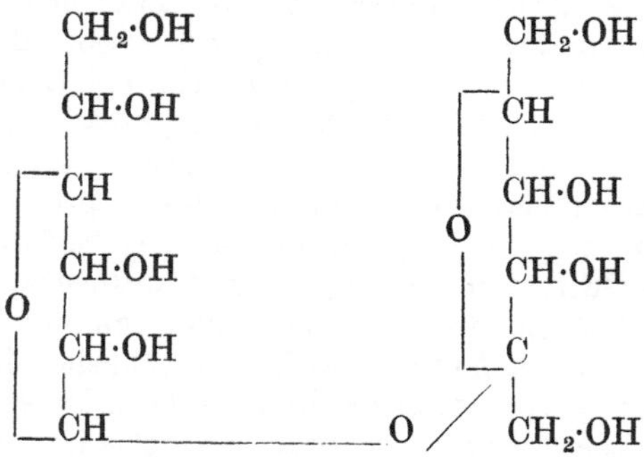

Common sugar belongs to a large class of chemical compounds called "carbohydrates" from the fact that their composition may be expressed by a formula in which there is just the proportion of hydrogen present sufficient to form water with the oxygen. Thus the composition formula of sugar is

$$C_{12}H_{22}O_{11}$$

in which the number of hydrogen atoms is double that of the oxygen or $11H_2O$.

Nearly all these substances are important constituents of food, and they may be divided into two main groups, namely the crystalline sweet substances known as sugars, and the non-crystalline and nearly tasteless constituents of many vegetable tissues such as starch, gum, and cellulose. Both these groups include a great many members or distinct individuals from the chemical point of view, and the interest of the subject has been greatly increased within the last few years by the success which has attended their investigation, more especially by Professor

Emil Fischer. Not only have many of the natural carbohydrates been made by synthetical laboratory processes, but a large number of purely artificial sugars have been built up which have no existence in nature, or at any rate have not hitherto been discovered. The constitutional formula for common sugar, which harmonises with its synthesis, has been represented as shown in the diagram on the opposite page.

This looks rather formidable, but a glance will show that it is essentially made up of two chains each consisting of six atoms of carbon, with which are associated four hydroxyl groups, HO, and associated together by the agency of an atom of oxygen which links them together.

Before proceeding to enquire how this has been determined it may be of interest to examine why sugar possesses a sweet taste and is so soluble in water. This appears to be connected with the presence of the numerous hydroxyl groups attached directly to carbon atoms, which also are linked to one or two atoms of hydrogen. Many compounds having a more or less sweet taste and ready solubility are thus constituted. They are possessed of the chemical characters exhibited by common ethyl-alcohol $CH_3{\cdot}CH_2OH$, though not its intoxicating properties. Thus :—

Ethylene-alcohol or glycol $HO{\cdot}CH_2{\cdot}CH_2{\cdot}OH$.
Propylene glycol $CH_3{\cdot}CH{\cdot}OH{\cdot}CH_2{\cdot}OH$.
Glycerine or glycerol $CH_2{\cdot}OH{\cdot}CH{\cdot}OH{\cdot}CH_2{\cdot}OH$.
Mannite or Mannitol

$$CH_2{\cdot}OH{\cdot}CH{\cdot}OH{\cdot}CH{\cdot}OH{\cdot}CH{\cdot}OH{\cdot}CH{\cdot}OHCH_2{\cdot}OH.$$

All these substances have a decided sweet taste, though inferior to that of common sugar, and it will be found when the constitution of some of the other sugars is explained that they also contain an abundance of this constituent, hydroxyl, HO. It will be observed, however, that when hydroxyl is associated with carbon which is charged with oxygen instead of hydrogen its function changes and it gives rise to acid.

Thus glycol is sweet in virtue of the hydroxyl and the hydrogen combined with the carbon, but when the two hydrogen atoms are each replaced by one atom of oxygen oxalic acid is the result, $HO{\cdot}CO{\cdot}CO{\cdot}OH$. The same gathering of oxygen and hydroxyl together, which is usually called carboxyl, is characteristic of hundreds of known acids.

The whole story of the sugars is too technical to be presented in any approach to completeness before the general reader. It will be sufficient if we endeavour to trace the transformations of a few of the more important members of the group, and if possible give some idea of the views entertained by modern chemists as to their constitution.

The origin of cane sugar has been already sufficiently described. Another sugar which has the same composition but very different properties is contained in milk to the extent of nearly 5 per cent. Milk sugar or *lactose* is obtained by evaporating clear whey to a syrup, when on standing the sugar crystallises in hard crusts and crystals which contain $C_{12}H_{22}O_{11}$ united with one molecule of water of crystallisation.

Maltose is another sugar having the same formula which is produced by the action of malt extract, containing *diastase*, on starch, and hence is formed in the preliminary stages of the fermentation of beer or grain spirit. Both these substances are less sweet than common sugar, but all three agree in undergoing a change under the influence of a small quantity of an acid which results in each breaking up into two sugars of a simpler type, the molecule of which contains only six atoms of carbon, and which are called glucoses or *saccharoses*.

Disaccharose.	Saccharoses.
Cane sugar	Glucose Fructose[1]
$C_{12}H_{22}O_{11}+H_2O=$	$C_6H_{12}O_6+C_6H_{12}O_6$
Milk sugar	Glucose Galactose
$C_{12}H_{22}O_{11}+H_2O=$	$C_6H_{12}O_6+C_6H_{12}O_6$
Maltose	Glucose
$C_{12}H_{22}O_{11}+H_2O=$	$2C_6H_{12}O_6$

Glucose or grape sugar occurs in ripe grapes and raisins, and is manufactured by heating starch with water till liquefied, and boiling the solution with a small quantity of sulphuric acid. After complete conversion of the starch the liquid is neutralised by adding a slight excess of chalk, filtering, and after passing through charcoal, evaporating the solution to crystallising consistency. Glucose is manufactured for use in preserving fruit and jam, and in the production of beer and so-called "malt" vinegar.

Fructose or fruit sugar is associated with glucose in many fruits. Like glucose it ferments in the presence of yeast and

[1] This change occurs when sugar is cooked with sour fruit.

resembles glucose in many properties, differing from it in its action on polarised light. Glucose rotates the polarised ray to the right and is, hence, frequently called dextrose, while fructose rotates to the left and is sometimes called lævulose in reference to this fact. For reasons to be explained later these terms are liable to confusion. Fructose behaves in many respects as if possessed of the constitution known as ketonic.

As the result of recent researches, especially by E. F. Armstrong, chemical opinion as to the constitution of glucose has undergone a serious modification. It is now believed that there are two varieties of glucose containing the same constituents attached to the fundamental carbon atoms in the same order, but disposed differently in space so that in the main their properties are very close together, and they are mutually convertible the one into the other. The formulæ by which these sugars are now represented are shown below.

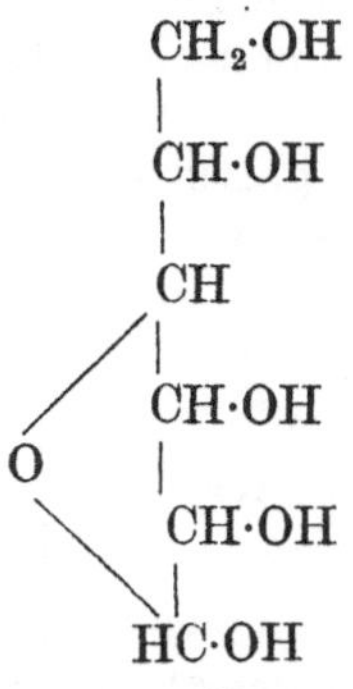

α. Glucose.

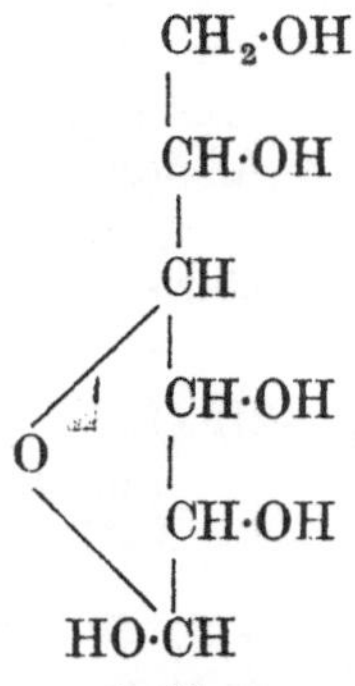

β. Glucose.

Here it will be observed an attempt is made to indicate the nature of this difference, though it must be understood that such formulæ pretend not in the slightest degree to afford a pictorial representation of the molecules (see Chapter XII).

The number of distinct saccharoses now known is very large. Some of the more important sugars of this class occurring in nature have been mentioned as derived from vegetable sources. Perhaps it ought also to be stated that glucose occurs in small quantity in the blood and tissues of the higher animals, where it is found as the result of changes in a peculiar compound known as glycogen which occurs in the liver. In books on physiology glycogen is often referred to as "animal starch," and while it differs from starch in solubility and other properties it appears that under the influence of the same reagents it will, like starch, break up into molecules of glucose. This sugar is excreted in large quantity in the disease known as diabetes. Glucose resembles an aldehyde in many reactions and until recently was classed with those compounds.

Glucoses are widely diffused in plants in the form of compounds called "glucosides," many of which are of great practical importance as well as scientific interest.

The glucosides may be regarded as analogous to fats in constitution, being made up of glucose and some complex, alcoholic or acid, the residues of which in the molecule of the glucoside only need the addition of the elements of water to enable them to separate and become independent. Thus a very familiar case is that of amygdaline, a white crystalline substance occurring in the bitter almond. When crushed with water the bitter almond, previously almost odourless, immediately emits a characteristic smell. This is due to the action of an enzyme occurring in the tissues of the almond itself, which in the presence of water causes the decomposition of the amygdaline in the manner expressed as follows :—

Amygdalin
$C_{20}H_{27}NO_{11}+2H_2O=$

Benzaldehyde		Hydrocyanic Acid		Glucose
C_7H_6O	+	HCN	+	$2C_6H_{12}O_6$

Among the very numerous natural glucosides the following may be mentioned as possessing great physiological importance in the economy of the plant or yielding products of practical interest and utility.

Arbutin, a bitter crystalline substance obtained from the leaves of the bear-berry (*Arctostaphylos uva-ursi*), a small shrubby plant belonging to the heath tribe, is used in medicine. Hydrolysed by emulsin it yields glucose and hydroquinone which has antiseptic properties. Arbutin occurs in the leaves of various species of pear.

Phloridzin is a somewhat similar substance found in the root bark of apple, pear, cherry, and plum.

Salicin is a bitter crystalline substance extracted from the bark of the willow or from poplar buds and is used in medicine.

When hydrolysed by emulsin it yields glucose and saligenin (*o* Hydroxybenzyl alcohol). The latter is oxidised to salicylic acid, a very important remedy for rheumatic affections, though now made almost exclusively by synthesis from phenol.

Coniferin is a glucoside which is present in the cambium sap of various fir trees. When hydrolysed by emulsin it yields glucose and coniferyl alcohol, $C_6H_3(OH)(OCH_3) \cdot CH : CH \cdot CH_2OH$, which by careful oxidation yields vanillin, $C_6H_3(OH)(OCH_3) \cdot CHO$, the fragrant constituent of the vanilla pod.

The production of indigo from the plant has already been described (see Dyes), and it was explained that the blue colouring matter does not occur ready formed in the juice, but is deposited as the result of a kind of fermentation. It is present in the form of indican, a glucoside which splits up on hydrolysis into glucose and indoxyl, from which by oxidation in contact with air indigo-blue is produced.

Amygdalin, described above, is not the only glucoside which yields prussic acid. Researches within recent years have shown that glucosides which, on hydrolysis by their own enzymes, yield this poisonous product are more widely spread than was formerly supposed. Thus the Lotus arabicus, a small leguminous plant resembling a vetch, which grows abundantly in the valley of the Nile and is used as fodder, contains a yellow crystalline glucoside together with an enzyme, and when moistened with water and crushed the leaves of the plant evolve prussic acid. Hence the plant is very poisonous to cattle, and the effect is most marked in the young plant up to the period of seeding. Some other plants eaten by cattle contain similar substances and require to be used with caution.

Tannin, which is a very widely diffused vegetable principle which constitutes the astringent agent in many barks, woods, leaves, and other parts of plants, is also a glucoside. It yields gallic acid and glucose when hydrolysed. Tannins are not only protective to the plant, but like other glucosides provide a store of reserve material to be used when the plant requires help in the development of buds or leaves or in the ripening of fruit.

Starch and other carbohydrates serve a similar purpose.

It would be unsuitable to these pages to enter into the necessarily long story of the successive steps by which, already long ago, the general character and chemical constitution of the sugars have been unfolded. The history of the researches on the subject dates back to the period, now sixty years ago, when chemical constitution began to be understood, and of course the facts and deductions from them are recounted in all the principal textbooks of organic chemistry. But the steps forward which resulted from Emil Fischer's work twenty years ago represent one of the great triumphs of synthetical chemistry, of which the first stages were indicated in a recent chapter (p. 334), and of which a further development will have to be reported further on.

The first step to be taken in the study of such a class of compounds as the sugars is to discover a method by which the several compounds may be discriminated and recognised even in the presence of one another. Nearly all these compounds are characterised by rotating the plane of polarisation of a ray of polarised light, and this serves as an indication that the molecule contains one or more atoms of carbon in the condition which has been described as asymmetric (p. 215). This alone would be insufficient, but Fischer was successful in finding a chemical reagent, phenyl-hydrazine, with which all the saccharoses unite and with an excess of it interact to produce a characteristic compound, called an osazone, by which each sugar may be separated, purified, and identified.

The first formed product of the union of glucose, for example, with phenylhydrazine $C_6H_5NH{\cdot}NH_2$, is a soluble compound called a hydrazone which is represented as follows :—

$$\begin{array}{l} CH_2{\cdot}OH \\ | \\ CH{\cdot}OH \\ | \\ CH{\cdot}OH \\ | \\ CH{\cdot}OH \\ | \\ CH{\cdot}OH \\ | \\ CH : N{\cdot}NHC_6H_5 \end{array}$$

water being formed at the same time.

In the presence of a larger quantity of the reagent a yellow, almost insoluble precipitate of the osazone is thrown down, and this may be collected, examined under the microscope, and its

Emil Fischer

To face page 420.

melting-point ascertained. The osazone is represented by the following formula :—

$$\begin{array}{l} CH_2{\cdot}OH \\ | \\ CH{\cdot}OH \\ | \\ CH{\cdot}OH \\ | \\ CH{\cdot}OH \\ | \\ C : N{\cdot}NHC_6H_5 \\ | \\ CH : N{\cdot}NHC_6H_5 \end{array}$$

Fructose undergoes exactly similar changes and yields an osazone identical with the compound obtained from glucose. The osazones when heated with a little hydrochloric acid give up their phenzlhydrazine, and the resulting compound (called an osone) by the action of zinc dust and glacial acetic acid, takes up two atoms of hydrogen and is converted back into a saccharose. When the original sugar, like glucose, belonged to the aldehyde type of sugar, the product of this change will be a sugar of the keto type ; if, for example, the process starts with glucose the decomposition product of the osazone will be fructose.

These saccharoses are readily convertible the one into the other, sometimes by mere dissolution in a neutral solvent like water, or by prolonged contact with solutions of ammonia or other alkali. Ordinary glucose, for example, when freshly dissolved in water rotates the plane of polarisation to the right, and its specific rotatory power is represented by 105°·2, but if kept for a few hours this falls to 52°·6. This change is attributed to the formation of an equilibrium mixture of three isomeric forms of glucose.

The methods by which the synthesis of the natural sugars, as well as of others, containing respectively less or more carbon in the fundamental chain can only be indicated by one or two examples. Of these the most interesting perhaps are those which start from very simple compounds, the synthetical formation of which was long ago established.

The first case which naturally presents itself is the sugar which has been called *acrose*, and was one of the first produced synthetically by Fischer. Acrolein and its dibromide have long been known; by treating the latter with baryta a condensation of two molecules into one is brought about, the bromine being eliminated; $2C_3H_4OBr_2+2Ba(OH)_2=C_6H_{12}O_6+2BaBr_2$.

From the sugar thus produced others can be obtained as the result of internal changes brought about by methods of which the general nature has already been indicated. α Acrose appears to be identical with inactive fructose.

To build up a higher saccharose from one containing a smaller number of atoms of carbon a well-known method has been made use of which consists in adding on the elements of hydrogen cyanide to the lower sugar and then splitting off the nitrogen from the resulting compound. Thus glucose $C_6H_{12}O_6$ may be made to yield by a series of steps the heptose or glucoheptose $C_7H_{14}O_7$.

On the other hand, glucose may be deprived of one atom of carbon and the chain shortened down to five atoms by a series of operations which have a certain relation to the above.

By the action of hydroxylamine glucose, like other aldehydic compounds, yields a substance called an oxime, and from this the elements of water may be removed. By further operations the elements of hydrogen cyanide are eliminated. The principal stages are shown below:—

CH_2OH	$CH_2{\cdot}OH$	$CH_2{\cdot}OH$	$CH_2{\cdot}OH$
$(CH{\cdot}OH)_3$	$(CH{\cdot}OH)_3$	$(CH{\cdot}OH)_3$	$(CH{\cdot}OH)_3$
$CH{\cdot}OH$	$CH{\cdot}OH$	$CH{\cdot}OH$	CHO
CHO	CH : NOH	CN	
Glucose	Glucose oxime	Gluco-nitril	Arabinose

By such means then the monosaccharose group has been extended so as to include members containing from two to nine atoms of carbon with constitutional differences corresponding to the glucose or aldehydic type and the fructose or ketonic type. The actual number of individuals recognised and characterised is, however, very much larger than might be inferred from such a simple statement. To realise the great extent and the complicated nature of this field of enquiry it is necessary to remember that the majority of these substances rotate the plane of the polarised ray, and that this property is traced to the presence in each molecule of several asymmetric atoms of carbon. The general nature of this conception has been already explained, but the case of the sugars requires a little closer examination.

Imagine a single atom of carbon united with four other atoms or groups of atoms all different. These may follow one another, projected on the plane of the paper, in either of the two ways shown below, that is in the direction in which the hands of a clock move or the reverse.

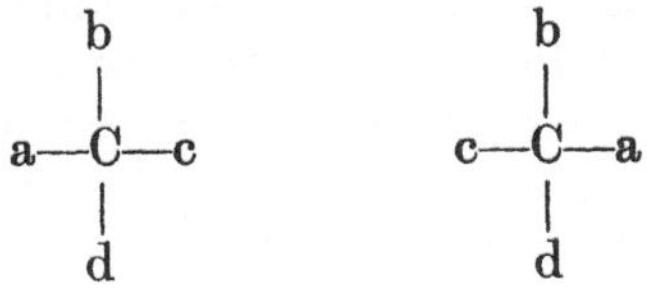

It is obvious that these two arrangements are such that the one is a mirror-image of the other. If but one atom of carbon in this condition is present in the compound there may be two varieties of the same which agree together in all chemical characteristics, and, generally speaking, in all physical characters except one, and that is in their action on the polarised ray. One will rotate the plane to the right, the other to the left through the same number of degrees. But if a pair of compounds, both being alcohols or acids or aldehyds or ketones, etc., contain two asymmetric atoms, then there will be a pair of stereoisomerides, one dextro- and one lævo-rotatory, while there will be a third substance which will be optically inactive. Now as glucose contains four asymmetric carbon atoms, there must be, according to the hypothesis, eight stereoisomeric forms with right-handed rotation and eight others with equal left-handed rotation. Most of these compounds are actually known, though only a small proportion of them occur in nature.

Though practically identical in respect to chemical reactions a very important difference is observed in their physiological relations, for they are acted upon very differently by yeast and other ferments. Thus the two components of invert sugar, glucose, and fructose are fermented into alcohol and carbonic acid at different rates, and while *d* glucose is fermentable *l* glucose, its optical isomeride remains unaffected. In fact, it is in general only the dextroforms among the hexoses which are capable of entering into fermentation. The majority of the saccharoses containing less and more carbon are unfermentable. These facts are difficult to account for, except on the assumption that there is a relation of spatial structure between the sugars and the enzymes which appear to be the agents in the fermentive process which is analogous to the interaction of lock and key. There is considerable evidence that in the first stages of fermentation there is production of compounds into which the molecule of the sugar enters, and, in fact, a hexose phosphate—$C_6H_{10}O_4(H_2PO_4)_2$—has actually been isolated.

The whole process, however, is still imperfectly understood and will continue to furnish problems for research. Changes of a kind similar to those which go on in fermentation probably occur in the tissues of both animals and plants and the subject is one which is of great importance in physiology.

CHAPTER XXIX

PROTEINS OR ALBUMINOUS SUBSTANCES

THESE are slimy, glutinous, or gelatinous substances which form the basis of the animal body, and which also occur in the juices of vegetables.

Chemists and physiologists between them have been so busy that they are able to distinguish and characterise some fifty substances which are gathered under this name. Some of these are familiar enough in white of egg, in the serum of blood, in milk, in animal cartilage, bone, and horn, in the gluten of wheat flour, and in many seeds such as peas, beans, and almonds. They are all, as found in nature, mixtures of highly complex compounds containing the elements carbon, hydrogen, nitrogen, oxygen, usually sulphur, sometimes also phosphorus and iron.

The study of these compounds has been proceeding, it may be said, long before systematic chemistry began, for some of the old chemists two hundred years ago learned to recognise a chemical difference between substances of this kind which are particularly characteristic of animal matters, and the materials of which wood and vegetable tissues chiefly consist. When the former are subjected to heat, and are destructively distilled they yield a fœtid liquid which smells ammoniacal and acts like an alkali on test paper. Wood and vegetable matter, on the other hand, give an acid distillate. But such insight as the chemical physiologist now enjoys into the constitution of the proteins is the result of researches which have been accomplished almost entirely within the last twenty years.

The large number of apparently distinct compounds which, presenting as they do few of the characters by which pure chemical individuals are usually recognised, renders the investigation of the proteins one of the most difficult tasks undertaken by the chemist. An attempt has been made to classify these substances, and provisionally we may accept the scheme which has been brought forward, with the assurance that, while temporarily useful, the results of further work will lead perhaps to new classes, certainly to some modification in those already recognised. The scheme set forth below results from the work of a Committee of the American Society of Biological Chemists (1908). It includes vegetable as well as animal proteins.

I. The Simple Proteins.
- *a.* Albumins.
- *b.* Globulins.
- *c.* Glutelins.
- *d.* Prolamins.
- *e.* Scleroproteins or albuminoids.
- *f.* Histones.
- *g.* Protamines.

II. Conjugated Proteins.
- *a.* Nucleoproteins.
- *b.* Glycoproteins.
- *c.* Phosphoproteins.
- *d.* Hæmoglobins.
- *e.* Lecithoproteins.

III. Derived Proteins.

1. Primary Protein Derivatives.
   - *a.* Proteans.
   - *b.* Metaproteins.
   - *c.* Coagulated Proteins.

2. Secondary Protein Derivatives.
   a. Proteoses.
   b. Peptones.
   c. Peptides.

The *simple proteins* and the *conjugated proteins* are all substances which are supposed to exist in the tissues and juices of animals and vegetables. They are separated by taking advantage of their solubility or insolubility in saline solutions, such as aqueous ammonium sulphate or sodium chloride or in alcohol of different strengths. They are all substances which present more or less decidedly an " amphoteric " character, that is they have the power of uniting either with acids or with bases owing to the presence in them of carboxylic groups, –CO·OH, or amino groups, $-NH_2$. The molecular weight is not known accurately, but is certainly very great. The composition of serum albumin, which may be regarded as typical, and so far as percentages of the elements are concerned, is closely similar to other albumins. It is as follows :—[1]

Carbon	53·08	per cent.
Hydrogen	7·10	,,
Nitrogen	15·93	,,
Sulphur	1·90	,,
Oxygen	21·99	,,

From various considerations a formula has been calculated,

$$C_{450}H_{720}N_{116}S_6O_{140}$$

which corresponds to a molecular weight 10,176. Very little importance can be attached to such expressions, they only serve to indicate a recognition of the highly complicated character of the molecule. The albumins are for the most part colloid substances, though many of them have been observed naturally or obtained by laboratory processes in a crystalline state.

Many of these compounds are coagulated by heat and are precipitated from solution in water, in the form familiar enough in white of egg, as a kind of network of films and fibres which are not reconvertible into the soluble form. The albumins are also precipitated by the addition to the aqueous solution of the strong

[1] Michel, quoted in *Mann's Proteids*, p. 327.

mineral acids and by the salts of practically all the heavy metals.

The precipitate in each case consists of a compound or mixture of compounds of the albumin with the acid or salt used.

A curious point about the albumins is the fact that they are optically active and that they all rotate the polarised ray to the left. In the case of the sugars it has already been mentioned that the most physiologically active and fermentable rotate the ray to the right.

Conjugated proteins consist of one or more molecules of albumin associated with some other substance of a different nature such as sugar. They do not always contain sulphur, but phosphorus is in many of these compounds a prominent constituent. Several, such as hæmoglobin, the red colouring matter of blood, contain iron.

The *derived proteins* represent the first stages in the process of degradation which the proteins, simple or conjugate, undergo in contact with almost any reagent. Indeed the process of extraction which involves any change in the composition of the fluid in which the natural albumin is normally found probably produces some modification. The solubility is affected even by the comparatively simple operation of diffusion by which the sodium chloride or other saline component of the albuminous fluid is removed. Contact with acids causes incipient hydrolysis, and the same effect is induced by alkalis, by metallic salts and probably also by the enzymes concerned in the process of digestion in the stomach. The products, called albumoses and peptones, are very nearly allied to the albumins and answer to many of the chemical tests which are supposed to characterise those bodies. They are also of high molecular weight. More severe treatment with hydrolysing agents leads to a break up of these complicated molecules, and the products are for the most part comparatively simple. It is of course significant that these products of hydrolysis consist principally of substances generally known as amino-acids.

One of the first cases of the kind was observed by Braconnot nearly one hundred years ago in the production of what was then called " gelatine sugar " or the " sugar of glue," glycocoll (from γλμκυς, sweet ; κόλλα, glue) in allusion to its sweet taste. This substance which is now called glycine is amino-acetic acid $CH_2NH_2{\cdot}CO{\cdot}OH$. Two similar substances, namely alanine or amino-propionic acid $CH_3 \cdot CHNH_2 \cdot CO \cdot OH$ and leucine or amino-caproic acid, $CH_3 \cdot CH_2 \cdot CH_2 \cdot CH_2 \cdot CHNH_2 \cdot CO \cdot OH$, have also been obtained as common products of this kind of change. Tyrosine has also long been recognised among these products; it is phenyl-*p*-hydroxy-*a* amino-propionic acid

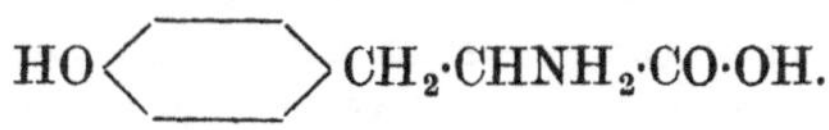

Other more complicated substances are also found among the products of disintegration of proteins, but they almost all partake of the same character and behave as amino-acids.

These facts furnished the principal clue to the mystery of the constitution of albuminoid substances generally, and led to the remarkable achievements in the direction of their synthesis which we owe chiefly to Emil Fischer and his school. The assumption is that the peptones are constituted of amino-acid groups, and that the number of such residues joined together in the molecule of the protein is relatively small. This view has been practically established by the production of a number of compounds called by Fischer, *polypeptides*, which in many respects resemble the natural peptones, and there seems every reason to expect that further efforts on similar lines will lead to the synthesis of some substances identical with a natural albumin.

The simplest of the peptides is glycyl-glycine which, containing the residues of two amino-acid groups, is called a dipeptide. Others containing three, four, or many such residues are named respectively tri-, tetra-, and poly-peptides.

Glycyl-glycine is obtained as follows: *a* amino-esters, when heated, part with the elements of alcohol and a closed-chain anhydride is formed, thus:—

Two molecules of glycine ethyl ester

$$2(NH_2 \cdot CH_2 \cdot CO \cdot OC_2H_5) \text{ or } \left\{ \begin{array}{l} NH_2 \cdot CH_2 \cdot CO \cdot OC_2H_5 \\ C_2H_5O \cdot CO \cdot CH_2 \cdot NH_2 \end{array} \right.$$

yield, by loss of H from the NH_2 in one molecule and C_2H_5O from the other molecule,

$$\underset{\text{anhydride}}{NH{<}\begin{array}{c} CH_2\text{—}CO \\ CO\text{—}CH_2 \end{array}{>}NH} + \underset{\text{alcohol}}{2C_2H_5OH}$$

When the anhydride is warmed with hydrochloric or hydrobromic acid or with dilute alkali, the elements of water are taken up and the dipeptide results; thus

$$NH<{CO—CH_2 \atop CH_2—CO}>NH + H_2O$$

$$= {HO·CO—CH_2 \atop NH_2·CH_2—CO}>NH \text{ Glycyl-glycine.}$$

A step by which the higher stages of condensation are reached consists in first forming a compound of the lower peptide with the chloride of an acid radicle, such as acetyl chloride, containing an atom of halogen, e.g. chloracetyl chloride $CH_2Cl·CO·Cl$. In such case HCl is eliminated and a compound is formed which, by the action of ammonia, exchanges Cl for NH_2 and the polypeptide results. Thus glycyl-glycine gives chloracetyl-glycyl-glycine

$$CH_2Cl·CO·NH·CH_2·CO·NH·CH_2·CO·OH$$

which yields diglycyl-glycine (a tripeptide)

$$CH_2NH_2·CO·NH·CH_2·CO·NH·CH_2·CO·OH$$

Proceeding on similar lines more complex molecules are built up by making use of other amino-acids, such as aspartic acid

$$HO·CO·CH_2·CHNH_2·CO·OH$$

the acid derived from asparagine, a crystalline, soluble amide which occurs commonly in plants. Aspartic acid is also found among the products of the degradation of proteins.

More than one hundred of the artificial polypeptides have been produced by similar methods. Fischer has described the preparation of an octo decapeptide, derived from fifteen molecules of glycine and three molecules of *l*-leucine, which in its external properties closely resembles many natural proteins. Thus penta glycyl glycine was allowed to react with *d*-bromo-*iso* capronyl diglycyl-glycyl chloride and the product so obtained was treated with ammonia with formation of *l*-leucyl octo glycyl glycine. By a repetition of this series of reactions the octo decapeptide, *l*-leucyl triglycyl—*l*-leucyl triglycyl—*l*-leucyl octo glycyl glycine is produced. The formula of this is given on the following page.

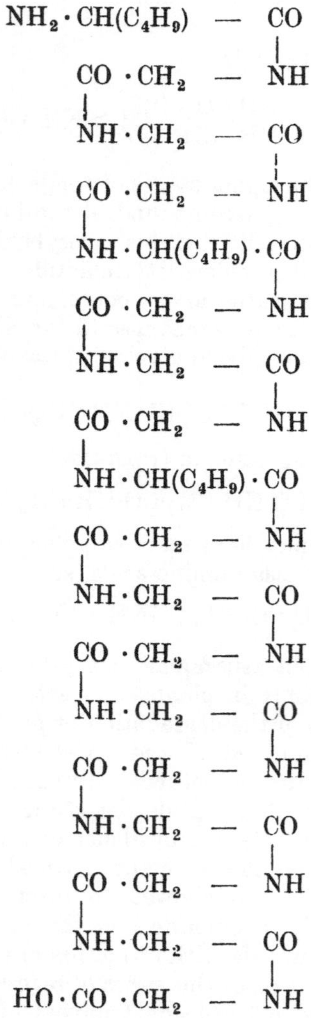
NH₂·CH(C₄H₉) — CO
CO ·CH₂ — NH
NH·CH₂ — CO
CO ·CH₂ — NH
NH·CH(C₄H₉)·CO
CO ·CH₂ — NH
NH·CH₂ — CO
CO ·CH₂ — NH
NH·CH(C₄H₉)·CO
CO ·CH₂ — NH
NH·CH₂ — CO
CO ·CH₂ — NH
NH·CH₂ — CO
CO ·CH₂ — NH
NH·CH₂ — CO
CO ·CH₂ — NH
NH·CH₂ — CO
HO·CO ·CH₂ — NH

This array of symbols, formidable as it appears, is evidently only that of an amino-acid of which the amino group is at one end of a long chain, while the carboxyl group is at the other.

Adopting the usual hypothesis as to the configuration of the carbon, and probably the nitrogen atom, the formula in space would be represented by a spiral.

The early days of organic chemistry were associated almost exclusively with the study of medicine and biology. This was natural owing to the readiness with which definite crystalline principles were obtainable by easy processes from plants and from animal matters. Nearly all the operations undertaken down to the beginning of last century had some practical utilitarian object, such as the production of dyes or tanning materials, the processes of fermentation, the distillation of spirits, the examination and preparation of drugs. It was only in the hands of a few of the older chemists that organic matters were examined with the primary object of ascertaining something as to their nature. When, however, such a man as the Swedish Scheele (1742–1786) entered the field discoveries were made so rapidly that it appears almost surprising that his example was not followed more freely by others. There were, however, two great difficulties in the way, the one was the lack of methods by which the composition of such compounds could be accurately determined, the other was the absence of the fundamental ideas which would enable the results of such analysis to be interpreted. The latter was provided in due time by the application of Dalton's Atomic Theory (1808), but it was much later before the methods of organic analysis were devised and perfected by Liebig.

But even before the advent of the Atomic Theory many definite substances had been extracted from natural sources. Citric, tartaric, malic, lactic acids were known, urea and uric acid, asparagine, morphine from opium, and glycerine by saponification of fats, beside hydrocyanic acid and many other carbon compounds. These, however, were isolated examples of naturally occurring principles, and their relations to one another or to the animal or vegetable sources from which they were extracted or to any system of things had yet to be discovered. One of the first important steps was taken when in 1828, Wöhler found that the inorganic salt ammonium cyanate by mere evaporation of its aqueous solution was transformed into urea by a rearrangement of its elements without loss or gain of any constituent. Changes of this kind now excite very little surprise. Scores of metamorphoses of a similar character are now known, and in the years which have elapsed since Wöhler's time hundreds of chemists have been busily at work. It is estimated that about 130,000 distinct carbon compounds are now known. These great advances, including the remarkable department of stereo-chemistry (Chap. XII), which has sprung into existence within the last forty years, are the natural outcome of the pursuit of knowledge for its own sake. Similar advances have been recorded in other branches of science on the recognition of the same principle. But the time has arrived long ago when the systematisation of the vast mass of knowledge acquired concerning the origin, synthesis, and properties of carbon compounds may be turned to account in the endeavour on the part of physiologists to penetrate some of the mysteries connected with living things. This has become imperatively necessary not only in the practice of medicine, to which the organic chemist has supplied so many now indispensable medicaments, but for the better understanding of the processes which go on in the body, in health as well as in disease. Nor is this knowledge applicable only to the human and animal body. Vegetable physiology brings to the agriculturist a practical assistance which enables him more successfully to cultivate his land, to understand the application of manures and the rotation of crops, and generally to gather from the soil a larger yield of food stuffs of all kinds.

But in order that all these applications of chemical knowledge may be accomplished the knowledge itself must be stored up and made accessible. To this end the chemist must study carefully and record fully the composition and properties both physical and chemical of a very large number of compounds.

It has already been pointed out in the chapter on sugars that not only is the proportion of carbon, hydrogen, and oxygen present in these compounds a matter of importance, but the number of atoms which enter into the molecule and the configuration of the molecule, that is the arrangement of the atoms in space, determine the part which these substances play in the vegetable and animal economy. Two sugars may have exactly the same composition and molecular weight and yet the one will have a more decided sweet taste than the other, the one will be resolved into alcohol and carbonic acid by the action of yeast, while the other will not. But though the chemist can in the laboratory build up from the elements carbon, hydrogen, and oxygen, compounds identical with those which are provided for us by nature in the plant, the physiologist cannot as yet form more than rude conjectures as to how they are generated in the plant.

The first step in the construction of the tissues of the higher plants consists in the decomposition of carbon dioxide derived from the air in the green parts of leaves and stems under the stimulus of the sun's rays. But how this comes about no one knows, and there is difference of opinion even as to the nature of the first product of the fixation of the carbon and rejection of the oxygen which is known to go on. It is probably a sugar, and the preponderance of evidence appears to be in favour of the view that it is the comparatively complex disaccharose, common sugar. The case is still more difficult when an attempt is made to conceive how protein matters are formed which require the co-operation of nitrogen, sulphur, and phosphorus derived from mineral compounds of these elements brought up from the soil. It is the first step which causes the chief difficulty; once a molecule is formed of the kind which has been described as a protein, even of the very simplest kind, the attachment of carbon dioxide to its amino-constituent or of ammonia to its carboxylic constituent may conceivably lead to accumulations of carbon and of nitrogen which may lead to the formation of the complex albuminous matters which are so intimately concerned in the affairs of the living tissue.

Beside the sugars fats also play an important part even in the vegetable economy. Nearly all the oils used for food or other practical purposes are derived from plants, and more especially from the fruits (palm and olive oils for example) and seeds (coconut, cotton-seed, linseed, etc.). The composition and constitution of fixed fats and oils is well known, but the problem of their production in nature is far from being solved.

Carbohydrates furnish the raw material in all probability, but how they are converted into fatty acids is quite unknown, though the resolution of a sugar into glycerin is readily conceivable. On the other hand the digestion of fats in the stomach is a subject on which physiologists are not agreed. And to refer these changes to the action of enzymes is only to substitute one kind of mystery for another, inasmuch as the precise nature of enzymes as already explained is unknown and likely to remain a mere subject for conjecture for a long time to come.

Emil Fischer has expressed his views very clearly in the Faraday Memorial Lecture given in 1907, in accordance with custom, in the lecture theatre of the Royal Institution in London. "Of the numerous attempts to unravel the constitution of the proteins by analytical means," he said, "the only method which has given useful results hitherto is that of hydrolysis. Hydrolysis can be effected by acids or by alkalis and also by digestive enzymes; the products as is well known, besides ammonia, are albumoses, peptones, and ultimately amino-acids. The wide range of variation in composition of these amino-acids is illustrated in the following table :—

Glycine	(Braconnot, 1820)
Alanine	(Schützenberger, Weyl, 1888)
Valine	(v. Gorup-Besanez, 1856)
Leucine	(Proust, 1818; Braconnot, 1820)
iso Leucine	(F. Ehrlich, 1903)
Phenyl-alanine	(E. Schulze and Barbieri, 1881)
Serine	(Cramer, 1865)
Tyrosine	(Liebig, 1846)
Aspartic acid	(Plisson, 1827)
Glutamic acid	(Ritthausen, 1866)
Proline	(E. Fischer, 1901)
Oxyproline	(E. Fischer, 1902)
Ornithine	(M. Jaffé, 1877)
Lysine	(E. Drechsel, 1889)
Arginine	(E. Schulze and E. Steiger, 1886)
Histidine	(A. Kossel, 1896)
Tryptophane	(Hopkins and Cole, 1901)
Diaminotrihydroxy-dodecanoic acid	(E. Fischer and E. Abderhalden, Skraup, 1904)
Crystine	(Wollaston, 1810; K. A. H. Mörner, 1899)

"In this table are included all the substances hitherto prepared from the proteins, the existence of which is established, with a short reference to their discovery.[1] . . .

[1] This was in 1907. Since that date further progress has been made in building up polypeptide molecules. Among the important steps may be mentioned the introduction of tyrosine groups into the molecule. Tyrosine is of frequent occurrence in natural proteins, and it appears now that the difference between natural and synthetic polypeptides consists less in the number of the

"The nineteen amino-acids in the table are the chief hydrolytic cleavage products of the proteins and those which are generally met with. . . . The proportions in which the various amino-acids are obtained from the different proteins vary very considerably. In some cases they are altogether lacking, as may be proved by application of the definite tests for tyrosine, tryptophane, or glycine; but it is worthy of note that, as a rule, the amino-acids referred to as isolated from the mixtures produced by subjecting albuminous substances to hydrolysis all occur almost without exception; especially is this true of the important proteins which play the chief part in animal or vegetable metabolism; so that the conclusion must be drawn that none of them can be dispensed with in organic life. With the exception of diaminotrihydroxydodecanoic acid they have all been so thoroughly investigated that their structure is well established. The majority also have been synthesised, proof of their structure having, in fact, been given in this way. Only oxyproline, histidine, and diaminotrihydroxydodecanoic acid remain still to be synthesised.

"With the exception of glycine all the amino-acids derived from natural sources are optically active; but when prepared by ordinary synthetic methods, as is well known, they are obtained in the first instance in the racemic form. The resolution of the racemoids into their optically active components has been effected quite recently in most cases. Asparagine, however, which is closely related to aspartic acid, had been resolved into the two active forms by recrystallising the inactive synthetic product from water, and separating the two constituents mechanically. Moreover, in the case of some other amino-acids, for example leucine, the antipode of the natural form had been obtained by partially fermenting the synthetic product with moulds. The complete synthesis of the active amino-acids which are obtained from natural sources was first accomplished by the method I introduced based upon the use of the acyl derivatives. The method has been applied with success to the majority of the synthetic products; its extension to the remaining cases, proline, lysine, tryptophane, and cystine, is not likely to be attended with any difficulties. The synthetical

amino groups than in the conjunction in the natural substances of groups of several different kinds. The further study of the proteins tends to show that the intricacy of the subject increases with progress of knowledge.

results are summarised in the following table, in which the inactive products are marked *d l*, and the natural active products are recorded separately :—

Glycine	(Perkin and Duppa, 1858)
Alanine *d l*	(A. Strecker, 1850)
,, *d*	(E. Fischer, 1899)
Valine *d l*	(Fittig and Clark, 1866)
,, *d*	(E. Fischer, 1906)
Leucine *d l*	(Limpricht, 1855 ; E. Schulze and Likiernik, 1885)
,, *l*	(E. Fischer, 1900)
iso Leucine *d l*	(Bouveault and Loquin, 1905)
,, ,, *d*	(Loquin, 1907)
Phenyl-alanine *d l*	(Erlenmeyer and Lipp, 1883)
,, ,, *l*	(Fischer and Schöller, 1907)
Serine *d l*	(Fischer and Leuchs, 1902)
,, *l*	(Fischer and Jacobs, 1906)
Tyrosine *d l*	(Erlenmeyer and Lipp, 1883)
,, *l*	(Fischer, 1900)
Aspartic acid *d l*	(Dessaignes, 1850)
,, ,, *l*	(Piutti, 1887)
Glutamic acid *d l*	(L. Wolff, 1890)
,, ,, *d*	(Fischer, 1899)
Proline *d l*	(R. Willstätter, 1900)
Ornithine *d l*	(Fischer, 1900)
,, *d*	(Sörensen, 1905)
Arginine active ; partial synthesis from ornithine	(Schulze and Winterstein, 1899)
Lysine *d l*	(Fischer and Weigert, 1902)
Tryptophane *d l*	(A. Ellinger and Flamand, 1907)
Cystine *d l*	(Erlenmeyer, jun., 1903) "

There can be little doubt that any association of amino-acids could be brought about by application of the existing methods. But to deal with the whole of the proteins will be a gigantic task, and after all it may turn out that the natural proteins do not occur singly, but quite possibly are generated in the living tissue two or more together, and that metabolic changes in the body involve the transformation of several of these complex compounds simultaneously.

There is evidently much work yet to be undertaken by the chemist and physiologist.

The study of nutrition, at any rate from the chemical side, has already made some progress, but the proteins of the animal body are very numerous and very diverse, and it is at present uncertain how many even of those which are known are necessary for the accomplishment of healthy normal changes in digestion, and in the building up of new tissue. It is interesting to learn from the researches of Abderhalden that a mixture of amino-acids alone without polypeptide is capable of maintaining life, that is to say that the complex proteins of food are not indispensable, and an animal can be kept in health when supplied with the products of their hydrolysis, combined only with suitable amounts of pure fat, starch, cane sugar, and the necessary inorganic salts. One reservation only is necessary here, and that is that a minute amount of the mysterious substance or substances belonging to the class of agents called *hormones* must be added to the food. What these compounds are is unknown, except that they are not proteins and are usually soluble in alcohol and ether. Normally these substances are secreted in the body by special glands, such as the thyroid, or in digestion from the salivary glands.

In some of the experiments which have been made on animals an alcoholic extract of fresh milk is the form in which this non-nitrogenous material has been supplied. In any case the quantity required is not more than perhaps 1 per cent of the food given.

Among the amino-acids which are produced by hydrolysis of proteins, tryptophane seems to play a special and peculiar part. A supply of this substance in the food was shown to be necessary some years ago, and Dr. Gowland Hopkins has described experiments in which the absence of this compound from the diet was immediately followed by a falling off of the body weight of the animal under observation. Tryptophane has the formula :—

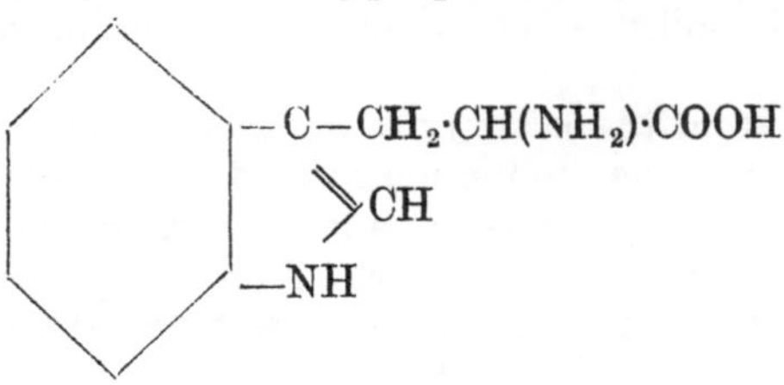

This includes the indol ring :—

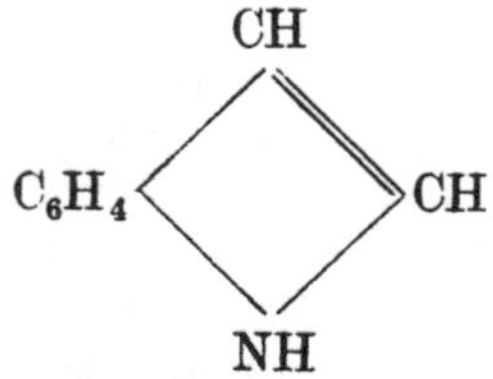

which is characteristic of several excretory products, and which the animal body does not seem fitted to produce from other materials. This seems to explain why gelatine is of such very inferior value as food, for on hydrolysis gelatine yields neither tyrosine nor tryptophane.

Those who desire a more extensive review of the advances which have been made in the knowledge of protein food stuffs cannot do better than read in the Transactions of the Chemical Society the report of a lecture given on May 18th last by Dr. F. Gowland Hopkins, F.R.S., entitled "Newer Standpoints in the Study of Nutrition."

Even at the present early stage of the development of the subject some practical lessons may be learned. Among these it is evident that one kind of protein-containing food cannot be substituted for another indiscriminately.

To borrow the language of Saint Paul: "all flesh is not the same flesh; but there is one kind of flesh of men, another flesh of beasts, another of fishes, and another of birds." This is true in strict biochemical sense, and it is no less true that a diet which will maintain one person in health would not be suitable for every individual of the same race.

Of late years a number of materials under fanciful names have been offered to the public as possessing extraordinary nutritive qualities. The phospho-proteins are specially valuable in promoting the growth of young animals, and as they are present in milk, preparations made from curd have a certain value.

Yeast from the brewery is another material which is turned to account in the production of food. Dry yeast contains about half its weight of proteins, and without preparation it may be used advantageously as a cattle food. A product made from it by a patented process closely resembles extract of meat, and may be used in a similar way.

CHAPTER XXX

NATURAL COLOURS

THE vegetation which clothes so large a part of the earth's surface, though differing so much in form, height, structure, and general character from the tropics to the limits of temperate zones, agrees in one character, namely, the colour of the foliage when young and active. This green colour extends to all except the lichens which cover some rocks, the fungi, of which the reproductive apparatus is alone conspicuous, and some few algæ. The green colour is in practically all cases due to the presence of a substance named *chlorophyll*, literally the green of the leaf (χλωρός grass green, φύλλον a leaf). This is deposited in the form of irregularly shaped granules contained within the cells of the leaf and other green parts of a plant. By viewing under the microscope a thin section of any such part it will be seen that the chlorophyll grains are chiefly found in the cells which lie just beneath the surface of the leaf. As a rule this substance is only formed in the presence of daylight, and its production is greatly accelerated by direct sunlight. When a plant is allowed to grow in the dark the chlorophyll is suppressed and the result is the familiar "blanching" which is commonly practised by the gardener. The production of the green colouring matter is intimately associated with the development of many characteristic vegetable principles, such as essential oils, bitter and flavouring substances, as may be noticed in the blanched stems of celery, sea-kale, leek, etc. Its relation to the production of starch is of the utmost importance in vegetable physiology, as starch is usually first formed within the chlorophyll granules and is probably formed by them or with their indispensable assistance directly from atmospheric carbon dioxide. It appears from experiments made by Pfeffer long ago that when plants are kept in an atmosphere entirely deprived of carbon dioxide they form no starch, even in strong sunlight.

It is not, however, in the explanation of how starch is produced in the plant nor in the origin of chlorophyll that advances of special interest have been accomplished within recent years, but in the practical solution of the question as to the composition and constitution of this green pigment. The problem is not new to chemists. For the greater part of a century attempts have been renewed many times to devise a method by which chlorophyll can be extracted from green leaves without altering it in some degree by the action of the solvents employed for the purpose. Dr. Schunck of Manchester, more than twenty years ago, supposed that he had succeeded in isolating pure chlorophyll, and he prepared from it a series of crystalline derivatives. By the action of alcoholic potash he obtained a compound which he called phyllotaonine and a derivative of this named phylloporphyrin. The latter was found to have an absorption spectrum nearly allied to that of hæmoporphyrin, a derivative of the colouring matter of blood. In these researches he was joined by the Polish professor, Marchlewski. The formula they proposed was, however, not confirmed by the later researches of Willstätter, to whom we owe the greater part of the knowledge we now possess of the composition and products of decomposition of this important substance.

Chlorophyll is a salt containing magnesium, and it is represented by the rather complex formula $C_{55}H_{72}O_6N_4Mg$.

Chlorophyll as it exists in the plant is an amorphous substance, the crystalline chlorophyll which has been described being now known to be a product of decomposition. Chlorophyll is the methylphytol ester of a tribasic acid to which the name chlorophyllin has been given with the formula :—

$$C_{31}H_{29}N_3Mg \left\{ \begin{array}{l} CO{\cdot}OH \\ CO{\cdot}OH \\ CO{\cdot}NH \end{array} \right.$$

The full formula of chlorophyll is therefore :—

$$C_{31}H_{29}N_3Mg \left\{ \begin{array}{l} CO{\cdot}OCH_3 \\ CO{\cdot}OC_{20}H_{39} \\ CO{\cdot}NH \end{array} \right.$$

Phytol is an unsaturated alcohol, $C_{20}H_{39}OH$, which in the isolated state is a colourless oil which boils at a high temperature, and any attempt to distil it, except in a vacuum, leads to its destruction.

The colour appears to be connected with the complex nitrogenous nucleus, and the magnesium is supposed to be united in a peculiar manner with the four nitrogen atoms. Treatment of chlorophyll with alkaline reagents fails to disturb the magnesium, but it is extracted and removed by the action of acid liquids. The nitrogen atoms are believed to exist in the molecule of chlorophyll in the form of closed chains constituted in the same manner as in the compound known to the chemist as pyrrol :—

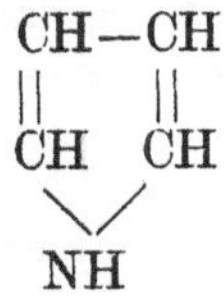

In the derivatives of chlorophyll three or four of these hydrogen atoms are replaced by methyl, CH_3 or ethyl, C_2H_5.

One of the most interesting facts in connection with this enquiry is the discovery that hæmoglobin, the red colouring matter of blood, yields by chemical decomposition compounds having the same fundamental structure.

There is thus a near relationship between hæmoglobin and chlorophyll, with one important difference. It has already been mentioned that chlorophyll contains magnesium attached to the nitrogen atoms. Hæmoglobin contains iron in a similar position. The presence of small quantities of a metal as an essential constituent of these colouring matters is a point of considerable interest, and though of much smaller importance a very curious instance is found in the wing feathers of certain birds which contain not iron but copper. The red colour exhibited by a number of African birds, called Turacos or Plantain-eaters, was examined by the late Sir Arthur Church in 1869, and found to be due to a pigment which he called turacine, which contained some 8 per cent of copper bound up with a nitrogenous structure.

It appears that turacine is actually a cupriferous derivative of hæmatoporphyrin which may be regarded as essentially the colouring matter of blood deprived of its iron. How the birds acquire the copper found in their feathers is not clear, but the same remark would apply to the minute quantity of other elements found in the tissues of animals and plants, the fluorine, for instance, found in the bones and teeth. The papers relating to turacin make no mention of the blood of the birds in whose feathers this red colouring matter is found; presumably the blood contains only iron as in all other cases.

The only question which appears open to doubt is whether the chlorophylls obtained from different plants are absolutely identical, and whether the chlorophyll obtained from any one plant is not a mixture of two or more closely similar compounds. Supposing this to be the case the constitution of the several modifications is of the same type, and the differences may arise from mere differences of arrangement of the constituent atoms in space. Molecules so complicated as those of chlorophyll are open to many possibilities of this kind.

In view of the general prevalence of chlorophyll and the change of the green bud into the coloured flower as well as the frequent tendency to reversion of coloured parts to the green state, it might be supposed that the various bright hues of flowers were produced by some kind of chemical transformation of chlorophyll. This idea was certainly accepted at one time, but it appears to be without foundation.

Many vegetable colouring matters were extensively used as dyes before the discovery of the coal-tar colours, and a few still retain their position. It is only necessary to remind the reader that, notwithstanding the advent of the very numerous synthetic dyes, natural indigo, logwood, safflower, and madder colours are still used to some extent in the dye house, and that the colouring matters of the damask rose, the red poppy, turmeric root, litmus, and red cabbage are employed in other ways. The diversity of natural colouring matters is a very interesting fact. In many cases it is probably true that the accumulation of colouring matter, especially in the coverings of the flower, is connected with a definite advantage to the plant. The bright coloured corolla, often associated with the secretions of essential oil which fills the surrounding air with perfume, brings the visits of insects which in many cases play an essential part in the process of fertilisation. On the other hand, there are many coloured substances secreted in the inner parts, in woods, like the barberry and logwood, in roots, like the turmeric and rhubarb, in bark, as in the quercitron (*Quercus tinctoria*), in berries, such as those of buckthorn and various other species of Rhamnus. In such cases the advantage, if any, must be of a different kind. It seems probable that, in many cases at any rate, the colouring matters thus deep-seated are like many of the other chemical compounds found in the tissues of plants, merely waste products concomitant with those which must be regarded as essential.

A plant requires, for example, to manufacture the chemical compound cellulose of which the membranous walls of its cells and vessels are composed, and in wood these are thickened and strengthened by encrusting deposits of similar material.

Cellulose is a carbohydrate and may conceivably be formed by condensation from other carbohydrates, the sugars and starch, which are among the first, if not the very first, compounds built up from the elements of water and carbon taken from the carbon dioxide of the air. But these synthetic operations are only affected through the agency of the complex mixture of nitrogenous substances contained in the protoplasm, which are themselves constantly in process of formation and decomposition.

In these highly complicated chemical changes by-products are doubtless formed, and unless they are utilised for some purpose in the plant itself they may accumulate in the tissues and may even obstruct the processes of growth. Such deposits are sometimes mineral, as in the deposits of phosphate of lime in teak, sometimes resinous, as in the heart wood of many trees, sometimes in the form of alkaloids, such as quinine, which are found in the bark. These by-products may be compared to the chips and shavings which collect in a carpenter's shop, but which have no relation to the form or purpose of the object which occupies his labour.

The blue, violet, and red pigments which are extracted from fruits, flowers, and many leaves are called *anthocyanins*. These exist in the flowers in the form of glucosides, and probably, therefore, serve to some extent as store of nutriment available during the process of fertilisation and development of the ovary. The colouring component, which is associated with the glucose or other sugar, is called an *anthocyanidin*, and it exhibits a certain degree of basic character as it combines with the elements of acids. These colouring matters contain no nitrogen, but only carbon, hydrogen, and oxygen, and, unlike chlorophyll, no magnesium or other metal. The anthocyan pigments are frequently mixed with a yellow substance, and as a result of this mixture some of the brown effects seen in such flowers as the dark wallflower are due. This is easily reconcilable with the existence of pure yellow varieties and a purple kind of wallflower in which the pigments are separate. The reds, purples, and blues of flowers seem in most cases to be attributable to the same substance, the tint varying according to the degree of acidity of the plant juice, acidity producing from the blue a red shade. On the other hand, the action of alkaline solutions is to produce from any of them a green colour. These colouring matters appear to be formed by oxidation from colourless substances resident in the substance of the petal. If a red or purple stock, for example, is immersed in strong spirit of wine the colour is dissolved out rapidly and the petals become colourless. If the decolourised petals are then placed in water at room temperature they begin almost at once to regain their colour, and in a quarter of an hour each petal is found to have recovered its original colour, and in the same parts of the petal which exhibited the colour, so that a white striped purple variety reproduces faithfully the purple and white pattern of the original.

It appears probable that there is a genetic connection between the yellow pigments present in many flowers and the red and blue colouring matters. It is stated by Dr. Everest (*Proc. Royal Soc.*, 1914, p. 326), that under the influence of reducing agents (magnesium and hydrochloric acid preferably), the former yield red substances which answer to the tests and characters of the anthocyanins. Research on this fascinating subject appears to be proceeding in the right direction, and as the constitution of the yellow substance is almost agreed upon by the chemists who have been occupied with the question the constitution of the anthocyanins will probably soon become clear. This does not mean that chemists or botanists are likely to be able to tell us how the plants produce their decorative pigments; that is another question which will probably take a long time to solve.

An investigation by Professor Willstätter of two yellow substances which occur together with chlorophyll in green leaves has led to some curious results. One of these, identical with the yellow colouring matter of the carrot and in reference to that fact named *carrotene*, is a hydrocarbon with the molecular formula $C_{40}H_{56}$. It crystallises in copper coloured leaflets which transmit a red light, but it absorbs oxygen readily and is converted into a colourless substance.

Xanthophyll, which accompanies chlorophyll in the alcoholic extract of leaves, is similar to carrotene in appearance, but is yellow. It has the formula $C_{40}H_{56}O_2$. It behaves neither as an acid, an alcohol, nor ketone. It absorbs oxygen readily, and is thereby converted into a colourless compound $C_{40}H_{56}O_{18}$.

The story of all these natural dyes, with which earth's garment of vegetation is tinted in such rich variety, would tell, if it could be completely and truly read, how much there is still to learn of nature's secrets. The chemist is just beginning in the twentieth century to find out the mere composition of a few of these dye-stuffs contained in the minute cellular laboratories of the plant, but the materials out of which they are formed and the process by which they are elaborated are alike unknown. From remote antiquity it has been known that solar energy is the source or, at any rate, a condition of all life. It must seem strange, therefore, that the agent which more than any other controls the chemical changes which result in the production of vegetable colours should be the last to engage systematic enquiry.

The action of light in producing decomposition in certain metallic compounds has been recognised, and the result is the art of photography. But the student of organic chemical changes has not until recently given much attention to the changes which light brings about, and as yet there is little to report. The field is wide and the enquiry will be difficult, for it will be necessary to know the relative effects of light of different colours, that is of different degrees of refrangibility, and in the first instance it would be well to operate on pure substances alone or two together. The greater part of the photo-chemical observations hitherto recorded are oxidations or reductions which have been obtained in other ways.

A few cases of polymerisation, that is condensation of two or more molecules into one, almost completes the list of known changes induced by light, and apart from the extension of photographic processes it is evident that a wide and fertile field is open to the industry of future generations. Out of such investigations may come, and probably will come, a more intimate knowledge of the way in which the plant does its work.

Before leaving the subject of natural colouring matters we may return to a brief survey of some colouring matters of animal origin. Of these the red substance in blood corpuscles is obviously the most important. This exists in the blood in two conditions according as it is taken from the arteries or from the veins, and especially after asphyxiation. In the former it is called oxyhæmoglobin, and consists of hæmoglobin combined loosely with oxygen. This power of entering into union with various gases is characteristic of hæmoglobin, and in at least one case, namely, carbonic oxide, this ought to be borne in mind as it serves to explain the dangerously poisonous effects of breathing an atmosphere containing even a small quantity of that gas.

Oxyhæmoglobin is crystallisable, and the form of the crystals differs in the blood of different animals. Hæmoglobin and its oxidised form both consist of a compound of a protein and a coloured substance called hæmatin, which is said to have the formula $C_{32}H_{30}N_4FeO_4$. If blood is treated with acids the iron is removed from the red colouring matter, and a new substance called hæmatoporphyrin $C_{16}H_{18}N_2O_3$ results. The views of the experts engaged in this interesting problem being still unsettled it is not advisable to attempt in these pages a display of the constitutional formulæ attributed to these compounds. The only point which appears clearly established is the identity of the ultimate pyrrolic products of vigorous oxidation or reduction obtained from the green matter of the leaf, and the red colouring matter of blood. This close similarity of chemical structure has led to speculations as to the changes which must have come about in the early stages of organic evolution. It may be supposed that the common colouring matter prevailing in the protozoa as well as all the early algæ and other organisms living in water was a substance essentially the same as the chlorophyll now prevalent. It must therefore be inferred that at some time, when worms or other creatures of distinctly animal characteristics began to appear, the provision of much iron in the soil or water inhabited by them led to a modification in the composition of the protoplasmic material of their tissues, and the change may have been further promoted by the exclusion of light from their muddy or earthy habitations. Whatever may have been the change of conditions which led to the change of composition from magnesian chlorophyll to ferruginous hæmatin the retention of the same fundamental atomic structure in the molecules of these two substances now so widely separated in function as well as in colour can only be regarded as strongly indicative of a common origin. No other colouring matter found in animal matters has the same importance and interest as hæmoglobin, and in nearly all cases our knowledge of their composition and properties amounts to very little. The plumage of many birds, the hair of man and many animals, and the skin of large sections of the human race contain a dark or black pigment, but it may be safely said that nothing important is known of the composition of this substance or mixture of substances, or whether, for example, the wool of the negro and the feather of the rook owe their blackness to the same or to different pigments. But the difficulty of investigations of this kind will be sufficiently illustrated by reference to one other pigment of animal origin, and that a familiar one.

Everyone is acquainted with the beautiful water-colour paint carmine, and the red essence of cochineal, which is used in cookery for colouring jellies, etc. The red substance in this case is derived from the body of the cochineal insect (*Coccus cacti*) which is cultivated in Mexico and in the Canary Islands, collected, dried, and sent into commerce in the form of small silver grey masses about a quarter of the size of small peas. These are the bodies of the females, the males being furnished with wings and have but a very brief existence after development from the larva. From 1909 to 1913, according to the statement of the Board of Trade for 1913, the amount of cochineal imported into this country almost entirely from the Canaries was as follows :—

1787—1269—1692—1623—1401 cwts.

This, though a large quantity, is a great falling off from the amount imported in the days when the British army was clothed in scarlet and cochineal was the dye. The red colouring matter has been the subject of experiment since 1813 when it was analysed by Pelletier and Caventou who assigned to it a formula which included the element nitrogen. Arppe and Warren de la Rue examined it again thirty years later and showed that it did not contain nitrogen. Since that time the carminic acid, so named by the last-mentioned chemists, has had half a dozen different formulæ attributed to it, of which the one selected by Hlasiwetz and Grabowski in 1867 approximates nearest to the truth. Corrected by Professor A. G. Perkin and Mr. C. R. Wilson it appears that the expression $C_{11}H_{12}O_6$, or the double of this, $C_{22}H_{24}O_{12}$, certainly represents its composition. Carminic acid is a crystalline compound and yields well-defined salts, and the question as to the order in which its constituent elements are united together in the molecule should not be difficult to answer. The synthetic production even of this colouring matter should be a practical matter for the manufacturer. Whether it would take a permanent place among the numerous red dyes would then be chiefly a question of cost.

CHAPTER XXXI

ENZYMES

THE process by which grape and other saccharine vegetable juices are converted after much frothing, turbidity, and ultimate clarification, into an intoxicating drink is as old as the history of mankind. But the true nature of the change which goes on was established only after a long controversy, in which the representatives of a purely mechanical or physical theory originated by Liebig were finally defeated by Pasteur, who established the dependence of ordinary alcoholic fermentation on the action of the living yeast cell.

This was fifty years ago, but since that day investigations into the phenomena of fermentation have reached a new stage, in which attention is concentrated on the agents by which the cell accomplishes its own growth and development at the same time that it brings about chemical changes in the surrounding medium. These agents are called *enzymes*, a term which was brought into use so recently as 1878 by the German physiologist W. Kühne.

Enzymes are not organisms like moulds or bacteria, but may be described as unorganised, colloidal, nitrogenous substances universally present in living animal and vegetable tissues and, though lifeless themselves, are at present producible only from living matter. They are distinguished by a remarkable *catalytic* action on carbon compounds, especially carbohydrates, fats, and proteins. Their actions are in some cases selective, but not always, and they are coagulated and rendered inactive by a temperature below that of boiling water.

The general nature of enzyme action will be understood if a few individual cases are described.

One of the earliest to be recognised and one of the most important of these substances is *diastase*. Early in the nineteenth century it became known that an aqueous extract of malt possessed the power of changing soluble starch very rapidly into dextrin and sugar. Payen and Persoz attempted in 1833 to isolate the active constituent of malt, but with no great success. Forty years later O'Sullivan described a method which was as follows : finely ground pale barley malt was mixed with sufficient water just to cover it, and after three or four hours the extract was separated by means of a filter press. The clear solution was then mixed with strong alcohol as long as a precipitate was formed, and the latter was collected, washed with alcohol, pressed between a cloth, and dried *in vacuo* over sulphuric acid. Prepared in this way diastase is a white powder, easily soluble in water, and possessing the activity of malt extract. It is, however, far from being a pure substance, as it contains a considerable percentage of mineral matter (chiefly phosphate) which is left as ash when the diastase is burnt.

Calculated on the ash-free substance the results of analysis by two different observers are as follows :—

	(Lintner)		(Szilágyi)
Carbon . . .	46·66	..	46·80
Hydrogen . . .	7·35	..	7·44
Nitrogen . . .	10·42	..	9·98
Sulphur . . .	1·12	..	1·14
Oxygen . . .	34·45	..	34·64

It will be evident on comparison with the analysis on page 426 that this is not the composition of an albumin. It should be added that other analyses have led to different proportions of the elements.

The characteristic property of diastase is its power of converting starch under suitable conditions of solution and temperature into a mixture of maltose and dextrin :—

$$5(C_{12}H_{20}O_{10})_{20}+80H_2O=80C_{12}H_{22}O_{11}+(C_{12}H_{20}O_{10})_{20}$$

Soluble starch[1] Maltose Stable dextrin

Diastase, or a substance resembling it in its action on starch, appears to be widely distributed in the vegetable kingdom in leaves and other organs, and it occurs in the grain of all cereals whether raw or germinated.

Another interesting case is that of the bitter almond. It must have been known as long as the almond itself that this seed when dry or crushed is without peculiar odour. When strongly pressed both sweet and bitter almonds yield a considerable quantity of a bland fatty oil, which may be eaten as food or used for making soap. If the bitter almond is pounded and mixed with water, a characteristic familiar aromatic odour is developed, which is due

[1] This the formula attributed to soluble starch, which is the first, or one of the earliest products, of the degradation of natural starch by acids.

to the production of an essential oil which can be distilled off in steam, and is sold as a perfume and flavouring essence. This substance is benzoic aldehyde, $C_6H_5{\cdot}CHO$, and it is well known that in the crude state it is very poisonous, owing to the presence of prussic acid, which accompanies it. These facts were explained by Liebig and Wöhler in 1837. If the cake of bitter almond from which the fixed oil has been squeezed out is exhausted with boiling alcohol a crystalline substance, amygdalin, can be procured from the solution. This is inodorous and almost tasteless, but if mixed with a small quantity of the pulp of *sweet* almonds and water the essential oil is at once developed :—

$$\underset{\text{Amygdalin}}{C_{20}H_{27}NO_{11}}+2H_2O=\underset{\text{Benz-aldehyde}}{C_7H_6O}+\underset{\text{Hydro-cyanic acid}}{HCN}+\underset{\text{Glucose}}{2C_6H_{12}O_6}$$

The bitter and the sweet almond both contain an enzyme, or rather a pair of enzymes (amydalase and prunase) commonly known under the collective term *emulsin*, which together break up the amygdalin in the manner shown in the equation. It would appear in such a case that the amygdalin, the glucoside, and the enzyme are contained in separate cells within the seed. The seeds of many other fruits of the same natural order as the almond, namely peach, apricot, cherry, etc., as well as the leaves of the common cherry laurel yield essential oil in a similar way.

Another familiar instance of enzyme action is afforded by common mustard. The domestic mustard flour is destitute of pungency so long as it is dry, but when mixed with water the odour of the essential oil becomes apparent almost immediately. Both black and white mustard seed, like the almond, yield by expression a quantity of a fixed fatty oil which has nothing to do with the pungency. The black mustard seed contains a glucoside called sinigrin or potassium myronate associated with an enzyme called *myrosin* contained in a separate system of cells. When crushed with water the glucoside is broken up into glucose, mustard oil (allyl isothiocyanate), and potassium hydrogen sulphate :—

$$\underset{\text{Potassium myronate}}{C_{10}H_{16}NS_2O_9K}+H_2O=\underset{\text{Glucose}}{C_6H_{12}O_6}+\underset{\text{Allyl isothio-cyanate}}{C_3H_5NCS}+\underset{\text{Potassium hydrogen sulphate}}{KHSO_4}$$

A glance at each of the three equations given will show that the resolution of the glucoside is due to the addition of the elements of water. This is effected by the agency of the enzyme, which in these cases acts as a hydrolysing agent, the chemical compounds which result being the same as those which are commonly produced by dilute acids or alkalis, though of course the *modus operandi* must be different. A hydrolytic action is in fact brought about by the majority of enzymes, but they differ from inorganic hydrolytic agents in the fact that many of them do not carry the process so far as acids do, and also that enzymic action is very often specific. In this last respect, however, they do not seem to differ essentially from some inorganic catalysts and some of them, emulsin for example, act on a great variety of substances. It would serve no useful purpose in this place to attempt an enumeration of the enzymes mentioned in chemical literature until more has been learnt concerning their composition and the range of their activities. Some of the enzymes which are known to hydrolyse the chief glucosides have already been mentioned in connection with sugar. Two or three may be added which are produced in the animal body and are concerned in the processes of digestion. Ptyalin, secreted by the salivary glands, changes cooked starch, as in food, into maltose and dextrin. Then there is trypsin, which is secreted by the pancreas and causes the degradation of proteins and their derivatives giving rise to amino-acids and the simpler polypeptides.

Pepsin is contained in the gastric secretion, and papain from the juice of the Carica Papaya or Papaw tree is said to have the property of making meat tender. Pepsin acts best in an acid medium such as the gastric juice which contains 0·2 per cent of hydrochloric acid. Trypsin, on the other hand, works best in an alkaline solution such as the pancreatic juice, which also contains several other enzymes.

Among the latter must be mentioned lipase, which splits up fatty matters into glycerin and fatty acid. Other enzymes are secreted by the liver, the kidneys, and the mucous lining of the intestines; their action generally is hydrolytic.

But all enzymes are not hydrolytic in their action. A different class is represented by rennet, which is prepared from the lining of the stomach of the calf and is used for curdling milk in the manufacture of cheese. The clotting of blood is brought about by a similar agent originating under certain conditions in the blood itself and called thrombin.

Yet another class of enzymes possess the power of effecting rapid oxidation, and in reference to this property are called oxydases. The process of oxidation in the tissues is one of great importance, for example in connection with respiration, but the mode of action and origin of this class of enzymes is still very obscure.

The whole subject of enzyme action is so difficult and its systematic study has been undertaken so comparatively recently that only a few generalisations have as yet been recognised.

As to the origin of enzymes it appears that, while they all originate in living protoplasm, they do not exhibit in the early stages of existence the characteristic catalytic actions which later they exercise. In this preparatory condition the substance is called a zymogen; thus pepsin is formed from pepsinogen, trypsin from trypsinogen, etc.

It appears also that in many cases, of which one example, emulsin, has already been mentioned, two enzymes habitually act together to produce the characteristic change, hydrolytic or other, which either enzyme separately is incapable of bringing about or can at most carry partly into effect. An interesting case of this kind has been described by Dr. A. Harden of the Lister Institute, in the study of yeast juice. It was discovered in 1897 by E. Buchner that the cell of the yeast as a whole is not necessary to alcoholic fermentation. By grinding and subsequent filtration a juice may be separated which introduced into a solution of sugar sets up this change. Harden's experiments have since shown that the fermentation of glucose and fructose by yeast juice is dependent not only on the enzyme, zymase, but requires the presence of another substance the nature of which is obscure, but which can be separated by dialysis and withstands the temperature of boiling without destruction of its activity. A phosphate is in addition always necessary. These two substances the enzyme and the co-enzyme are incapable separately of causing alcoholic fermentation.

The co-operation of two colloidal agents in the production of a given effect suggests that there may be bodies which are not mutually helpful, but on the contrary antagonistic. The question why the blood does not clot in the veins, why the stomach is not itself dissolved by the digestive juices which it generates from its own surface require an answer. This is partly supplied by the hypothesis of anti-enzymes which neutralise the hydrolysing or other action of the enzymes. Thus an anti-thrombin is supposed to prevent the coagulation of the blood while in the vessels by the fibrin ferment or enzyme. Similarly it is assumed that there are antipepsin and antitrypsin which check the action of the pepsin and trypsin in the stomach and intestines. There appears to be, however, some considerable differences of opinion among experts on this question and evidently further investigation is necessary. The relation of antitoxins to toxins is probably of the same character, and the production of "immunity" in respect to certain diseases results from the development in the body itself of some protective substance, or the injection into the body of a serum prepared in the tissues of another animal.

In 1898 the first case of reversible enzyme action resulting in the synthesis of a disaccharose was discovered by Dr. A. Croft Hill. Having observed that the hydrolysis of maltose by the action of the enzyme maltose in yeast was incomplete, he found that starting from glucose alone in strong solution a disaccharose was produced. The substance thus formed by the union of two molecules of glucose was originally supposed to be maltose, but, appears to be isomaltose, a sugar obtained by Fischer by the condensing action of strong acids on glucose.

$$2C_6H_{12}O_6 \rightleftarrows C_{12}H_{22}O_{11}+H_2O.$$

The reversed arrows indicate that the change may proceed in either direction according to the conditions of the experiment. This is in accordance with a very common form of chemical change in which three substances together attain a condition of equilibrium which is disturbed on changing the temperature or altering the proportion of any one of the substances present.

The principle is very important in connection with chemical or biochemical reactions, for it must be borne in mind that in the great majority of cases such a change tends to slacken or to be stopped altogether if the products of the change are allowed to accumulate. Removal of such products occurs when a gas escapes, or the solution becomes diluted, or any acid or alkali formed is neutralised. In such cases there is no accumulation because the products are removed from the sphere of action or from a condition of activity.

Other cases of synthetical formation of sugars have been

observed since 1898, and the reversibility of enzymic action is now well recognised.

The mechanism of enzyme action has received a great deal of attention. How do these complex substances do their work? That in so many cases they are selective in their attitude toward the carbohydrate, protein, or other substance with which they are in contact suggested to Emil Fischer long ago the analogy of the lock and key. When an enzyme finds itself in the presence of two glucoses, for example, having the same composition, molecular weight, and general character, but differing from each other only in "configuration" as indicated by their optical properties, the conclusion seems irresistible that the enzyme is able to fit itself into the body of the one and not into that of the other. This seems to imply that the preliminary to action is a state of union between the enzyme and the "substrate," and the only remaining question is whether this combination is of a chemical nature, or a physical or mechanical nature. In the former case definite proportions would be expected to interact or combine. Evidence in this direction is, however, unsatisfactory.

In the latter case the phenomena of "adsorption" are referred to in which the amount of combination is dependent, as in the case of other colloids, wholly on the extent of surface. Surface combination is in many cases of a very intimate nature and quite as difficult to dissociate as many a true chemical compound.

Mention has been made in previous chapters of the tenacity with which air adheres to the surface of glass, of the withdrawal from solution of various substances, such as iodine, bromine, and organic colouring matters by contact with charcoal, of the staining of fibres of cotton, wool, hair, or silk, all colloids, by dyes.

That this adsorption is in many cases selective is shown by the following facts. If a sheet of fine filter paper is wetted with a drop of solution of a salt of silver, lead, or mercury moderately concentrated, and the spot after a few minutes is exposed to contact with sulphuretted hydrogen gas, a dark stain due to the formation of the sulphide of the metal is produced in the centre which is surrounded by a wide ring of pure water. To produce a corresponding effect with a solution of copper, nickel, or cobalt, a much more dilute solution must be used. On the other hand, a solution of a cadmium salt, even very dilute, when exposed to the gas gives a yellow stain of sulphide which extends quite to the edge of the blot. If a mixture of a cadmium salt with one of silver, lead, or mercury is treated in the same way a patch of black sulphide in the centre is surrounded by a yellow ring, showing that the metallic ions travel to different distances and spread themselves to different extents over the surface of the paper fibre, and that they all behave differently from water.[1]

Similar observations have been made on salts of barium and calcium as compared with those of potassium.[2]

The whole subject of enzyme action is, however, still under investigation by a considerable number of chemists and physiologists; in fact it forms part of a new and extensive department of organic chemistry which is usually designated "biochemistry" in allusion to its close association with the phenomena of living beings, vegetable or animal. Any survey of the phenomena exhibited by enzymes cannot fail to excite wonder at the powerful action of these complex and sensitive agents. They are capable of bringing about changes which can be effected by ordinary chemical agents, such as strong acids or alkalis, only under circumstances of considerable concentration or high temperature. The hydrolysis of a fat for example can be accomplished either by boiling with alkali, when a soap is produced, or by steam, heated considerably beyond the boiling point of water, when glycerin and a fatty acid are produced. The enzyme lipase can do its work without sensible rise of temperature, and remains active after all is over. Invertase (from yeast), according to O'Sullivan and Tompson, can change 100,000 times its weight of cane sugar into glucose and fructose, and can still go on producing inversion, an effect enormously in excess of that produced by sulphuric acid at the same temperature.

In fact the metaphor already made use of to illustrate the specific action of enzymes may be extended a little in order to emphasise the contrast between the operation of these substances and the accomplishment of the same chemical change by ordinary chemical agents. For suppose it is desired to enter a house the action of the former may be compared to the simple and peaceful process of inserting the right key into the lock, the action of the latter would be more nearly represented by breaking down the door.

The effects described here are nearly all brought about by the

[1] T. Bayley, *Trans. Chem. Soc.*, 1878, p. 304.
[2] Schönbein, *Poggendorff's Ann.*, 1861, p. 275.

enzymes separated from the living tissue in which they were generated. It is usually assumed that their activity and mode of action is the same in the parent tissue, but this assumption, however probable, is incapable of strict verification. In any case it is a matter for further research to what extent the processes of absorption and assimilation, of growth and development are wholly dependent on these catalytic processes, or are at least partly the result of still more complex changes wrought by the living protoplasm itself.

CHAPTER XXXII

ORGANIC CHEMISTRY

In previous chapters an account has been given of some of the more important constituents of animals and vegetables and of the definite products of secretion. The study of such substances as sugar or colouring matters was formerly called, perhaps not altogether improperly, organic chemistry, as an abbreviated expression meaning the chemistry of organised beings. Such compounds as those mentioned and many others were supposed to be producible only by living things through the agency of what was called *vital force.*

The expression "organic" chemistry has become established by long usage, and it seems impossible to get rid of it, but it should be remembered that the customary application to the chemical history of all the multitudinous hydrocarbons, alcohols, acids, bases, sugars, etc., etc., is not intended to imply that there is any difference in the fundamental principles of *organic* and *inorganic* or mineral chemistry. A very large number of the known definite organic compounds, in which carbon is the characteristic central element, can now be produced by purely artificial processes in the chemical laboratory, independently of any operation in which plant or animal life is concerned. A considerable number of examples have been mentioned in the earlier pages of the book, and especially in the recent chapters. The history of organic synthesis may be said to have begun in 1828 when Wöhler observed the change of ammonium cyanate into urea and, though not ended, may be said to have culminated in the methods by which Emil Fischer has built up the complex molecules of some of the proteins.

So far as animal tissues are concerned the problem of their constitution and how they are wasted and renewed is, from one point of view, simpler than the corresponding problem presented by plants, inasmuch as animals find the materials they require in their food, ready formed in the vegetable matter from which directly or indirectly they derive nourishment. No animal is known to assimilate any element except oxygen from the air, or to build up fats, carbohydrates, or proteins from such simple materials as carbon dioxide, water, and ammonia. The case of the plant is very different. The forest of timber trees equally with the humble moss or lichen with which their trunks are clothed derives the whole of the carbon which is the chief component of wood, leaves, and fruit from the carbon dioxide of the air. And this gas is found in the atmosphere to the extent of only 3 to 4 parts in 10,000 under ordinary conditions. How is this accomplished ? That is the great problem on which chemists and physiologists have been more or less engaged since chemistry and physiology began. It was proved by experiment two hundred years ago that a willow tree planted in a tub of pure sand and watered with rain water grew and flourished without limit (it actually weighed 60 pounds at the end of the experiment), and though the result seemed to prove, at that time, that vegetable matter consisted only of water it was shown by later experiments made in the first instance by Priestley, and subsequently confirmed by many other chemists, including Sir Humphry Davy, that the growth is entirely due to the use of carbon derived from the carbon dioxide of the atmosphere. In Davy's Lectures on Agricultural Chemistry given in 1813 he describes experiments in which, first a square of turf and in another case a branch of vine, was exposed to an atmosphere containing a large quantity of carbonic acid. Under the influence of sunshine the carbonic acid disappeared and was replaced by oxygen.

It is unnecessary to cite the further experiments of the older observers, but among the many researches in connection with this question which have been recorded during more recent times, there is none more important than those of Dr. Horace T. Brown published during the years 1893 to 1905 on the nature of the substances formed in the leaf during exposure to sunlight and the rate of decomposition of the atmospheric carbon dioxide. In order that this constituent of the air may come into contact with the active surface it is necessary for it to penetrate into the interior through the small openings, called stomata, which exist for the most part on the lower surface of the leaf, and these minute pores when fully open do not exceed 1 or 2 per cent of the total area of the leaf surface. The astonishing result has been arrived at that in a fully active leaf the atmospheric carbon dioxide is taken up at least fifty times as fast as it would have passed into a series of small openings of equal size with the stomata, if these had been filled with a strong solution of caustic alkali.

The changes which go on in the interior of the leaf are associated in a mysterious way with the green colouring matter, chlorophyll, which plays the part of a sieve or filter of the sun's rays, stopping some and allowing others to proceed.

From the researches of C. A. Timiriazeff, Professor of Botany in the University of Moscow, it has been shown that the reduction of carbon dioxide as well as the production of starch is due to the rays which are absorbed by the chlorophyll. As the chlorophyll transmits chiefly green light the part which is stopped lies chiefly in the red, and on passing through a spectroscope the light which is transmitted is seen to have a dark band between the lines B and C of the solar spectrum. The blue and violet rays produce very little effect.

It is, however, not known with certainty what compound is the first result of the decomposition of the carbon dioxide, though it is commonly assumed that formaldehyde is the initial product. Its formation can be expressed by the simple equation

$$CO_2+H_2O=CH_2O+O_2.$$

This accounts for the elimination of the equal volume of oxygen which is known to be the other product. Formaldehyde is inimical to living organisms, and if it accumulated to any appreciable extent in a living leaf would speedily put an end to all vital processes within; in other words, it would kill the protoplasm by which it is supposed to have been produced. But while the formation and existence of formaldehyde in minute quantity appears to have been demonstrated in leaves exposed to light,[1] the greater part of it disappears immediately in consequence of condensation into some kind of carbohydrate, it may be a form of glucose, or, as appears more probable, starch, which remains stored up for use in the growth of the plant. It is not

[1] Usher and Priestley, *Proc. Roy. Soc.*, 77, B, 369.

necessary here to attempt to follow the transformations of the starch thus deposited in order that it may become available as food for the plant. It is only necessary to state that it is obvious that grains of starch being insoluble in water, this substance must be changed into something which is soluble, so that it may pass by diffusion from cell to cell in the tissue. It has been shown that this soluble compound is a sugar, and that it is formed from the starch by the action of an enzyme associated with the living protoplasm.[1]

Observations of this kind have led to various attempts to effect the change by which, from carbonic acid and water, formaldehyde and oxygen are formed, without the aid of the living plant and by the use alone of inorganic materials, the necessary energy required to bring about the reaction being derived from the sun. This result is said to have been achieved by several workers, of whom the most recent, Professor Benjamin Moore and Mr. T. A. Webster, in a paper contained in the Proceedings[2] of the Royal Society, review and partly confirm the results of earlier workers in the same field.

The materials used in these researches consisted of an aqueous solution of carbon dioxide mixed with a solution of colloidal uranic hydroxide or ferric hydroxide (see Colloids). The mixed solutions were exposed to the rays of the sun for one or two days or more, and at the end of the exposure the liquid was distilled and tested, by methods known to be very sensitive, for the presence of formaldehyde.

Similar solutions kept in the dark gave no indication of the production of any such substance. Now as it has been shown repeatedly that formaldehyde in the presence of lime or of certain salts of inorganic nature and origin condenses somewhat readily into various sugars, here is a process by which it is conceivable that a so-called organic compound might be formed by a natural operation independently not only of living matter, but of the artificial conditions provided by the intervention of man.

For the essence of the process is the occurrence together of water, carbon dioxide, and some colloidal or other substance which may act as a catalyst in the presence of solar radiation.

[1] The reader who is interested in such questions should read Brown and Morris, "On the Chemistry and Physiology of Foliage Leaves," *Trans. Chem. Soc.*, 1893, pp. 604–677.

[2] *Proc. Roy. Soc.*, 87 B, 163 (1913).

In the plant the green colouring matter chlorophyll plays an important part, the exact nature of which is not yet understood. All that can be said at present is that it is not merely catalytic, and that the efficiency of the green matter in association with the living protoplasm in fixing carbonic acid is far greater than that of any combination of inorganic materials yet tried.

Researches of this kind have been connected, especially during recent years, with speculations as to the origin of life. Man finds himself in a world so full of miracles, and the daily spectacle is so familiar as almost to paralyse the faculty of wonder. Nevertheless the desire to form a theory or view as to how it all came about has been in all ages and among all peoples so urgent that in the absence of direct and positive knowledge mythology has always centred round a special act of creation. "In the beginning," when the earth was "void," that is empty, it was filled with all manner of beast and bird and creeping thing, and with the herb and tree which was to be their food. In no case is the nature of the act of creation revealed, or what would be called in modern language the physical or chemical acts or doings by which the water and the dry land were furnished with inhabitants.

Geology assures us that there was a time when the earth was at a temperature at which no living animal or vegetable could exist. There is abundant evidence that as it cooled down it gradually became clothed with a vegetation differing in form and structure from that which now covers its surface, and with a succession of animals of which the earliest were chiefly inhabitants of the water, while the latest of all included man himself.

From these facts and from the knowledge laboriously acquired, chiefly during the last century, concerning the forms, the structure, the habits, and mode of propagation of plants and of animals, the doctrine of organic evolution has arisen, and with slight variation of detail has been accepted by the whole civilised world. This doctrine teaches that the higher animals and plants possessing more specialised organs and internal structure arose from lower, less specialised forms by a process which involved what may be termed experimental trials by Nature, through the results of spontaneous variation and survival of the fittest. The imagination of the naturalist then travels back from mammal to bird and reptile, through fishes and crustacea to sponges and corals, till the animal can no longer be distinguished from the vegetable, and the lowly organism takes the form of an apparently structureless but ever moving mass of jelly.[1] If this is to be thought of as the primal form in which life resided "in the beginning," how did it receive the inspiration which differentiates it from a minute drop of white of egg or gelatinous silica or any similar mass of colloid ?

Present views seem to be divided between two opposite camps. On the one hand are found those who cling to the idea that life is a directive influence distinct from any ordinary physical forces, taken either singly or together. On the other hand are the advocates of the view that all the operations of living beings are the result of physical and chemical processes going on in the organism itself or in response to forces acting on it from the outside.

The adherents of the former view are frequently referred to by those who profess the opposite opinion as "vitalists" and the doctrine as "vitalism" with a tone which seems to imply that such an idea is obsolete. The vitalists do not attempt to explain what life is, but like the rest they are eager to account for the existence of living beings in this world of ours, and to do so must choose between the hypothesis of a special creation and

[1] I cannot refrain from quoting here the following humorous verses by a young poetess and student of philosophy, Constance Naden, who, after a brilliant career in the early days of the University of Birmingham, died on 29th December, 1889, at the early age of 31 years.

"We were a soft Amœba
  In ages past and gone,
Ere you were Queen of Sheba
  And I King Solomon.

Unorganed, undivided,
  We lived in happy sloth,
And all that you did I did,
  One dinner nourished both :

Till you incurred the odium
  Of fission and divorce—
A severed pseudopodium
  You strayed your lonely course.

When next we met together
  Our cycles to fulfil,
Each was a bag of leather,
  With stomach and with gill.

But our Ascidian morals
  Recalled that old mischance,
And we avoided quarrels
  By separate maintenance.

Long ages passed—our wishes
  Were fetterless and free,
For we were jolly fishes
  A-swimming in the sea.

We roamed by groves of coral,
  We watched the youngsters play,
The memory and the moral
  Had vanished quite away.

Next each became a reptile,
  With fangs to sting and slay :
No wiser ever crept, I'll
  Assert, deny who may.

But now, disdaining trammels
  Of scale and limbless coil,
Through every grade of mammals
  We passed with upward toil.

Till anthropoid and wary
  Appeared the parent ape,
And soon we grew less hairy,
  And soon began to drape.

So from that soft Amœba
  In ages past and gone,
You've grown the Queen of Sheba,
  And I, King Solomon."

(From Solomon Redivivus, 1886, "Evolutional Erotics," in the volume entitled *A Modern Apostle*. Kegan Paul, Trench and Co., 1887.)

the idea that life has existed, if not in this planet, elsewhere, and that it is as old as matter. To account for its appearance on the earth it would be necessary to make a further assumption. Either we must suppose with Lord Kelvin[1] the arrival of a "seed-bearing meteoric stone" from space outside our atmosphere, the result of the disruption of some other life-bearing planetary body in consequence of collision or otherwise. Or the hypothesis of *panspermia* may be accepted. This assumes that the minutest germs of some of the lowest organisms may be small enough to be carried through the cosmical spaces from one world to another, by the pressure of some radiant form of energy.

In either case there remains no problem for the chemist or physiologist to investigate as to the *origin* of life. This problem belongs to the field in which the advocates of the other view have long been at work.

The triumphs of synthetical chemistry during the last forty years which have resulted in the production not only of compounds like formaldehyde containing a small number of atoms and of simple constitution, but substances especially of the protein class containing a very large number of atoms and therefore consisting of large molecules, have encouraged the idea that by similar methods substances of still more complex type may be produced which will resemble the natural colloids or even be found identical with them.

The "Conclusions" added to the paper, already quoted, by Messrs. Moore and Webster contain the following passages :—

"Such a synthesis[2] occurring in nature probably forms the first step in the origin of life. . . .

"Without the presence of organic material when life was arising in the world, any continuance of life would be impossible.

"The process of evolution of simple organic substances having once begun, as now experimentally demonstrated, substances of more and more complex organic nature would arise from these with additional uptake of energy. Later organic colloids would be formed possessing meta-stable properties and these would begin to show the properties possessed by living matter of balanced equilibrium, and up and down energy transformations following variations in environment.

"There can be little question that such energy changes as are

[1] Address to British Association. [2] That of formaldehyde.

above described occur at present, and are leading always to fresh evolutions of more complex organic substances, and so towards life, and equally is it true that they must occur on any planet containing the necessary elements for the evolution of inorganic colloids and exposed to light energy under suitable conditions of environment."

Such statements deserve a close examination. As to the first there will be probably no difference of opinion, for it is obvious that any organism born into a world which contained no organic matter must forthwith perish. But the speculations set forth in the latter part of these conclusions require us to believe that if molecules become big enough they will consequently begin to show signs of life.

What would happen to a very large molecule as soon as it is formed can only be guessed at, and there is absolutely nothing but analogy for guide. It seems agreed that the atoms of elements which attain large dimensions become unstable and in their break-down show the phenomena of radio-activity. But when uranium or radium disintegrates there is, beside a great liberation of energy, the production of two or more substances which obey ordinary physical laws as gas or solid. There is no indication of a return of any part of them to their original state, there is no cycle of events. But, as Sir Humphry Davy is reported to have said, "the substitution of analogy for fact is the bane of natural philosophy."

Chemical synthesis has accomplished some wonderful things by well-known laboratory methods. These methods involve very commonly the use of high temperatures, caustic alkalis, strong acids, and solvents such as alcohol, ether, or acetone which, at any rate in a concentrated form, never appear among the constituents of either plant or animal. In fact the processes of the laboratory have not the remotest resemblance to those which must be assumed to go on in living tissue.

The chemist can take carbon and hydrogen and by the aid of a high temperature can make them unite together to produce ethylene. From ethylene he can build up tartaric acid by a succession of steps which, however, require the use of chlorine or bromine. The grape-vine also manufactures tartaric acid, but it uses neither a high temperature nor a halogen. And even in such a case as that of formaldehyde, described on a recent page, the absorption of carbon dioxide by the living tissue is about fifty times as rapid as its fixation by strong caustic potash, which is practically instantaneous.

The proposition that living matter is being generated afresh during every day of sunshine from mere mineral matter at the present time and through all the past ages of the world since the dawn of life is an assumption which can never be proved. It is also unnecessary if the object were only to account for the continuance of the efflorescence of living forms which cover the surface of the earth. The question to be met relates to the initial act or process by which life was first established on this globe, and to do this it is necessary to recall the conditions prevailing on the earth's surface when in the beginning the globe had cooled down sufficiently to allow of the formation not only of a solid crust, but the deposition of water in the liquid form and its retention at a temperature far below the boiling-point. The materials then available, beside the solid silicates, oxides, carbonates, etc., of the crust, would be water, gaseous oxygen, nitrogen, and carbon dioxide, with possibly small quantities of ammonia and sulphuretted hydrogen. Possibly some volatile compound of phosphorus might be formed among the multitudinous products of chemical changes going on, but this would be speedily removed from the atmosphere by oxidation and fixed in the solid crust in the form of phosphate. Among the products formed by the action of water on the silicates and other compounds containing metals there would doubtless be an ample supply of colloidal substances suspended in the waters or deposited in crevices of the crust, and some of these would doubtless be qualified to act as catalysts in promoting the formation of carbon compounds from the carbon dioxide which at that period would be found as a copious ingredient in the primeval atmosphere. At the time imagined the fixation of the carbon which afterwards took place by the action of vegetation and its replacement by an equal bulk of oxygen, had not begun. That carbon was afterwards withdrawn and stored up in the beds of coal and in the forests of living trees with which so large a part of the earth is clothed.

We have seen how, if Professor Moore's results are to be accepted as conclusive, formaldehyde and consequently some carbohydrate might be brought into existence by the co-operation of the atmosphere, the water, and some colloid constituent of the solid crust. There would thus be provided one ingredient in the dietary necessary for living organisms. But protoplasm is believed to consist of colloid material in which not only carbon, hydrogen and oxygen are elements, but nitrogen is an essential component. Phosphorus is also a constituent of some proteins, and indispensable at some stages of development. Now formaldehyde is a very active substance which readily enters into chemical reactions of all kinds, and in the presence of ammonia is converted into a definite basic substance, hexamethylene-tetramine, which under the various names hexamine, etc. has long been used in medicine. Whether the minute and highly diluted quantities presumably formed in the process described by Professor Moore would yield hexamethylene-tetramine by contact with a mixture of gases containing ammonia it is difficult to say, but for the sake of argument it may be assumed that *some* organic compound containing nitrogen would be produced. We may even go further and suppose that, by a series of changes the nature of which cannot now be even conjectured, a complex colloidal protein was actually formed. We may in the present state of knowledge safely enquire—What then? No chemist will be induced to believe that a pulpy mass of one or more aminoacids, no matter how complex or how associated with saline electrolytes, will cease to exhibit the characters which belong to chemical compounds in general, and acquire of its "own mere motion" the power of utilising and controlling energy supplied from external sources in such a way as to give rise to the cycle of events exhibited in every particle of living substance, from the amœba onwards.

Something has been made of a supposed resemblance between cell membranes and the curious forms which some of the very simplest organisms assume and the films and cavities formed by inorganic colloids in the process of drying, or when in contact with other matters in a different state of hydration. There is just as much resemblance in such case as is to be found between the forms of fossil plants in the more ancient rocks and the foliaceous tracery produced by frost on the pavements in winter. It has even been suggested that some of these impressions in the palæozoic rocks may after all have been left by frost, and not by the fronds of ancient ferns. No one, however, supposes that if it were so these forms represent the beginning of life.

But protoplasm cannot be thought of merely as a solution of mixed colloids and saline electrolytes. It must consist of aggregates or clusters of molecules of various dimensions and possessing a consistency which no solution could show, inasmuch as a solution would possess viscosity and cohesion equal in every direction. The amœba if merely a drop of colloid solution would, like a drop of any jelly, gradually melt away into the surrounding water by the operation of ordinary liquid diffusion. The amœba has extensibility and retractility, and therefore cannot be an ordinary solution.

There is another point which seems to have escaped discussion by biochemists. The skin which is formed on a warm colloid solution, say of glue, is produced first because it is that part of the liquid which is cooled most quickly, being exposed to the air. The process of solidification gradually extends through the entire mass, and the extent of the film is dependent on the size of the vessel and the extension of the liquid. It is therefore indefinite. But when a new cell is formed by partition or budding a limit is set to the extension of the membrane. It continues to grow till it reaches the average dimensions of the cells which compose that particular kind of tissue. Consider the case of a vegetable cell the wall of which is composed of cellulose. One molecule after another of cellulose is generated within and is added to pre-existent molecules and cemented to them by the operation of an unknown cause, probably not ordinary chemical attraction, or what is called cohesion, because of the limit which is set to the process. The cell remains always small and microscopic, only occasionally reaching such dimensions as to become visible to the unaided eye. Ordinary chemical and physical laws will not account for this phenomenon: no one can say as yet what it is that makes molecule stick tenaciously to molecule, forming so strong a continuous membrane, and what it is that puts a stop to the process when a sufficient extent of membrane has been produced. Neither can anyone yet say why, in a mass of cellular tissue in which cells all alike have been multiplying side by side under the same conditions, some of these cells suddenly take new forms and proceed to secrete new products, such as colouring matters, not previously found in them. If ordinary chemical and physical processes had the field to themselves, undisputed by that directive influence which is exercised by the vital principle, whatever that is, there could be no orderly arrangement in nature. Any living mass of cellular matter provided with the necessary temperature, moisture, and pabulum would develop into an indefinite mass the form of which would depend on the rate at which supplies were furnished or conditions favourable. For how can chemistry and physics explain heredity ? The seed of wheat contains within itself the incentive to produce a plant of the order of grasses, and no matter how it may be cultivated or neglected it never produces anything else.

Consider again the propagation of the animal races by the sexual process, and there can be no fear of contradiction in the statement that in the whole range of physical and chemical phenomena there is no ground for even a suggestion of an explanation. The mammalian ovum consists of a small cell about $\frac{1}{150}$ inch in diameter, while the spermatozoon is a far more minute body. The progeny which results from their interaction exhibits, more or less obviously, the characteristics of both the immediate parents or even of earlier generations. The bodily size, form, markings, and colours, as well as in the higher animals, the mental peculiarities of ancestors are reproduced. It may fairly be asked what chemical or physical property can be transmitted by any such process, or could conceivably be so stored up and utilised ?

Too much has been made of the curious observations by J. Loeb and others on the supposed fertilisation of the ova of sea urchins by immersion in solutions of sodium or magnesium salts, or by a stimulus provided by an electric current. These observations, even if not open to suspicion on account of the free diffusion of the spermatozoa of these and other creatures in the surrounding water, prove nothing of importance in relation to the question now under discussion, which is the initiation of organic living matter from inorganic lifeless material. The ovum contains within itself the potentialities of a new generation, and the stimulus necessary to bring them into operation may well be derived from various sources in the case of creatures so low in the scale and so little removed from forms which are habitually reproduced by subdivision.

This is not the place to pursue such a discussion further. It will be sufficient to add that those who accept the purely materialistic doctrine as to the origin of life have before them the necessity of establishing a vast number of facts before such doctrine can be made generally acceptable to the scientific world. The progress which has been made in the desired direction is far from being as yet a justification for the pronouncements which have within the last few years found their way into print and which have too much the air of being uttered *ex cathedra*.

The origin of life and hence the origin of mind constitute problems which it is safe to assert will occupy mankind for generations to come. Fascinating as they are, the further we penetrate the more perplexing these problems become, and it is open to those who look at them from the standpoint of the pure physicist or the pure chemist to hold the view that physiology, being the chief handmaid of medicine, would be rendering a greater service to humanity by devoting all her great powers to furthering that branch of science rather than attempting the solution of problems which have every appearance of being insoluble. At any rate, it is urgently desirable that any statement of the new views should be communicated to the public only when the fundamental facts have been established beyond controversy, and that the bio-chemical and physiological student will not allow his enthusiasm to colour his hypotheses independently of the light which can be cast upon them.

The scientific chemist has a large field to himself in which will be found problems as perplexing as those which are presented to the biologist. At present who can say what is chemical affinity or attraction, what is the proper measure of valency, what is the real nature of the relation of the elements to one another? The term "energy" is freely used, and the physicist speaks commonly of potential energy and distinguishes it from kinetic energy, but he cannot define energy, he can only measure it. "Energy," therefore, is in the same category as "life," but no one would deny its existence because he can no more say exactly what it is than he can define matter, space, or time.

The future of scientific chemistry will probably depend on the activity of research in two main directions. On the one hand, there will certainly be large additions to the long list of already known definite compounds, especially in the so-called "organic" division of compounds built up on a foundation of carbon as the characteristic element. In this direction little that is new in principle must be expected. But, on the other hand, developments of physical chemistry will doubtless lead to a better knowledge of the laws which regulate chemical change and which connect together chemical constitution and physical properties. The extension of this kind of knowledge will enable the chemists of the future to calculate in advance what will be the colour and crystalline form of any compound it is proposed to make, what its physiological properties will be, and therefore its use, if any, as a medicine. Some few steps have already been taken in this direction, but there is need for a much larger body of workers properly qualified not only by the possession of theoretical knowledge, but by sufficient laboratory experience to give any results they may arrive at the indispensable quality of being trustworthy. And something further is eminently desirable, and that is some organisation of the resources which are available for the collection of facts, and for performing the large amount of routine work necessary in providing data which may be made stepping-stones to further advances. There is frequent reference to State assistance in research. This is a difficult question. While there would be practical unanimity in the feeling that State assistance should be given in the form of money, there would probably be much difference of opinion as to the way in which it should be applied. State-aided or controlled institutions are apt to fall under the wheels of routine, and epoch-making discoveries are not likely to proceed from such establishments, but on this very ground they would be well fitted to carry out efficiently experimental work which has for its object the determination of physical constants and the provision of data of all kinds derived from exact observation.

The conduct of research into questions connected with special processes, materials, or patents connected with industry must be left to the manufacturers. So also must the question how far industrial research can be carried on in colleges and universities. The association of research with the teaching of advanced chemistry is a matter which concerns the professors. And it is to be hoped that in future the governors of these institutions will see to it that there is perpetual evidence of activity in this direction, though of the results of the work done or discoveries made they may not be qualified to act as judges.

Lastly, there is the real heaven-sent researcher endowed with that kind of inspired curiosity which drives him to labour for the mere love of it, provided only that it takes the form of putting questions to nature. All that can be hoped for is that he will not be hindered by artificial obstacles created by official stupidity, and that such assistance as money can give will not be beyond his reach. For it cannot be too often repeated that *pure science*—that is, the correct observation of fact and the establishment of "law"—stands ever in practical importance before *applied science*, which is invention. But this is a hard saying, and there are still too many people who believe that the true and only business of science is to find out useful things. Even Francis Bacon, in his famous fable of the "New Atlantis," seems to have taken this view, for in the Order or Society which he imagined under the name of "Solomon's House," he supposes only three members of the community set apart as "Interpreters of Nature," all the rest being occupied in drawing out of their discoveries things of use to mankind.

It is, however, only necessary to consider any application of science to useful purposes to perceive that such application became possible only at the end of a long series of observations, experiments, and arguments which occupied the labours of several generations of men. Each step forward is usually the result of some apparently trivial scrap of new knowledge acquired without regard to the question whether it is likely ever to be turned to any practical purpose.

Real progress comes from the pursuit of knowledge for its own sake.

APPENDIX

BRIEF BIOGRAPHICAL NOTES CONNECTED WITH THE PORTRAITS

IN ALPHABETICAL ORDER

PROFESSOR ARRHENIUS.
PROFESSOR BERTHELOT (died 1907).
SIR WILLIAM CROOKES.
MADAME CURIE.
PROFESSOR EMIL FISCHER.
PROFESSOR MENDELÉEFF (died 1907).
SIR WILLIAM PERKIN (died 1907).
PROFESSOR SIR WILLIAM RAMSAY (died 1916).
LORD RAYLEIGH.
PROFESSOR THEODORE RICHARDS.
PROFESSOR SIR JOSEPH J. THOMSON.

APPENDIX

BRIEF BIOGRAPHICAL NOTES CONNECTED WITH THE PORTRAITS

PROFESSOR SVANTE AUGUST ARRHENIUS is Director (since 1905) of the Physico-Chemical Department of the Nobel Institute in Stockholm.

He received his education at Upsala, and took his degree of Ph.D. at the University of that town in 1884. He began teaching physics as Privat-docent immediately after graduation, and held office successively as Teacher of Physics (1891), Professor of Physics (1895), and Rector (1897–1902) in the same University.

Professor Arrhenius is famous as the originator of the ionic dissociation theory, whereby the chemical properties of substances are connected with the electric conductivity of their solutions. He has also published a large amount of experimental work in support of his views and their application in chemistry and physiology.

In 1902 he was awarded the Davy Medal by the Royal Society, and in 1903 he received the Nobel Prize for Physics. He is a member of many academies and learned societies in Europe and America. In 1898 he was elected one of the forty Honorary Members of the Chemical Society, and in 1910 he became a Foreign Member of the Royal Society.

In 1914 Professor Arrhenius delivered the Faraday Memorial Lecture in the Royal Institution, where this lecture is given about once in three or four years, always by a very distinguished foreign chemist. He received at the same time the Faraday Medal from the Chemical Society.

Beside textbooks and works on Electro-Chemistry Arrhenius has published *Worlds in the Making* (1908), and *Life of the Universe* (1909).

PIERRE EUGÈNE MARCELLIN BERTHELOT was born in the heart of old Paris, in the Place de Grève, on October 25th, 1827. He died in Paris, March 18th, 1907.

The son of a physician, Dr. Jacques Martin Berthelot, he received a sound classical education and retained throughout life a love of ancient literature. After leaving school he completed a full medical course, but was ultimately led to adopt a purely scientific career. His first paper, on a simple method of demonstrating the liquefaction of gases, was presented to the Academy of Sciences in 1850.

Beginning with the humble appointment of lecture assistant to Balard, the discoverer of bromine, then Professor of Chemistry in the Collège de France, he became in succession Professor in the Ecole Supérieure de Pharmacie, Professor of Organic Chemistry in the Collège de France, a post which he held till his death, and, finally, he succeeded Pasteur as Perpetual Secretary of the Academy of Sciences in 1889. In 1900 he became one of the forty French Academicians.

Berthelot lectured before the Chemical Society in London, June 4th, 1863, "On the Synthesis of Organic Substances," and a few years later was elected an Honorary Member of the Society. He received a large number of marks of honour from various academies and learned societies. It is only necessary to mention here that he was elected a Foreign Fellow of the Royal Society in 1877 and received the Davy Medal in 1883. The Copley Medal, the highest distinction the Royal Society has to bestow, was awarded to him in 1900.

Berthelot was an extraordinarily prolific investigator and writer. The great variety and extent of the subjects he attacked are indicated by the following brief summary :—

1. Synthesis of fats and characterisation of glycerol as a polyhydric alcohol.
2. Synthesis of hydrocarbons, acids, etc.
3. Action of mass, and study of the law of equilibrium.
4. Thermochemical researches embodied in two large volumes entitled *Essai de Mécanique Chimique*.
5. Explosives, and especially the explosion wave in gases.
6. Fixation of atmospheric nitrogen and its relation to vegetation.
7. Study of Greek and Arabian alchemistic writings.

In November, 1901, Berthelot's seventy-fifth birthday and the jubilee of his first appointment in the Collège de France was celebrated by a public ceremonial at the Sorbonne, where he was received by the President of the Republic. It was in harmony with the public sentiment of respect that on the death of the great chemist, which followed on the same day that of his wife, a State procession with military escort, conveyed the bodies of husband and wife to final rest in the Pantheon. "They order these things better in France."

For details see the Memorial Lecture given by Professor H. B. Dixon to the Chemical Society, 23rd November, 1911 (*Trans. Chem. Soc.*, II, 1911, pp. 2353–2371).

Sir William Crookes, O.M., Foreign Secretary (1908–1912), and President of the Royal Society (1913–1915). He has received the Royal, Davy, and Copley Medals of the Royal Society.

Sir William Crookes received his scientific education in the Royal College of Chemistry, London, under Hofmann's professorship.

In 1851 he published his first paper, being then in his nineteenth year. By the use of the then new method of spectrum analysis he discovered in 1861 the metal thallium, which presents curious chemical features intermediate between those of potassium and lead. As a result of the study of certain peculiarities of attenuated gases he devised in 1875 the radiometer. Pursuing his investigation on the electric discharge he was led to announce a fourth state of matter (1879), and described the phenomena exhibited by "radiant matter" in 1881. These early investigations were the means of attracting the attention of other investigators to the subject, and provided a starting-point for the work of the Cambridge school which under Sir J. J. Thomson has in recent years achieved such wonderful discoveries. Crookes' views on the nature and origin of the elements have attracted much attention, and a condensed account of them is given in the text.

For the purpose of collecting information concerning the origin of the diamond he paid two visits to Kimberley, the second in 1905, when over seventy years of age.

Sir William Crookes was President of the Chemical Society in 1888 and 1889. He has also been President of the British Association and of the Institution of Electrical Engineers. He has received many honours, including Honorary Membership of the R. Accademia dei Lincei (Rome), and Corresponding Membership of the Institute of France (Academy of Sciences).

Madame Marie Curie, née Marie Sklodowska, was born at Warsaw in 1867, the daughter of Professor Sklodowski. She received her education first at the Lycée in her native city, afterward at the Sorbonne in Paris, where she first graduated as Licenciée ès Sciences Physiques, Licenciée ès Sciences Mathématiques, and ultimately as Docteur ès Sciences, on the publication of her thesis on Radio-active Substances. A full translation of this thesis was published in the *Chemical News* in 1903.

Marie Sklodowska married Professor Pierre Curie and worked with him on the radioactive properties of minerals. Madame Curie succeeded, after a protracted research, in separating from pitchblende salts of the element which has ever since been known as radium. She also determined its atomic weight and fixed its position among the elements.

Professor Curie was killed in a street accident in Paris in 1906. Madame Curie is now Professor of Physics in the Sorbonne.

PROFESSOR EMIL FISCHER is Ordinary Professor of Chemistry in the University of Berlin, and Director of the Laboratories in the Chemical Institute. He was appointed to this Chair on the death of Hofmann in 1892. Hofmann was the first Professor in the Royal College of Chemistry London, from its foundation in 1845 till his return to his own country in 1865.

Professor Fischer was born at Eus Kirchen (Rhenish Prussia) in 1852. He began his scientific studies in Strasburg.

After working for some years under von Baeyer in Munich, latterly as Extraordinary Professor in the analytical department of the University, he was appointed Ordinary Professor of Chemistry at Erlangen (1882).

Three years later he occupied the chair at Wurzburg till the call came which took him to Berlin. He has, of course, received all the distinctions which are in Germany naturally associated with the position he occupies in the premier university—Geheim-rat, Regierungs-rat—to which has been added the title "Excellenz." He became a member of the Munich Academy of Sciences in 1881, and of the Imperial Academy of Sciences in Berlin in 1893. He has also received from the Royal Society the Davy Medal, and from the Chemical Society the Faraday Medal on the occasion of his delivering the Faraday Lecture in 1907.

Fischer's researches may be grouped under four principal heads. First, in 1878, in conjunction with Otto Fischer, he established the nature of the rosaniline dyes as derivatives of triphenylmethane. Having then discovered phenyl hydrazine he applied this substance as a reagent in his masterly study of the sugars a few years later. Next the synthesis of uric acid and a number of closely allied compounds cleared up the confusion previously existing in this important group of substances. During nearly twenty years Fischer has been occupied with the study of the protein constituents of animal and vegetable tissues, on which he has thrown a flood of light by his synthesis of a number of complex amino-acids, some of which are described in the text.

DMITRI IVANOVITSCH MENDELÉEFF was the fourteenth and youngest child of his parents. The elder Mendeléeff was Director of the College at Tobolsk (Siberia), and here was born 27th January, 1834 (O.S.), the son who was to become so famous. The full story of his life and work has been told in the Memorial Lecture given to the Chemical Society by Sir William Tilden on 21st October, 1909, and printed in the *Transactions* of the Society for that year.

The subject of this notice was about 1861 appointed Professor of Chemistry in the Technological Institute at St. Petersburg (Petrograd). In 1866 he became Professor of General Chemistry in the University, the Chair of Organic Chemistry being occupied at the same time by Butlerow.

In 1890, in consequence of a difference with the authorities, he retired from his professorship. In 1893, however, he was appointed by M. Witte to the office of Director of the Bureau of Weights and Measures, which he retained till his death. This occurred on 20th January (O.S.), 1907, within a few days of his seventy-third birthday.

Mendeléeff's name is indissolubly connected with the evolution and final establishment of the principle of periodicity among the elements, the first recognition of which we owe to John A. R. Newlands, an English chemist. Mendeléeff's table of the elements was first drawn up in 1869, and in 1871 was modified so as to give it nearly the form in which it is to be found in every modern textbook of chemistry.

The German Professor Lothar Meyer, assisted toward the general acceptance of the principle by the publication in 1869 of his graphic demonstration of the relation between atomic weights and atomic volumes.

Mendeléeff alone was the first to foretell the properties of these *undiscovered* elements and to alter atomic weights in confidence in the validity of the law.

The famous manual entitled *Principles of Chemistry*, which bears Mendeléeff's name and of which several English editions have appeared, was essentially an exposition of the periodic scheme. It had many original features and will always be a landmark in the history of chemistry.

Mendeléeff's experimental researches related to the constitution of solutions, and the physical properties of gases. Another subject to which he gave great attention and on which he was consulted by the Russian Government, was the nature and origin of petroleum.

In 1882 the Royal Society conferred on Mendeléeff, jointly with Lothar Meyer, the Davy Medal. In 1883 the Chemical Society elected him an Honorary Member, and in 1889 it conferred on him the highest honour in its power to bestow, namely, the Faraday Lectureship, with which is associated the Faraday Medal. In 1890 he was elected a Foreign Member of the Royal Society which, in 1905, awarded him the Copley Medal. So far as England is concerned his services to science received full acknowledgment. It is all the more remarkable, therefore, that he never became a member of the Imperial Academy of Sciences in his own country.

WILLIAM HENRY PERKIN was born in London, 12th March, 1838, the youngest son of G. F. Perkin, a builder and contractor, who died in 1865.

After a few years at the City of London School, where he received his first notions of chemistry from the late Thomas Hall, one of the masters, he was allowed, at the early age of fifteen, to enter the Royal College of Chemistry as a student under Hofmann in the year 1853. At the end of his second year, so rapid was his progress, he was allowed to begin research, and the first subject attempted was the isolation from coal-tar and nitration of anthacene (then known as paranaphthalene).

Undaunted by want of success, for reasons which can now be appreciated, the boy was led to undertake other work in which better results were secured, and he was made a member of the teaching staff. In 1856, in the pursuit of an attempt to produce quinine synthetically he submitted to oxidation a specimen of commercial aniline, which as then manufactured consisted of aniline mixed with variable proportions of the toluidines. The result was a dark coloured precipitate from which, after further experiments, the famous dye, "Aniline Purple," or "Tyrian Purple," or, as it was called in France, "Mauve" was manufactured. This discovery and the extensive use of the purple, especially in France, was the starting-point of numerous investigations, out of which ultimately grew the great coal-tar colour industry.

Later, in 1868, Perkin resumed the study of anthracene, and introduced a process for the production from it of alizarin. Notwithstanding the invention of the sulphuric acid process by Caro, Graebe, and Liebermann at the same time, it was stated by Perkin himself (Hofmann Memorial Lecture, 1896) that up to the end of 1870 the Greenford Green Works (that is Perkin's) were the only works producing artificial alizarin.

From the first Perkin's heart was devoted to research for science's sake, and all through the period when he was occupied as a manufacturer of dye-stuffs he continued to pursue investigation in other directions without reference to industrial questions. His synthesis of coumarin in 1868, and later of cinnamic acid were among the results of a general method introduced by him and usually referred to as Perkin's reaction.

In 1874 Perkin retired from business as a manufacturer, and devoted himself wholly to scientific research. He built a new house at Sudbury on land adjoining that on which stood the house in which he had lived up to that time, and which was thereafter converted into a laboratory. Here many years were devoted to the work on Magnetic Rotation.

Perkin was elected into the Royal Society in 1866. He was

Secretary of the Chemical Society from 1869 to 1883, and President from 1883 to 1885. He was also President of the Society of Chemical Industry, and at the time of his death, which occurred on July 14th, 1907, he was President of the Society of Dyers and Colourists. Of course honours of all kinds were showered on him, and at the jubilee of the discovery of mauve, celebrated in 1906, many distinguished foreign chemists were present.

Sir William Ramsay, K.C.B., F.R.S., Ph.D., LL.D., D.Sc., etc. Emeritus Professor of Chemistry in University College, London.

Sir William Ramsay is famous all the world over as the discoverer of terrestrial helium and of neon, krypton, and xenon, the companions of argon in the atmosphere. He was associated with Lord Rayleigh in 1894, in the study of argon, the then newly discovered inert gas found in the air. Previously to this work he had published many papers on physico-chemical subjects of which perhaps the most important are those on the Molecular Surface Energy of Liquids. After the discovery of radium he, in conjunction with Professor Soddy, proved that the gas which escapes from radium is helium.

Sir William Ramsay received his scientific education at the universities of Glasgow and Tübingen. After teaching for some years in the university of Glasgow he was appointed in 1880 Professor of Chemistry in University College, Bristol, where he soon afterwards became Principal of the College. In 1887 he succeeded Williamson at University College, London, and retired from the Chair in 1912.

During the last twenty years a profusion of honours has been bestowed on the discoverer of the inert gases of the atmosphere. In recognition of the importance and interest of this discovery, and of the skill displayed in the investigation, he has been elected an honorary member of nearly every scientific academy in the world. In 1904 he was awarded the Nobel Prize for Chemistry. Sir William Ramsay has also taken a prominent place in connection with several of the scientific bodies in his own country and has occupied the position of President of the Chemical Society, the Society of Chemical Industry, the British Association, and of the International Congress of Applied Chemistry at the meeting held in London in 1909.

Ramsay is distinguished for his skill as a manipulator, which in the management of the difficult operations connected with the collection and measurement of the minute quantities of the emanation and gas from radium has contributed largely to his success in this work. He has also a remarkable command of foreign languages. At the International Congress in London the readiness with which he addressed the opening meeting in the four official languages successively, English, French, German, and Italian, attracted admiration. He has also lectured to large popular audiences in Berlin in German, and in Paris in French.

He was created K.C.B. in 1902. The Davy Medal was awarded to him by the Royal Society in 1895, and the Longstaff Medal by the Chemical Society in 1897.

P.S.—Since the foregoing lines were written, and while this book is in the press the news has reached the author that his great and distinguished friend is no more. To the great grief of a large circle of scientific friends and others it became known very early this year that he was suffering from a malignant disease for which there was no hope. He was released from further pain early in the morning of Sunday, 23rd July, 1916.

It is impossible in a few lines to estimate justly the immense importance of Ramsay's work. Some of the most interesting of his discoveries have been described in the foregoing pages, and it will perhaps be sufficient in this place to remind the reader that to have added an entire group of new elements to the periodic scheme is an achievement both unexpected and unparalleled.

The Rt. Hon. John William Strutt, LORD RAYLEIGH (3rd Baron), O.M., Past Secretary (1885–1896) and President of the Royal Society (1905–1908), Chancellor of the University of Cambridge and formerly Cavendish Professor of Physics (1879–1884), Professor of Natural Philosophy in the Royal Institution (1887–1905).

Lord Rayleigh was Senior Wrangler and Smith's Prizeman at Cambridge in 1865. He has received many honours of which it is only necessary to mention the Royal, Rumford, and Copley Medals of the Royal Society and the Faraday Medal from the Chemical Society. He was also Nobel Laureate in Physics, 1904.

Lord Rayleigh's first paper was published in the *Philosophical Magazine* in 1869, and since that time a stream of communications, chiefly to the Royal Society, has been poured forth on a great variety of physical subjects, in which a rare combination of high mathematical powers with great experimental skill is manifest. Chemistry is indebted to his work chiefly for the series of experimental investigations, begun about 1887, on the relative densities of the principal gases. This enquiry culminated in the discovery of argon in the atmosphere, which was announced in association with Professor Ramsay by Lord Rayleigh at the Oxford Meeting of the British Association in 1894.

PROFESSOR THEODORE WILLIAM RICHARDS, Professor of Chemistry, Harvard University, Cambridge (Mass.), U.S.A., and Director of the Wolcott Gibbs Memorial Laboratory.

Professor Richards, though still a young man, having been born in 1868, has accomplished a very large amount of chemical research. He is especially distinguished for his work in the revision of the atomic weights of upwards of twenty of the elements. During recent years he has occupied himself with the problems connected with the compressibility of elementary atoms, and has shown that this property, like other properties of elements, is periodic with atomic weight.

Professor Richards is a member of the National (American) Academy of Sciences and an Honorary Member of many European academies and institutions.

The Davy Medal was awarded to him by the Royal Society in 1910, the Faraday Medal, which is given to the Faraday Memorial Lecturer, by the Chemical Society in 1911, and the William Gibbs Medal in 1912. In 1915 he also received the Nobel Prize for Chemistry.

Sir Joseph John Thomson, O.M., Cavendish Professor of Experimental Physics, Cambridge, and Professor of Physics at the Royal Institution.

Sir Joseph Thomson was elected President of the Royal Society in 1915, in succession to Sir William Crookes, whose period of office, in consideration of his advanced age, extended to only two years.

Sir Joseph Thomson was educated at Owens College, Manchester, and Trinity College, Cambridge. He was Second Wrangler and Second Smith's Prizeman in 1880, and was forthwith elected a Fellow of his College. After working as a Lecturer at Trinity he succeeded Lord Rayleigh as Cavendish Professor of Physics in 1884, and has carried out in the Cavendish Laboratory the long series of investigations on the Discharge of Electricity through Gases, which have resulted in the remarkable discoveries of which some are described in previous pages.

Sir Joseph Thomson has been the recipient of many honours, including membership of the most important academies in Europe and America. He has also received the following medals and prizes in recognition of the importance of his discoveries; a Royal Medal in 1894, and the first Hughes Medal, 1902, both from the Royal Society, the Hodgkins Medal from the Smithsonian Institute, Washington (1902), the Copley Medal (1914), the highest honour in the power of the Royal Society to award. In 1906 he was awarded the Nobel Prize for physics. The Order of Merit, which was instituted in 1902, has been awarded among the rest to six men of science, of whom Lord Rayleigh, Sir William Crookes, and Sir Joseph Thomson are the survivors.

INDEX